Multivariate Applications in Substance Use Research

New Methods for New Questions

Multivariate Applications Book Series

The Multivariate Applications book series was developed to encourage the use of rigorous methodology in the study of meaningful scientific issues, and to describe the applications in easy to understand language. The series is sponsored by the Society of Multivariate Experimental Psychology, and welcomes methodological applications from a variety of disciplines, such as psychology, public health, sociology, education, and business. Books can be single authored, multiple authored, or edited. The ideal book for this series would take on one of several approaches: demonstrate the application of a variety of multivariate methods to a single, major area of research; describe a methodological procedure or framework that could be applied to a variety of research areas; or present a variety of perspectives on a controversial topic of interest to applied researchers.

There are currently four books in the series:

- *What if There Were No Significance Tests?*, co-edited by Lisa L. Harlow, Stanley A. Mulaik, and James H. Steiger (1997).
- *Structural Equation Modeling with LISREL, PRELIS, and SIMPLIS: Basic Concepts, Applications and Programming,* by Barbara M. Byrne (1998).
- *Multivariate Applications in Substance Use Research: New Methods for New Questions,* co-edited by Jennifer S. Rose, Laurie Chassin, Clark C. Presson, and Steven J. Sherman (2000).
- *Item Response Theory for Psychologists,* co-authored by Susan Embretson and Steve Reise [in press].

Interested persons should contact the editor, Lisa Harlow, at the University of Rhode Island, Department of Psychology, 10 Chafee Road, Suite 8, Kingston, RI 02881-0808; Phone: (401) 874-4242; Fax: (401) 874-5562; or e-mail: LHarlow@uri.edu. Information may also be obtained from one of the editorial board members: Leona Aiken (Arizona State University), Gwyneth Boodoo (Educational Testing Service), Barbara Byrne (University of Ottawa), Scott Maxwell (University of Notre Dame), David Rindskopf (City University of New York), or Steve West (Arizona State University).

Multivariate Applications in Substance Use Research

New Methods for New Questions

Edited by

Jennifer S. Rose
Indiana University

Laurie Chassin
Clark C. Presson
Arizona State University

Steven J. Sherman
Indiana University

LEA LAWRENCE ERLBAUM ASSOCIATES, PUBLISHERS
2000 Mahwah, New Jersey London

Lawrence Erlbaum Associates, Inc., Publishers
10 Industrial Avenue
Mahwah, NJ 07430

Cover design by Kathryn Houghtaling Lacey

Library of Congress Cataloging-in-Publication Data

Multivariate applications in substance use research : new methods for new questions / edited by Jennifer S. Rose ... [et al.].
p. cm.
Includes bibliographical references and index.
ISBN 0-8058-2942-3 (cloth : alk. paper)
1. Substance abuse—Research. 2. Multivariate analysis. 3. Substance abuse—Longitudinal studies—Statistical methods. I. Rose, Jennifer S.
HV5809.M84 1999
362.29'01'519535—dc21 99-28611
CIP

Books published by Lawrence Erlbaum Associates are printed on acid-free paper, and their bindings are chosen for strength and durability.

Printed in the United States of America
10 9 8 7 6 5 4 3 2 1

We dedicate this volume to
our graduate students
for dragging us, kicking and screaming,
into the 21st century;
to our colleagues in quantitative methods
for showing us the way;
to NICHD and NIDA
for supporting the journey;
and to our families
for allowing us the time to make the trip.

—Jennifer S. Rose
—Laurie Chassin
—Clark C. Presson
—Steven J. Sherman

Contents

Preface

This volume was inspired by a number of motivations. One was to share part of the adventure that we have had on the Indiana Smoking Project over the past 20 years. During this time, we have studied how it is that adolescents come to smoke, and how smoking relates to various aspects of people's lives as they develop through adolescence and enter and establish adult roles. In so doing, we have also learned a great deal about the factors involved in smoking initiation and transitions. In addition, we have learned a lot about large-scale research projects, and have come to appreciate the importance of dialog between researchers in the area of substance use and abuse and researchers who are developing new quantitative techniques. In studying the etiology of tobacco use, the questions that we have asked have turned out to be very complex, and, over the 20-year period of our collaboration, these questions often required us to seek out newly developing quantitative methods. For example, during this period we went from asking whether a particular variable prospectively predicted a change in smoking behavior one year later, to asking about predictors of trajectories of smoking behavior over six waves of data spanning adolescence to adulthood. We went from asking whether parental smoking influenced adolescent smoking to testing complex models of the multiple mechanisms where parental smoking might exert its influence. We went from examining psychosocial influences in general to examining the shared versus nonshared nature of these influences as they affected sibling similarity and dissimilarity within families. As these examples illustrate, the changes in the sophistication and complexity of our research agenda necessitated tremendous changes in the way that we approached and analyzed our data.

As with many researchers in the area of substance use, most of us were trained in research traditions that were simpler and, quantitatively speaking, decidedly old-fashioned. Without dating the time of our training, let us simply say the ideal study might well involve a 2 X 2 design, with 8 or 16 subjects in each cell, and many chi-square tests and ANOVAs. These tools are not bad ones, and for some questions they are still highly desirable, even the best available, but they are not necessarily optimal for addressing important questions that currently face the field of substance use research. In this book, we hope to identify some of these important questions and illustrate how some recent advances in quantitative methods can help to answer them.

It is worth taking a moment to remember the context of research concerning psychosocial research on substance use in the mid-1970s. At that time, the field could be characterized as primarily descriptive, based on cross-sectional correlational studies, with little attention to developing and testing models of processes. There were glimmers of what has come to pass since the early-1980s, with beginning interests in longitudinal studies, the introduction of path analysis, and time-series analyses. Indeed, the National Institute on Drug Abuse published a volume in 1976 to "make available basic information on some of the latest and most relevant techniques," and this volume introduced drug researchers to path analysis, multiple regression, and discrimination analysis (Bentler, Lettieri, & Austin, 1976)[1]. Although these important multivariate techniques were established and available in standard programs for statistical analysis, researchers were often not trained in these methods. Not only were there barriers in our lack of training, but there were limitations in the lack of resources for implementing these complex, multivariate analyses. This was, after all, the world of punched cards for both programs and (often) data. The SPSS run to read the data from our original survey, to compute scales, and to save a system file took two hours of CPU time on the Indiana University mainframe.

In the context of these limitations of training and technology, it is not surprising that the questions addressed by substance use research had to be relatively simple ones. Once more, sophisticated techniques became more widely available, and as more researchers became familiar with them, the field developed more complex theories and conceptualizations. As the field continued to develop, important research questions have taxed each generation of analytic techniques, and have helped spur the development of newer applications and techniques. This synergism, between the theoretical questions that researchers have asked and the quantitative techniques that are developed to answer those questions, has spurred innovation in both theory and method, and that is the motivation for this volume.

Today, more than 20 years after the NIDA monograph, we find ourselves in somewhat the same position as those earlier authors, in wanting this vol-

---

[1]Bentler, P. M., Lettieri, D. J., & Austin, G. A. (1976). *Data analysis strategies and designs for substance abuse research.* Rockville, MD: National Institute on Drug Abuse.

ume to introduce substance use researchers to some recent advances in quantitative methods and to show how these methods can be applied to substance use data. However, one difference today is that the field is asking very different kinds of questions. First, the perspective of the field has shifted from a predominantly cross-sectional one to a longitudinal one, such that substance use researchers are critically interested in measuring processes of change over time. Researchers in the area of substance use are interested in the natural history of substance use and abuse, and in describing and predicting substance use trajectories. Moreover, researchers are interested in how these trajectories are changed by preventive or treatment interventions. In the past, techniques have modeled change over time at the level of a group. However, recent advances have allowed for modeling not only group level trajectories, but changes within a person over time. In this volume, Curran introduces the reader to the use of latent curve methods for describing individual trajectories of adolescent alcohol use over time. He demonstrates how a risk factor (in this case, parent alcoholism) can be used to predict individual differences in the escalation of alcohol use over time. Moreover, he demonstrates how changes over time in one construct (in this case peer alcohol use) can be related to changes over time in substance use.

In modeling longitudinal substance use data, a common problem is the existence of missing data. In the past, this problem has often been overlooked as researchers focused only on subjects with complete data. More recently, random coefficient models have allowed researchers to use all available data to estimate trajectories, and other forms of data imputation have also been employed, (see e.g., Schafer, 1997)[2]. In this volume, Hedeker and Rose show that it is important not only to consider missing data, but that the pattern of missing data is important as well. Applying pattern mixture models to our cohort-sequential study of the natural history of tobacco use, these authors demonstrate how different patterns of missing data influence interpretation of results. Specifically, they show how missingness interacts with cohort and time in describing development trajectories of cigarette smoking from adolescence to adulthood. The chapter by Muthén also provides an overview of missing data techniques in longitudinal growth modeling.

Another complex problem in modeling longitudinal substance use data over time is that background variables may have different influences over time on the growth process. For example, a variable that has a significant influence on teenage substance use may become less influential as participants grow older. Furthermore, within a sample, individual trajectories are likely to belong to different subpopulations corresponding to several forms of normative and non-normative development. Muthén shows how these issues can be

[2]Schafer, J. L. (1997). *Analysis of incomplete multivariate data*. London: Chapman & Hall.

addressed in longitudinal growth modeling with latent variables using a sample of participants from the National Longitudinal Survey of Youth.

For substance use researchers, the study of change over time is made even more complex because theoretically, researchers conceptualize substance use as a series of stages, and different transitions along these stages may be affected by very different factors. These stages typically include initiation, escalation, maintenance, cessation, and relapse. Our approach to dealing with the issue of stages is to choose individuals who are at the same stage, and then prospectively predict a single transition. In this volume, Rose, Chassin, Presson, and Sherman illustrate this approach with our longitudinal smoking data, using logistic regression to prospectively predict smoking cessation. Another approach is to examine multiple transitions within a single model, and ask whether a predictor variable differentially influences movement across different transitions. In this volume, Hyatt and Collins demonstrate how latent transition analysis can identify different transitions across multiple substances, and show how a risk factor (in this case, parental permissiveness) influences movement across these transitions.

In addition to studying trajectories of substance use and abuse, substance use researchers also want to influence these trajectories using interventions. With regard to prevention, common intervention programs are multicomponent in nature, attempting to change factors such as information about substance use effects, peer norms, and so on. In these cases, researchers want to know not only if their prevention program has succeeded but why it succeeded—that is, through what mediating mechanisms the program had an effect. In this volume, MacKinnon used data from a steroid prevention program to demonstrate methods that allow researchers to assess complex mediational models of preventive effects. MacKinnon describes how researchers can assess an individual mediated effect within a multiple mediator model.

Multivariate models of substance use may not only take the form of multiple mediating mechanisms, but also, there may be interactions among multiple predictors or other higher order effects (e.g., quadratic effects of a predictor). In this volume, Wood introduces a method that simplifies the testing of such effects in latent variable models. Previously, the intensive, error-prone programming required was a barrier to testing these effects, and convergence problems were common. Using an example prospectively predicting heavy alcohol use from prior use and alcohol expectancies, Wood demonstrates a program that writes the code necessary to test these interactive latent variable models.

Substance use researchers have also become increasingly interested in the effects of context. Recent theories have recognized, more explicitly, that individuals are embedded in multiple social contexts, including families, classrooms, neighborhoods, and treatment groups. Thus, techniques are needed that allow us to take into account the effects of these contexts and interactions among multiple levels of context. In this volume, Rowe, Rodgers, and Gilson apply epidemic modeling techniques to understand smoking status as a function of the prevalence of smoking in an adolescent's environment. Khoo and Muthén use data from the National Longitudinal Survey of Youth to illustrate how data from siblings can be used to appropriately consider their clustering within families, and also, to incorporate models of sibling influence on alcohol use.

Finally, when substance use researchers attempt to evaluate the effects of interventions (both treatment and preventive interventions), issues of design must be considered along with issues of analysis. The chapter by Ullman, Stein, and Dukes takes an experimental design of high quality (the Solomon Four Group Design) and provides researchers with a new way to analyze the data (in this case, data from the DARE program) using latent variable models.

Although we hope that this volume will inspire interest among substance use researchers in the usefulness of these recent quantitative methods, we recognize that simply reading a book will not suffice to allow researchers to benefit from these new methodologies. One of the real dilemmas is the one that faced our smoking project 20 years ago—finding out what tools are available and making them work with one's own data—especially when the investigator does not initially know the techniques. Fully utilizing these complex methodologies calls for a new partnership between those interested in substance use and those interested in methods development, and this partnership is not always easy to achieve The partnership begins in the planning and design of the study itself, not when the data have already been gathered. Thus, our major goal in constructing this book is to help provide a forum for dialogue between methodologists, who are innovating and developing complex multivariate statistical methodologies, and substance use researchers, who often have rich data sets and complex theoretical models, yet who may not be familiar with available data analytic alternatives. Certainly, we know (from painful experience) that launching into a complex modeling enterprise requires more than a cookbook and a statistical package. It requires even more than a hope and a prayer. It requires an understanding of both the substantive questions and the methodology used to answer those questions. We hope that this volume will make both tasks more attainable.

ACKNOWLEDGMENTS

We thank the volume's external reviewers who generously volunteered their time to write detailed, thoughtful reviews. They are: Sanford Braver, Lisa Harlow, Robert Laforge, Robert MacCallum, Scott Maxwell, Steven Pitts, Michael Seltzer, and Alexander von Eye. We are especially grateful to our contributors, who went above and beyond the call of duty by not only dedicating enormous amounts of time to writing their chapters, but also by carefully reviewing the chapters of other contributors.

Thanks also are extended to Lisa Harlow, *Multivariate Applications* Series Editor, for encouraging us to prepare this volume and providing valuable direction and feedback throughout the process.

Finally, we thank NICHD and NIDA for support in preparation of this volume.

1

A Latent Curve Framework for the Study of Developmental Trajectories in Adolescent Substance Use

Patrick J. Curran
University of North Carolina, Chapel Hill

There is a great need to better link theoretical models and statistical models in studies of change over time, and this is especially evident in longitudinal studies of substance use and abuse. Historically, linking appropriate statistical models to developmental models has been difficult because the necessary statistical techniques had not yet been developed, or had been developed but were largely inaccessible to applied researchers. Although many of the conventional analytic techniques commonly used to test developmental research hypotheses perform exceptionally well given certain strict assumptions, these methods can be biased when applied under conditions in which the assumptions have been violated. These assumptions are often violated in longitudinal studies of substance use. A broad class of newly developed statistical techniques has recently become available that are much better suited for studying individual differences in development and change. One such approach is the structural equation modeling-based latent curve analysis. Latent curve analysis utilizes multiple indicator latent factors to estimate individual and group level components of growth over time and is the focus of the chapter. The chapter begins with a brief introduction to six research hypotheses about the developmental relation between adolescent substance use and peer substance use over time. A traditional cross-lagged model is used to evaluate these six research hypotheses using a large longitudinal data set, and the potential limitations of this modeling approach are discussed. An introduction is then provided to growth modeling in general and to latent curve analysis in particular. A series of latent curve models are then presented in detail that evaluate the six research questions of interest. The relative advantages and disadvantages of these models are discussed, and it is argued that latent curve analysis has much to offer the study of substance use over time.

There are a variety of theoretical models that have been developed to better understand the initiation and escalation of alcohol and drug use in children and adolescents (for a review, see Chassin, Presson, Sherman, & Curran, 1992; Maisto & Carey, 1985; Petraitis, Flay, & Miller, 1995; Sher, 1991). These models are typically used to generate specific research hypotheses that are amenable to empirical evaluation using some observed sample of observations and an appropriate statistical model. The sample results of the statistical models then provide some basis for making inferences back to the original theoretical model of interest. A critically important step in this process is the appropriate matching of the statistical model to the theoretical model that generated the research hypotheses to be evaluated. In empirical studies of adolescent substance use, the process of matching theoretical and statistical models is particularly difficult given the need to not only articulate theoretical causal mechanisms of influence, but to also clearly delineate theoretical expectations about individual differences in development and change over time. The selection of a statistical model that optimally evaluates the proposed research hypotheses directly depends upon the proper theoretical articulation of change over time.

For example, a developmental theory of substance use might generate a research hypothesis that describes the relation between characteristics of a school setting in the prediction of changes in adolescent alcohol use over time. Here, given that the behavior of multiple individuals within the school is of key interest, individual change in alcohol use over time might be best evaluated relative to all of the adolescents in the school. Change is thus conceptualized as the average difference between each individual within the group and the mean of that group over time. This type of change is sometimes referred to as *average group change*. Alternatively, a theory might generate a research hypothesis that describes the onset and developmental course of adolescent substance use during the transition from middle school to high school. Unlike the previous example, the change of interest here is *not* individual change relative to a larger group, but change within each individual adolescent. This type of change is sometimes referred to as *intraindividual,* or *within-person, change*. Finally, a theory might generate a research hypothesis that describes the relation between characteristics of the adolescent's peer group in the prediction of patterns of intraindividual change in adolescent substance use during the transition from middle school to high school. In this case, the theoretical model posits between-person(or *interindividual*) differences in peer group characteristics that are related to within-person (or *intraindividual*) change in substance use over time. This type of change might be considered a comprehensive model of interindividual differences in intraindividual development over time.

All three of these examples are valid questions about stability and change, but each approaches change from a slightly different perspective and thus each is able to make unique inferences about development over time. One of the many reasons why it is so important for developmental theories of substance use to clearly articulate what type of change is being posited is that the statistical model selected to empirically evaluate the research hypotheses will directly depend on precisely what type of change or developmental process is hypothesized to exist. If the theory fails to specify, or incorrectly specifies, the type of change that is under study, the statistical model selected to evaluate the research hypotheses may be inappropriate. The failure to properly match the statistical model to the theoretical model can significantly limit, if not wholly bias, the inferences that can be made about the theoretical model of interest (Curran, 1997).

There is a great need to better link theoretical models and statistical models in applied developmental research and studies of change over time, and this is especially evident in developmental studies of substance use. Historically, linking appropriate statistical models to developmental models has been difficult because the statistical techniques needed to evaluate the theoretical research hypotheses had not yet been developed, or had been developed but were largely inaccessible to applied researchers. Although many of the conventional analytic techniques historically used to test developmental research hypotheses perform exceptionally well given certain rather strict assumptions, these same methods can be severely biased when applied under research conditions in which the assumptions have been violated (Rogosa, 1995; Rogosa & Willett, 1985; Willett, 1988). These assumptions are commonly violated in the study of individual differences in developmental trajectories of substance use, and this can unnecessarily limit the utility of many conventional statistical methods when empirically examining the onset and escalation of substance use in children and adolescents.

A broad class of newly developed statistical techniques are becomingly increasingly available that are much better suited for studying individual differences in development and change. These recently developed methods are often referred to as *random coefficient models*, and the historical roots of these methods can be traced to the fields of biostatistics (e.g., Diggle, Liang, & Zeger, 1994; Laird & Ware, 1982; Rao, 1958), education (e.g., Bryk & Raudenbush, 1992; Burstein, 1980; Cronbach, 1976; Goldstein, 1987) and psychometrics (e.g., McArdle, 1988; McArdle & Epstein, 1987; Meredith & Tisak, 1984, 1990; Tucker, 1958). Latent curve analysis, a specific type of random coefficient model developed in the psychometric tradition, utilizes multiple indicator latent factors to estimate the fixed (or group level) and random (or individual level) components associated with individual differences in developmental trajectories over time. This is a powerful and flexible analytic technique that overcomes many of the challenges commonly encountered in the empirical study of adolescent substance use over time. Latent curve analysis is the primary focus of this chapter.

This chapter provides both a conceptual and a statistical introduction to latent curve analysis. This is accompanied by a detailed presentation of a series of latent curve models designed to study six research questions relating to adolescent and peer alcohol use previously addressed in Curran, Stice, & Chassin (1997). I begin with a brief introduction to three competing theoretical models of adolescent substance use relating to the role of peer group influences on individual behavioral development. Drawing from these three theories and from previous empirical research, I propose six specific research questions, followed by a description of the sample of adolescents and the measures that are used to empirically evaluate these six questions. I then present the findings from a more conventional longitudinal statistical model that serve both as a logical starting point for the analysis as well as providing a comparison to the later latent curve models. After describing the potential limitations of this conventional statistical model, I provide a general conceptual introduction to the modeling of individual differences in growth, followed by a more technical introduction specifically related to latent curve analysis. This is followed by the presentation of a series of detailed latent curve models designed to evaluate the six research questions. These models are constructed in increasing complexity, and each is explicitly tied back to the six proposed research questions. Finally, I conclude with a discussion of further extensions and potential limitations of the latent curve model and directions for future research.

A THEORETICAL MODEL OF ADOLESCENT SUBSTANCE USE: THE ROLE OF DEVIANT PEER AFFILIATIONS

Of the influences found to be related to adolescent substance use, peer substance use is consistently one of the strongest (Brook, Brook, Gordon, Whiteman, & Cohen, 1990; Hawkins, Catalano, & Miller, 1992). These findings have prompted many to conclude that peer substance use is a key proximal determinant of subsequent adolescent substance use (Oetting & Beauvais, 1986, 1987; Swaim, Oetting, Edwards, & Beauvais, 1989). This theoretical model is sometimes referred to as a *peer socialization* model such that adolescents who affiliate with substance-using friends are socialized to the behavior of the group and are thus more likely to use substances themselves. However, an alternative explanation is that adolescents who initiate illicit substance use as a function of other risk factors seek out a peer group that more closely matches their newly acquired behavior and attitudes (Farrell, 1994; Farrell & Danish, 1993). This theoretical model is sometimes referred to as a *peer selection* model such that adolescents who are already using substances are choosing to affiliate with a substance-using group. A third explanation combines these two processes such that adolescents tend to select friends who are similar to themselves in their substance use but are also susceptible to pressures of conformity from these same selected friends (Bauman & Ennett, 1994; Fisher & Bauman, 1988; Kandel, 1985). This theoretical model is sometimes referred to as a *bidirectional influences* model that combines elements of both the peer selection and the peer socialization models.

Previous empirical studies of the relation between peer substance use and adolescent substance use resulted in conflicting findings both in support and in refutation of each of the three theoretical models of peer influence (e.g., Bauman & Ennett, 1994; Farrell & Danish, 1993; Swaim et al., 1989). However, all of these previous studies used more conventional statistical models to evaluate rather complex theoretical questions about interindividual differences in intraindividual developmental trajectories of alcohol use over time. All of these previous studies provide important information about the relation between peer and adolescent substance use. However, one potential explanation for the discrepant findings may be related to a less than optimal match between the theoretical model of substance use and the statistical model used to evaluate the research hypotheses. The goal of this chapter is to try to clearly articulate the theoretically derived research questions and to select a statistical model that most closely corresponds to the theoretical model under consideration. A variety of latent curve models is used to explore these three competing models of peer influences in adolescent substance use.

The first step in the study of these competing theories is to articulate clearly the specific research questions of interest by drawing directly on the theoretical models as well as on previous empirical findings in this area. The six research questions of interest are:

Question 1 (Q1): What are the characteristics of individual differences in developmental growth trajectories of adolescent alcohol use over time?

Question 2 (Q2): Are individual differences in developmental growth trajectories of adolescent alcohol use systematically related to the child's age, gender, or parental alcoholism diagnosis?

Question 3 (Q3): What are the characteristics of individual differences in developmental growth trajectories of peer alcohol use over time?

Question 4 (Q4): Are individual differences in developmental growth trajectories of peer alcohol use systematically related to the child's age, gender or parental alcoholism diagnosis?

Question 5 (Q5): How are individual differences in developmental trajectories of adolescent alcohol use related to individual differences in developmental trajectories of peer alcohol use?

Question 6 (Q6): Are earlier levels of adolescent alcohol use predictive of later developmental trajectories in peer alcohol use, and are earlier levels of peer alcohol use predictive of later developmental trajectories of adolescent alcohol use?

Questions 1 and 3 are based on previous empirical epidemiological findings describing the developmental trajectories of alcohol use in adolescence (e.g., Schulenberg, O'Malley, Bachman, Wadsworth, & Johnston, 1996); these questions address intraindividual differences in the development of adolescent and peer alcohol use over time. Questions 2 and 4 are based on both empirical and theoretical results that relate individual differences in age, gender, and parental alcoholism diagnosis to adolescent alcohol use and peer affiliations (e.g., Chassin, Curran, Hussong, & Colder, 1996; Sher, 1991); these questions address interindividual differences in intraindividual development over time.

Finally, Questions 5 and 6 are drawn directly from the three competing models of peer influences that explore the interrelations between adolescent alcohol use and peer alcohol use (e.g., Bauman & Ennett, 1994; Farrell & Danish, 1993; Oetting & Beauvais, 1987). Like Questions 2 and 4, Questions 5 and 6 also address interindividual differences in intraindividual development over time. Taken together, these six empirically and theoretically derived research questions address the developmental course, correlates, and predictors of adolescent and peer alcohol use over time.

Description of Data Set

Data were drawn from the Adolescent and Family Development Project (AFDP) to evaluate the six proposed research questions. AFDP is a longitudinal study of a large sample of children and their parents. Approximately one half of the sample consisted of children from families in which one or both of the parents were diagnosed as alcoholic; the remainder of the sample consisted of demographically matched control children from families in which neither of the parents were diagnosed as alcoholic. AFDP was funded by a grant from the National Institute on Drug Abuse to Laurie Chassin and Manuel Barrera at Arizona State University. Details about the design, recruitment strategies, and measures are presented in Chassin, Rogosch, & Barrera (1991), Chassin, Barrera, Bech, & Kossak-Fuller (1992), and Curran et al. (1997).

Subjects. The total sample at the first wave of measure (Time 1) consisted of 454 adolescents aged 10.5 to 15.5 years ($M = 12.7$, $SD = 1.45$) and their parents. Children of alcoholics (COAs, $N = 246$) had at least one biological and custodial alcoholic parent, and controls (CONs, $N = 208$) had no biological or custodial alcoholic parents. Of these 454 adolescents, 449 (99%) were interviewed at Time 2, 445 (98%) were interviewed at Time 3, and 442 (97%) were interviewed at all three time points. Because the current research questions focused on the relation between changes in adolescent alcohol use and changes in peer alcohol use, 74 adolescents were not included because they reported no substance use themselves *and* no peer substance use at any time point (e.g., complete abstainers).[1] Finally, five observations were found to be extreme outliers and exerting an inappropriate influence on the statistical models, and were subsequently dropped (see Curran et al., 1997, for details). The final sample used for the current analyses consisted of 363 families. The average adolescent age at Time 1 was 12.9 years, 56% were COAs, 48% were female, 25% were Hispanic, and 75% were non-Hispanic Caucasian.

Measures. Demographic variables consisted of *child age* (measured in years at Time 1) and *child gender* (male coded 1, female coded 0). Lifetime DSM-III diagnoses of *parental alcohol abuse or dependence* was a dichotomous variable reflecting the presence (coded 1) or absence (coded 0) of alcoholism diagnosis in either or both parents. Adolescents reported how many of their friends drank alcohol occasionally (one item) and how many of their friends drank alcohol regularly (one item) over the previous 12 months. Response options ranged from 0 (*none*) to 5 (*all*), and a single *peer alcohol use* score was calculated by summing the two items. Adolescents self-reported their frequency of consumption of beer/wine and hard liquor (2 items), frequency of consumption of 5 or more drinks in a row (1 item), and frequency of getting drunk (1 item) in the past 12 months. Response options ranged from 0 (*not at all*) to 7 (*every day*). A single *adolescent alcohol use* score was calculated by summing the four items. Correlations, standard deviations, means, skewness and kurtosis for all variables are presented in Table 1.1.

Conventional Fixed Effects Model

Prior to estimating the latent curve models, it is often helpful to begin by examining the data using more traditional analytical methods. This provides important insights into the general characteristics of the data, and also allows for a comparison between the conventional analysis and the latent curve models. To test the six research questions, I first used an auto-regressive, cross-lagged (ARCL) panel model (Dwyer, 1983). The

[1]These complete abstainers were not included in the analyses given the focus on studying the inter-relations among changes in adolescent alcohol use and changes in peer alcohol use, and this subgroup reported no use on either construct at any time point. For comparative purposes, all models were re-estimated including the complete abstainers. Although there were mean and variance differences in the growth trajectories, there were no differences in the substantive conclusions compared to the analysis without the abstainers.

TABLE 1.1

Means, Standard Deviations, Univariate Skewness and Zero-Order Correlations for all Predictor and Criterion Variables

	1.	*2.*	*3.*	*4.*	*5.*	*6.*	*7.*	*8.*	*9.*
1. Time 1 Adolescent Alcohol Use	1.00								
2. Time 2 Adolescent Alcohol Use	.68	1.00							
3. Time 3 Adolescent Alcohol Use	.50	.68	1.00						
4. Time 1 Peer Alcohol Use	.66	.58	.46	1.00					
5. Time 2 Peer Alcohol Use	.53	.65	.52	.65	1.00				
6. Time 3 Peer Alcohol Use	.38	.51	.61	.48	.58	1.00			
7. Age	.31	.31	.23	.48	.44	.29	1.00		
8. Gender	.01	.06	.10	-.09	-.02	-.13	.14	1.00	
9. COA Status	.12	.17	.20	.09	.12	.16	.12	-.15	1.00
Mean	1.36	2.12	3.18	1.33	1.75	2.48	12.91	0.52	0.56
Standard Deviation	2.81	3.98	4.79	1.74	1.82	2.01	1.40	0.50	0.49
Skewness	3.23	2.73	2.07	1.57	1.16	0.66	-.28	-.06	-.26

Note: Absolute values of correlations greater than $r = .19$ represent a per-test $p < .002$ and a familywise $p < .10$. Statistics based on $N = 363$.

most common form of the ARCL model is a fixed-effects, regression-based linear model for studying stability and change in a construct over time. These models can be estimated using a series of multiple regression analyses, but more typically ARCL models are estimated using a structural equation modeling (SEM) framework. The ARCL model estimates the stability of a construct over time by regressing later measures of a construct onto earlier measures of the same construct. For example, Time 1 alcohol use predicts Time 2 alcohol use, and Time 2 alcohol use in turn predicts Time 3 alcohol use. These across-time/within-construct predictions are often referred to as *stability coefficients*, or simply *stabilities*. Once these stabilities are estimated, then earlier measures of one construct are used to predict later measures of another construct. For example, Time 1 peer alcohol use predicts Time 2 adolescent alcohol use, above and beyond the effects of Time 1 adolescent alcohol use. This parameterization provides a test of whether earlier peer alcohol use predicts later adolescent alcohol use, above and beyond the effects of earlier alcohol use. In principal, ARCL models can be extended to include multiple time points across multiple constructs.

The current ARCL model was built in five progressive steps (see Curran et al., 1997, for details). First, the Time 1 measures of adolescent alcohol use and peer alcohol use were regressed on adolescent age, gender, and parent alcoholism diagnosis. Second, the stabilities between time-adjacent measures of adolescent alcohol use and peer alcohol use were estimated. Third, the errors of equations (or disturbances) of the repeated measures were correlated within-time/across-construct (e.g., Time 1 adolescent use was correlated with Time 1 peer use, etc.). Fourth, the four *cross-lagged* effects were estimated (the across-time/across-construct regression parameters). Each subsequent step was associated with a significant improvement in model fit assessed using nested chi-square tests. Finally, Lagrange multipliers suggested the post hoc inclusion of a direct effect from age to Time 2 peer alcohol use and from gender to Time 3 peer alcohol use. The final ARCL model was found to fit the observed data well ($\chi^2(14, N = 363) = 29.3$, $p = .01$; Comparative Fit Index [CFI] = .99; Tucker–Lewis Index [TLI] = .97; Root mean square error of approximation [RMSEA] = .055 with a 90% confidence interval [90CI] = .03, .08). The final model is shown in Fig. 1.1 and the EQS (Bentler, 1997) computer code is presented in Appendix A.

The ARCL model resulted in large and positive stability coefficients for both adolescent alcohol use and peer alcohol use suggesting that the average adolescent standing relative to the group mean on either construct at an earlier time point tended to be quite similar to the average adolescent standing relative to the group mean at a later time point. That is, on average, adolescents reporting alcohol use above the mean at Time 1 tended to report alcohol use above the mean at Time 2, and adolescents reporting alcohol use above the mean Time 2 tended to report alcohol use above the mean at Time 3. Note that, although research questions *Q1* and *Q3* addressed issues of *individual* change in developmental trajectories of adolescent and peer alcohol use, the ARCL model analyzed change relative to the *group* average over time. Because of this, it is quite difficult to make any meaningful inferences about research questions *Q1* and *Q3* based on these ARCL results.

Research questions *Q2* and *Q4* addressed the issue of systematic individual differences in developmental change in adolescent and peer alcohol use as a function of child age, gender, and parental alcoholism diagnosis. As with the previous questions, clear inferences about Questions 2 and 4 are difficult to make based on the ARCL model results. For example, older adolescents and COAs reported significantly ($p < .05$)

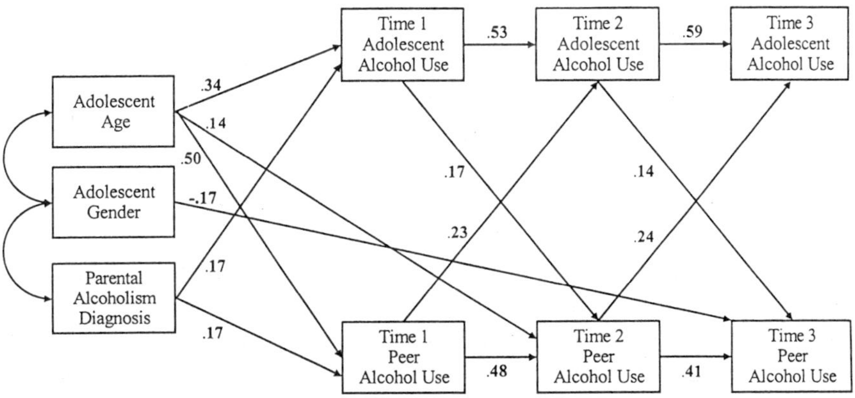

Fig. 1.1. Conventional auto-regressive cross-lagged panel model.
Note: Model fit was $\chi^2(14, N = 363) = 29.3$, $p = .01$, TLI = .97, CFI = .99, RMSEA = .055. Parameter estimates are standardized and were all significant ($p < .05$). Covariances among the within time disturbances of the repeated measures are not shown.

higher Time 1 adolescent alcohol use and higher Time 1 peer alcohol use, older adolescents reported higher Time 2 peer alcohol use, and males reported lower Time 3 peer alcohol use. Again these ARCL inferences are made with reference to average relative standing at a specific time point. So, parental alcoholism diagnosis was positively related to Time 1 adolescent and peer alcohol use, but in this modeling framework could not be used to predict continuous changes in alcohol use across Times 1, 2, and 3. This again severely limits our ability to draw meaningful inferences about the hypothesized research questions from the ARCL model results.

Finally, research question *Q5* inquired about the correlation between individual differences in developmental trajectories of adolescent and peer alcohol use, and research question *Q6* explored the existence of systematic individual differences in developmental trajectories as a function of earlier levels of one construct predicting later changes in the other construct. Unfortunately, the ARCL model allows no inferences to be made about question *Q5* in that this model is not estimating characteristics of individual growth, but instead of average relative standing across time. With respect to *Q6*, the ARCL model resulted in large and significant prospective bidirectional effects between adolescent alcohol use and peer alcohol use. That is, earlier peer use positively predicted later adolescent use, above and beyond earlier adolescent use; similarly, earlier adolescent use positively predicted later peer use, above and beyond earlier peer use. These prospective bidirectional paths were significant across both time lags. Again, because these prospective paths represent average relative standing between two discrete time points, and not continuous change across all three time points, these results only partially inform research question *Q6*.

In sum, the results from the ARCL model suggest that there is significant stability of adolescent alcohol use and peer alcohol use over time, that age, gender, and parental alcoholism diagnosis are related to time specific measures of alcohol use, and that there is support for a prospective bidirectional relation between adolescent and peer alcohol use such that earlier measures of one construct predicted later measures on the other construct. However, when comparing how developmental change was conceptualized by the three competing theories with how change is actually being evaluated in the statistical model, there seems to be a poor match between the theoretical and the statistical models of interest. To better understand this mismatch, we need to more closely consider the behavior of the ARCL model under conditions of systematic development over time. Rogosa (1988, 1995), Rogosa & Willett (1985), and Willett (1988) present excellent discussions of the general assumptions and related limitations of these types of models, so I will not review these issues fully here. I will, however, comment on several of the more problematic issues relating to the particular research questions at hand.

First, note that a single parameter characterizes the stability of each construct over two time points for *all* subjects in the sample. For example, the standardized regression parameter estimate predicting Time 2 alcohol use from Time 1 alcohol use for the current sample is $\hat{\beta} = .53$; this single value is meant to capture the relation between Time 1 and Time 2 alcohol use for all $N = 363$ subjects in the sample. Characterizing this prospective relation using a single value for all subjects is likely not feasible given the expected heterogeneity in change within the sample, and this certainly does not tell us much about the characteristics of growth in alcohol use over time. We will see that it is often extremely important to allow parameter values describing change over time to vary freely *across* individuals instead of having one value fixed for all individuals.

Second, the ARCL model calculates individual change between two time points relative to the group, *not* relative to the individual. This can be of profound importance depending upon how a particular theory construes change over time. For example, the average alcohol use for all $N = 363$ adolescents at Time 1 and Time 2 was 1.4 and 2.1, respectively. So, as a group, average alcohol use increased by .70 between Time 1 and Time 2. Say that Adolescent 1 reported alcohol use of 1.75 at both the first and the second time points. Given that the group mean increased .7 units over the two time points, but the individual reported equal rates of use at both time points, conventional statistical models would consider this individual to be making a *negative* contribution to the average covariance between the two time points. This is because Adolescent 1 reported alcohol use *above* the mean at Time 1, but *below* the mean at Time 2, even though their *own* alcohol use did not change. In contrast, say that Adolescent 2 reported alcohol use of .7 at Time 1 and 1.4 at Time 2. Although reporting a .70 increase in alcohol use between the two time points, this individual would subsequently make a contribution of *perfect stability* to the average covariance between the two time points because Adolescent 2 increased at a rate *equal* to that reported by the overall group. So, although Adolescent 2 reported an increase in their own alcohol use between the two time points, conventional statistical models consider Adolescent 2's alcohol use to be stable over time. As can be seen, it is of great importance for the researcher to articulate clearly whether research questions of interest relate to change *relative to the group*, or change *relative to the individual*. Conventional statistical models typically construe change relative to the group, but developmental theory is often interested in change relative to the individual.

A final limitation of the ARCL model is that we often assume from a theoretical perspective that the onset and escalation of alcohol use tend to follow a rather continuous developmental trajectory over time, particularly during childhood and adolescence. That is, substance use is typically characterized by some point of initiation in early adolescence, followed by an accelerated rate of escalation in middle to late adolescence, peaking and gradual decrease in use in the early 20s, and, finally, the development of some stable level of use in the middle or late 20s (Muthén & Muthén, in press). Given the existence of such a developmental trajectory, examining growth as a series of time adjacent comparisons (Time 1 vs. 2, Time 2 vs. 3) is artificially carving up the continuous trajectory into a series of two-timepoint snapshots of change (Willett, 1988). This further highlights the problem of using a statistical model to test a research hypothesis that does not optimally correspond to the theoretical model that gave rise to the research hypothesis.

Because of these limitations, and several others not detailed here, conventional approaches to longitudinal data analysis tend not to be well suited for studying interindividual differences in intraindividual developmental change over time. Let's now consider an alternative analytic approach that might allow for stronger empirical tests of our developmental research hypotheses.

Models of Individual Growth

How can we work toward overcoming the limitations of the more conventional statistical models? One solution is to approach the question of development and change from an entirely different perspective. The ARCL model predicted Time 3 alcohol use from Time 2 alcohol use, and Time 2 alcohol use from Time 1 alcohol use. Say instead that we were to use the three observed repeated measures of alcohol use to estimate a *single* underlying true growth trajectory for each person across all three time points. An example of this is displayed in Fig. 1.2 for a single subject drawn from the sample of $N = 363$.

Instead of examining the time-adjacent relations of the alcohol use measures between Times 1 and 2, and Times 2 and 3 (as was done in the ARCL model), we have estimated a single continuous line that "best fits" the three observed measures. This line of best fit can be thought of in much the same way we would fit a regression line to a scatter plot of data points between two measures. However, in this case, the x variable is time (where x equals 0, 1, and 2), the y variable is alcohol use, and we consider one subject at a time. We consider this line of best fit as an estimate of the *individual's true growth trajectory* of alcohol use over time.

Whereas three repeated measures were used to estimate the growth trajectory, the trajectory is completely characterized by just two pieces of information: a starting point (or intercept, which we will refer to as α_i to denote the α value specific to subject i) and a rate of change over time (or slope, which we will refer to as β_i to denote the β value specific to the same subject i). The three observed repeated alcohol use measures

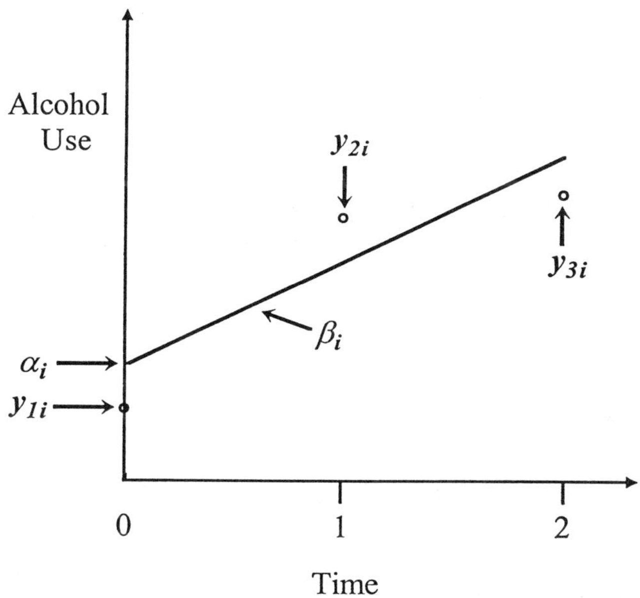

Fig. 1.2. Hypothetical growth trajectory fit to three observed measures of alcohol use. Note: α represents the intercept of the trajectory and β represents the individual slope of the trajectory.

for subject i can now be expressed as an additive function of the individual intercept and slope estimates such that:

$$y_{it} = \alpha_i + \beta_i a_t + \varepsilon_{it} \quad (1.1)$$

where y_{it} represents the observed alcohol use measure for person i at timepoint t, α_i represents the true intercept of the growth trajectory for person i, β_i represents the true slope of the growth trajectory for person i, a_t represents the value of time at timepoint t (e.g., 0, 1, or 2), and ε_{it} represents the time specific residual for person i at time t. Equation 1.1 is sometimes referred to as the *Level 1,* or *within person* model.

Say that we now fit trajectories for a sample of four individuals as is shown in Fig 1.3.

Now α_i and β_i from Equation 1.1 are random variables, and variation in these variables across individuals can be expressed as

$$\alpha_i = \mu_\alpha + u_{\alpha i} \quad (1.2a)$$

$$\beta_i = \mu_\beta + u_{\beta i} \quad (1.2b)$$

Equations 1.2a and 1.2b are sometimes referred to as the *Level 2,* or the *between person,* model. These equations highlight that the individual intercept and slope values describing the developmental trajectories are now the dependent variables of most interest to the model, and not the individual observed repeated observations of y over time.

The Level 2 model (Eqns. 1.2a and 1.2b) can be substituted into the Level 1 model (Equation 1.1) to result in a combined model

$$y_{it} = (\mu_\alpha + u_{\alpha i}) + a_t\,(u_\beta + \mu_{\beta i}) + \varepsilon_{it} \quad (1.3)$$

and simple rearrangement of terms results in

$$y_{it} = (\mu_\alpha + a_t\mu_\beta) + (u_{\alpha i} + a_t u_{\beta i} + \varepsilon_{it}) \quad (1.4)$$

Equation 1.4 clarifies that the observed repeated measures of y can be expressed as an additive combination of a *fixed* component of growth (e.g., $\mu_\alpha + a_t\mu_\beta$) and a *random* component of growth (e.g., $u_{\alpha i} + a_t u_{\beta i} + \varepsilon_{it}$). These parameters will combine to provide important information about our research questions of interest. The two fixed effects of growth are

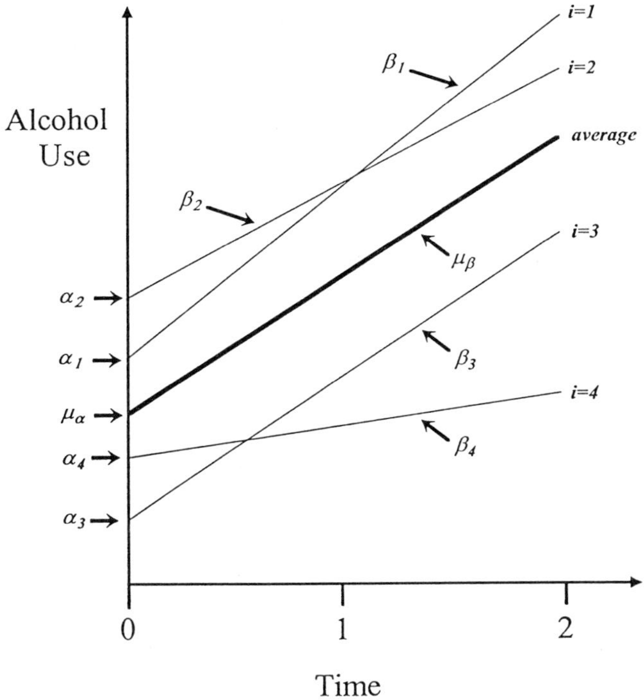

Fig. 1.3. Four fitted growth trajectories.
Note: Light lines represent individual trajectories and the heavy line represents the mean growth; α represents the individual intercepts, β represents the individual slopes, μ_α represents the mean intercept, and μ_β represents the mean slope.

μ_α (the mean intercept of the growth trajectory) and μ_β (the mean rate of change of the growth trajectory). The random effects of growth include the variance around the mean intercept ($var(u_{\alpha i}) = \sigma^2_\alpha$), the variance around the mean slope ($var(u_{\beta i}) = \sigma^2_\beta$), the covariance between the intercept and slope ($cov(u_{\alpha_i}, u_{\beta_i}) = \sigma_{\alpha\beta}$), and the time specific residual variance ($var(\varepsilon_{it}) = \sigma^2_\varepsilon$).

The model described in Eqn. 1.4 is sometimes referred to as an *unconditional* growth model (Bryk & Raudenbush, 1987); that is, we have estimated the mean starting point and mean rate of change of the developmental trajectories (the fixed effects) as well as the individual variability in both the starting point and rate of change (the random effects). Although this unconditional model allows for estimation of the general characteristics of individual differences in growth, we have made no attempt to *predict* these individual differences. We do not know what factors might differentiate children who report drinking at a high rate at the initial assessment period from those who do not, or what factors might differentiate children who report steep increases in alcohol use over time from children who do not. These are questions of *inter*individual, or *between person*, differences in *intra*individual change over time. To explore what factors are predictive of individual differences in starting point and rates of change over time, we must move to a *conditional* growth model.

The conditional growth model incorporates Equation 1.1 as stated above, but Equations 1.2a and 2b are expanded to include explanatory variables such that

$$\alpha_i = \mu_\alpha + \gamma_1 x_{1i} + \gamma_2 x_{2i} + u_{\alpha i} \quad (1.5a)$$

$$\beta_i = \mu_\beta + \gamma_3 x_{1i} + \gamma_4 x_{2i} + u_{\beta i} \quad (1.5b)$$

where x_{1i} and x_{2i} are two explanatory variables of interest (say gender and parental alcoholism diagnosis), and γ_1, γ_2, γ_3, and γ_4 are fixed regression parameters that relate the explanatory variables to the intercept and slope components of growth. (These are fixed parameters because they do not vary across individual). Whereas the unconditional model allowed us to examine general characteristics of individual differences in development over time (needed to examine questions *Q1*, *Q3*, and *Q5*), the conditional model allows us to predict individual differences in development as a function of our set of explanatory variables (needed to examine questions *Q2*, *Q4*, and *Q6*).

Summary. Using our six theoretically and empirically derived research questions as a guide, we conceptually know precisely what type of information we would like to extract from our observed data. First, we would like to use the observed repeated measures of adolescent and peer alcohol use to estimate an underlying true growth trajectory for each individual in the sample. Second, we would like to compute summary statistics for these individual trajectories across the entire sample to better understand characteristics of individual differences in growth over time. Third, we would like to incorporate explanatory variables to examine predictors of individual differences in growth over time. Finally, we would like to build a multivariate model that will estimate the interrelations between developmental trajectories of adolescent alcohol use and developmental trajectories of peer alcohol use over time.

Although we can clearly articulate the types of analytical models we require to optimally evaluate our research hypotheses, in the past it has been quite difficult to statistically estimate these types of models in an appropriate way. Fortunately, there are a variety of statistical modeling techniques that have recently become available that allow for the estimation of the fixed and random components of growth. Recent developments include latent curve analysis, hierarchical linear models, general estimating equations, general mixed models, and Bayesian growth models. A thorough comparison of these various techniques is beyond the scope of this chapter (but see MacCallum, Kim, Malarkey, & Kiecolt-Glaser, 1997, and Willett & Sayer, 1994, for excellent comparisons of several of these modeling techniques). I will focus explicitly on the use of latent curve analysis.

Latent Curve Analysis

Structural equation modeling (SEM) is a powerful analytic framework that allows for the simultaneous estimation of the relations between a set of observed variables and a smaller set of underlying latent constructs (the confirmatory factor or measurement model), as well as the relations among the latent constructs themselves (the structural model); (Bentler, 1980, 1983; Jöreskog, 1971a, 1971b; Jöreskog & Sörbom,1979). Latent curve analysis (LCA) is a highly structured type of structural equation model. Whereas SEM typically (but not necessarily) incorporates only information about the covariance structure of the observed measures, LCA uses information about both the covariance structure and the mean structure of the observed measures. Conceptually, LCA considers the observed repeated measures over time to be fallible indicators of an unobserved true growth trajectory. The observed repeated measures are thus used to define one or more underlying latent growth factors using the SEM framework. The fixed and random components of the growth trajectory are then estimated via the means and variances of the latent growth factors. These latent growth factors can then be regressed on one or more exogenous variables to examine predictors of individual differences in change over time. The technical developments of latent curve analysis are presented in McArdle (1988, 1989, 1991), McArdle &

Epstein (1987), Meredith & Tisak (1984, 1990), and Muthén (1991, 1993). Applied substantive examples of latent curve models include Curran, Muthén, & Harford (1998), Curran et al. (1997), Duncan, Duncan, Alpert & Hops (1997), MacCallum et al. (1997), and Stoolmiller (1994).

We will now use the latent curve models to estimate the fixed and random components of growth described above in order to address our six research questions of interest. Recall that the first research question (*Q1*) seeks to describe the characteristics of intra-individual differences in developmental trajectories in adolescent alcohol use. We will begin with the estimation of an unconditional growth model in which we examine the characteristics of the growth process prior to incorporating explanatory variables to predict growth. Using the structural modeling framework, two latent factors are defined to represent the intercept and slope of the alcohol use growth trajectory. The factor loadings relating the three observed alcohol use measures to the intercept factor are all fixed to 1.0 to define the starting point of the alcohol use growth trajectory. The factor loadings relating the observed repeated measures to the slope factor are a combination of fixed or free loadings that best capture the functional form of the growth trajectory over the three time points. The initial approach is typically to fix the factor loadings to 0, 1, and 2 to represent straight-line growth.[2]

The means of the two latent factors are freely estimated, but the intercepts of the observed repeated measures are fixed to zero. The observed mean structure among the repeated measures is thus reproduced completely through the means of the latent factors. The sample estimate of the mean of the latent intercept factor ($\hat{\mu}_\alpha$, denoted by a "hat" to indicate a sample estimate of a population parameter) represents the mean initial status of the alcohol use growth trajectory; the sample estimate of the variance of the latent intercept factor ($\hat{\sigma}^2_\alpha$) represents the

[2]One difficult issue to resolve in latent curve modeling is how to best define the factor loadings relating the observed repeated measures to the latent slope factor. In this example, I have fixed these loadings to values of 0, 1, and 2 to define linear growth. One advantage of the latent curve approach is that the adequacy of these loadings can be formally tested using nested group comparisons. However, given that only three assessment periods were available here, linear growth was the most complicated functional form of growth that could be estimated. A minimum of four time points is necessary to test higher order functions (e.g., quadratic polynomials). There are several approaches that can be used to define different growth functions, but a full discussion of these issues are beyond the scope of this chapter. McArdle (1989, 1991) and Willett & Sayer (1994) provide excellent discussions of these issues.

individual variability in initial levels of use. Similarly, the sample estimate of the mean of the latent slope factor ($\hat{\mu}_\beta$) represents the mean slope of the alcohol use trajectory, and the sample estimate of the variance of the latent slope factor ($\hat{\sigma}^2_\beta$) represents individual variability in rates of change in alcohol use over time.

Research Question *Q1*: Characteristics of Individual Differences in Developmental Trajectories of Adolescent Alcohol Use

To examine research question *Q1*, an unconditional latent curve model was estimated for adolescent alcohol use. Model 1 is presented in Fig. 1.4 and the computer code is presented in Appendix B.[3] To estimate this model, the three repeated measures of alcohol use were defined to load on both the latent intercept factor and the latent slope factor by fixing the factor loadings relating the observed measures to the underlying latent factors to prespecified values. The factor loadings relating the three adolescent alcohol use measures were fixed to 1.0 on the intercept factor, and 0, 1.0, and 2.0 on the slope factor. The means of the two latent growth factors (not shown in Fig. 1.4) were freely estimated whereas the means of the observed repeated measures were not. Finally, variances were estimated for both the observed measures and for the two latent growth factors. It is the variance in the growth factors that infers the presence or absence of individual differences in growth over time. The model shown in Fig. 1.4 fit the observed data well ($\chi^2(1, N = 363) = 2.1$, $p = .15$, TLI = .99, CFI = .99, RMSEA = .055 with 90CI = 0, .16). Together, these various measures suggest a close fit of the model to the observed data.

Once we determined that there is an adequate overall fit of the model to the observed data, we can now evaluate the eight estimated model parameters: two means (one for the intercept factor and one for the slope factor), five variances (three error variances for the repeated measures and two variances for the growth factors), and one covariance (between the intercept and slope factor). First, the means for the

[3]Although I used EQS to estimate all of the latent curve models here, any SEM software package that can estimate both mean and covariance structures can equivalently be used to estimate the latent curve models (e.g., Amos, CALIS, Mplxs, LISREL, Mx, etc.).

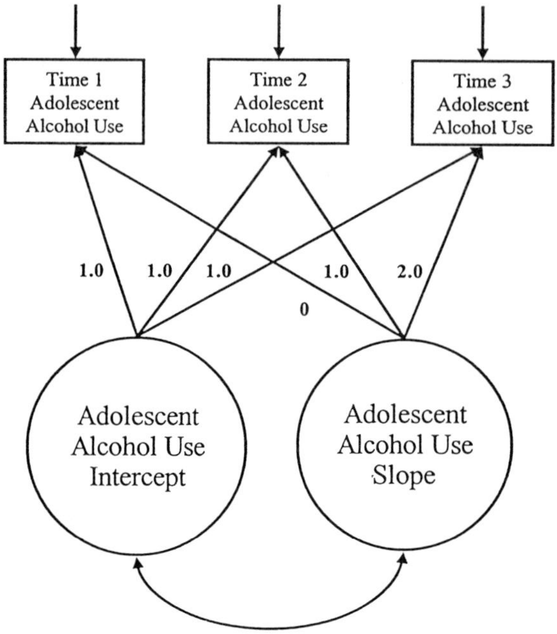

Fig. 1.4. Unconditional latent curve model for three repeated measures of adolescent alcohol use.
Note: Model fit was $\chi^2(1, N = 363) = 2.1$, $p = .15$, TLI = .99, CFI = .99, RMSEA = .055. Factor loadings are fixed to shown values; estimated parameters include five variances (one for each repeated measure, one for each latent factor), one covariance (between the two latent factors), and two means (one for each latent factor).

intercept and slope factors were estimated to be $\hat{\mu}_\alpha = 1.36$ and $\hat{\mu}_\beta = .88$, respectively, both of which significantly differed from zero ($p < .05$). This indicated that the estimated mean starting point of adolescent alcohol use was 1.36, and the estimated mean rate of true change was .88 units per year. Next, there were significant variance estimates for both the intercept ($\hat{\sigma}^2_\alpha = 7.9$) and slope ($\hat{\sigma}^2_\beta = 3.2$) factors, indicating that there was substantial individual variability around both the mean starting point and mean rate of change over time. Finally, there was a marginally significant ($p = .10$) estimate for the covariance between the intercept and slope factors ($\hat{\sigma}_{\alpha\beta} = -.54$)that was standardized to a correlation of $r = -.11$, indicating that adolescents reporting higher initial levels of alcohol use tended to report slightly smaller rates of increase in alcohol use over time.

In sum, the unconditional growth model indicated that the group reported a significant amount of alcohol use at the initial starting point, the group reported significant increases in alcohol use over the three time points, and there was significant individual variability in both the starting point and rates of change over time. These results are critical information needed to fully evaluate research question *Q1*. Note the greater depth and dynamism of information resulting from the latent curve results relative to those obtained from the ARCL stability coefficients estimated from the very same data. This reflects the additional information that can be extracted from the observed data using a growth modeling approach to the study of change over time.

Research Question *Q2*: Predicting Individual Differences in Developmental Trajectories of Adolescent Alcohol Use

Given the presence of variability in intraindividual differences in the developmental growth trajectories of adolescent substance use (that is, the significant variance estimates in the intercept and slope factors), we are now interested in trying to model this variance using additional explanatory variables. This allows us to explore predictors of individual developmental trajectories as was posited in research question *Q2*. To model the observed variability in the intercept and slope factors, we regress the two latent growth factors on three exogenous variables: age, gender, and parental alcoholism diagnosis. Model 2 is presented in Fig. 1.5 and the computer code is presented in Appendix C.

This conditional growth model was estimated and was found to fit the data quite well ($\chi^2(4, N = 363) = 5.5$, $p = .24$, TLI = .99, CFI = .99, RMSEA = .033 with 90CI = 0, .09). Of key interest here are the sample estimates of the regression parameters linking the three exogenous variables to the two growth factors because these estimates capture potential predictability of individual differences in development as a function of the explanatory variables. Examination of significant ($p < .05$) *z*-ratios indicated that older children and children of alcoholics reported higher initial levels of alcohol use and that older children, boys, and children of alcoholics reported steeper increases in alcohol use over time. Finally, even with the inclusion of the three explanatory variables, there remained significant variability in both the intercept and slope

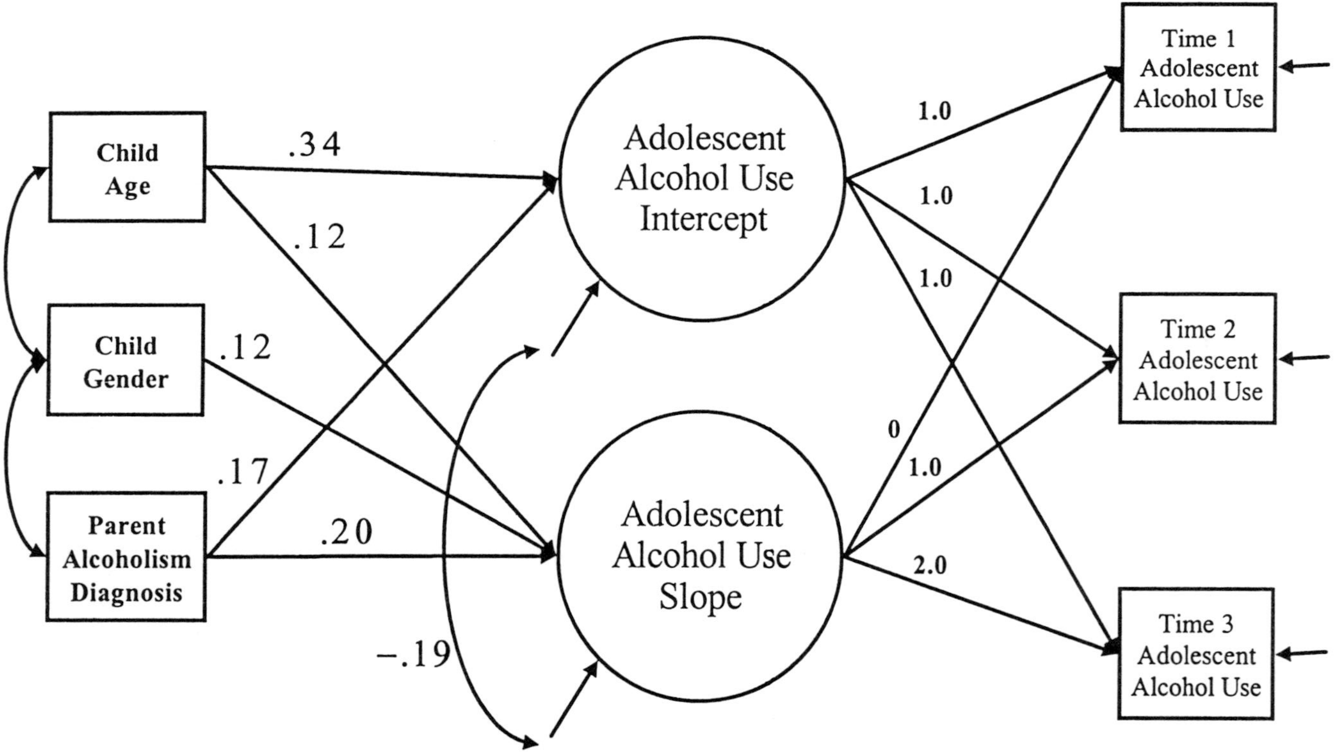

Fig. 1.5. Conditional latent curve model for adolescent alcohol use regressed on adolescent age, gender, and parental alcoholism diagnosis. Note: Model fit was $\chi^2(4, N = 363) = 5.5$, $p = .24$, TLI = .99, CFI = .99, RMSEA = .033. Only significant paths ($p < .05$) are shown. All parameter estimates are standardized.

factors, indicating that additional explanatory variables are necessary to fully model the observed individual differences in growth over time.

Thus far, we have learned a great deal about the characteristics of growth in adolescent alcohol use over the three time points, and in the prediction of growth in alcohol use from age, gender, and parental alcoholism diagnosis. Recall that one of the key goals of these analyses is to explore the relation of the child's peer's alcohol use and the child's own alcohol use (research questions *Q5* and *Q6*). An initial step might be to simply add the Time 1 measure of peer alcohol use to the conditional growth model containing the three exogenous variables we just estimated. Model 2 could be easily modified to incorporate Time 1 peer use as a fourth exogenous variable, and then the intercept and growth factors could be regressed on Time 1 peer use. However, our theories predict that peer use itself is developing systematically over time, so considering just the Time 1 measure of peer use in the presence of systematic change in peer use over time could be unnecessarily limiting. We thus must also examine growth in peer use prior to moving toward exploring research question *Q5*.

Research Question *Q3*: Characteristics of Individual Differences in Developmental Trajectories of Peer Alcohol Use

We need to retrace our steps and examine an unconditional growth model for peer alcohol use, which is identical in form to Model 1 but with the three peer alcohol use measures instead of the three adolescent alcohol use measures. This model was estimated and fit the observed data well ($\chi^2(1, N = 363) = 4.5$, $p = .03$, TLI = .97, CFI = .99, RMSEA = .09 with 90CI = .02, .19). A significant mean estimate for the intercept factor ($\hat{\mu}_\alpha = 1.3$) reflected that the estimated mean starting point for peer alcohol use was 1.3; a significant mean estimate of the slope factor ($\hat{\mu}_\beta = .55$) reflected that the estimated mean rate of change for peer alcohol use was .55 units per year. Further, large and significant variance estimates in both the intercept factor ($\hat{\sigma}^2_\alpha = 2.5$) and slope factor ($\hat{\sigma}^2_\beta = .40$) reflected the presence of individual variability in both mean starting point and mean rate of change over time. Finally, a significant correlation of r = .38 was found between intercept and slope, indicating that adolescents with lower initial levels of peer use tended to report steeper increases in peer use over time. Now we can extend this model to examine potential predictors of interindividual differences in these intraindividual developmental trajectories.

Research Question *Q4*: Predicting Individual Differences in Developmental Trajectories of Peer Alcohol Use

To evaluate question *Q4* that posited predictors of individual differences in peer alcohol use, the unconditional peer use model was regressed on the three measures of age, gender and parental alcoholism diagnosis. This model fit the observed data well ($\chi^2(4, N = 363) = 12.2, p = .02$, TLI = .94, CFI = .98, RMSEA = .08 with 90CI = .03, .13); significant parameter estimates indicated that older children and children of alcoholics reported higher initial levels of peer alcohol use, and that younger children reported steeper increases in peer alcohol use over time. As was found with adolescent alcohol use, significant variability remained in both the intercept and slope factors after inclusion of the three explanatory variables suggesting additional variables are needed to fully understand individual differences in growth in peer alcohol use over time.

Research Question *Q5* & *Q6*: Multivariate Relations Between Individual Differences in Developmental Trajectories of Adolescent Alcohol Use and Developmental Trajectories of Peer Alcohol Use

The four latent curve models presented previously are considered *univariate* given that change in only one construct was considered, even though multiple measures were taken on that one construct. We now create a *multivariate* latent curve model by simultaneously estimating the two unconditional univariate growth models of adolescent alcohol use and peer alcohol use, and regressing both growth processes on the exogenous measures of age, gender, and parental alcoholism diagnosis. This allows us to examine research questions *Q5* (the relation between developmental trajectories of adolescent and peer alcohol use over time) and *Q6* (earlier levels on one construct predicting later developmental change on the other construct). It is important to note that the more complicated these growth models become, the greater variety of ways there are to approach the model building strategy. There is not necessarily a right or wrong way to build and probe these models, but whatever steps are taken should be clearly articulated, well justified, and closely guided by the theoretical questions of interest. I follow these guidelines in the construction of Model 3.

I define the covariance structure among the latent growth factors such th factor of adolescent alcohol use is regressed on the intercept factor of peer alcohol use, and the slope factor of peer alcohol use is regressed on the intercept factor of adolescent alcohol use. This model represents the initial a priori multivariate growth model. Lagrange multipliers are then examined for potential estimation of additional parameters in the prediction of the two slope factors. See MacCallum (1986) and Kaplan (1990) for thorough discussions of post hoc model modifications, and Curran et al. (1997) for further details about the model-building process.

Model 3 was estimated and found to fit the observed data moderately well ($\chi^2(16, N = 363) = 36.9$, $p = .001$, TLI = .97, CFI = .98, RMSEA = .06 with 90CI = .03, .08). Examination of significant Lagrange multipliers ($p < .01$) suggested the addition of two parameters: gender and COA status predicting the adolescent alcohol use slope factor. The inclusion of these two regression parameters were associated with a significant improvement in model fit and the resulting model fit the data well ($\chi^2(14, N = 363) = 25.4$, $p = .03$, TLI = .98, CFI = .98, RMSEA = .04 with 90CI = .01, .08; see Fig. 1.6 for the model diagram and Appendix D for the computer code). Results indicated that older children and children of alcoholics reported both higher initial levels of adolescent alcohol use and higher initial levels of peer alcohol use compared to younger children and children of nonalcoholics. In addition, boys and children of alcoholics reported steeper increases in adolescent alcohol use over time compared to girls and children of nonalcoholics. After the inclusion of the three explanatory variables, there remained large and positive correlations between both the intercept factors and both the slope factors. This suggests that a child's initial status on adolescent alcohol use was quite similar to his or her initial status on peer alcohol use, and a child's rate of change on adolescent alcohol use was quite similar to his or her rate of change on peer alcohol use. These findings provide insight into research question *Q5* and suggest that there is a rather high degree of similarity between individual differences in developmental trajectories of adolescent alcohol use and developmental trajectories of peer alcohol use over the three time points.

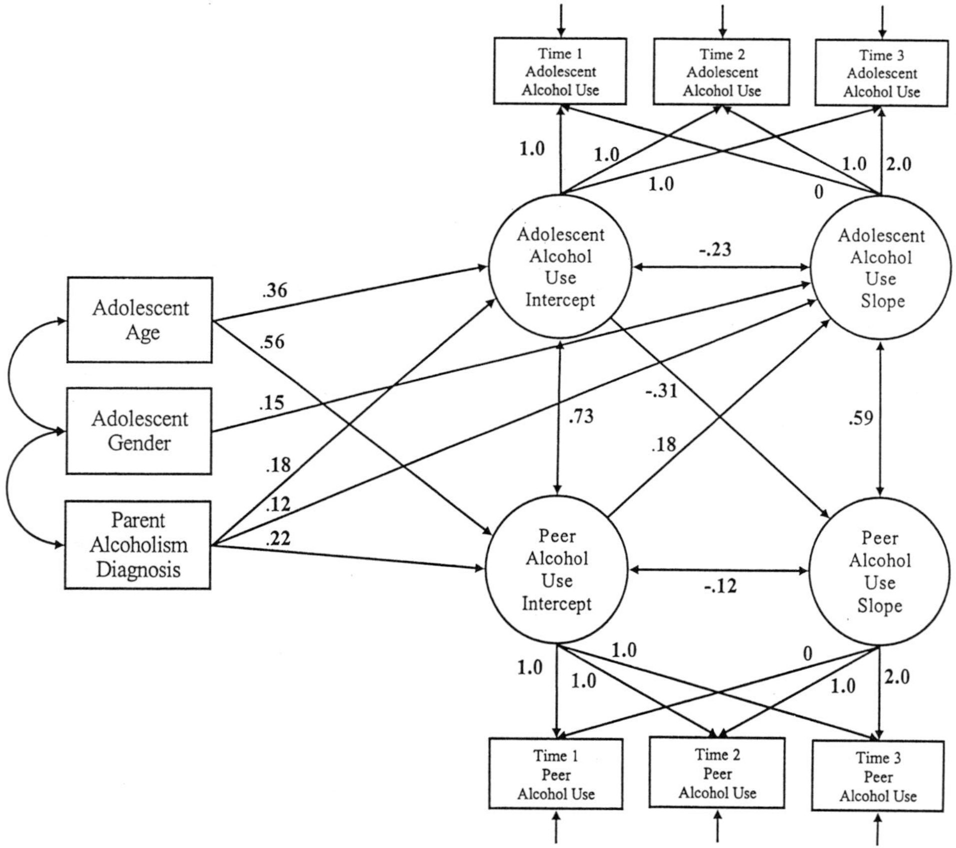

Fig. 1.6. Final multivariate conditional growth model between adolescent alcohol use and peer alcohol use regressed on adolescent age, gender, and parental alcoholism diagnosis.
Note: Model fit was $\chi^2(14, N = 363) = 25.4$, $p = .03$, TLI = .98, CFI = .98, RMSEA = .04. Only significant paths ($p < .05$) are shown. All parameter estimates are standardized. Covariances among the within time disturbances of the repeated measures are not shown.

The regression parameters between the intercept factor of adolescent alcohol use and the slope factor of peer alcohol use, and between the intercept factor of peer alcohol use and the slope factor of adolescent alcohol use allow us to examine the final research question *Q6* that inquired about the relation between earlier levels of one construct predicting later developmental trajectories of the other construct. Both of these regression parameters were found to significantly differ from zero ($p < .05$). The peer alcohol use intercept factor positively predicted the adolescent alcohol use slope factor indicating that higher initial levels of peer alcohol use were associated with steeper growth trajectories in adolescent alcohol use over the following 2 years. In comparison, the adolescent alcohol use intercept factor *negatively* predicted the peer alcohol use slope factor, indicating that higher initial levels of adolescent alcohol use was associated with *less* steep (but still positively increasing) growth trajectories in peer alcohol use over the following 2 years. It is extremely important to note that this negative relation in no way indicates *decreases* in peer alcohol use over time; instead, this implies that higher initial levels of adolescent alcohol use are associated with *smaller* rates of positive change in peer alcohol use.

To better understand these complex relations, it is helpful to probe the model implied growth trajectories on one construct as a function of initial status on the other construct. Conceptually, this is done in much the same way as we might probe a two-way interaction in multiple regression (e.g., Aiken & West, 1991). Although these techniques are well worked out for the regression model, it is not so clear how to best accomplish this for the latent curve model, and further work is needed in this area. The method I used here incorporates the final parameter estimates from the latent curve model to compute the model implied growth trajectories as a function of the predictor variables. The goal here is to compute the mean of the slope factor on one construct as a function of varying values of the intercept factor on the other construct. There is no strict definition for what is meant by varying values, but here I probed this relation using one standard deviation above and below the mean of each intercept factor. That is, I first used the final parameter estimates to provide the average model implied growth trajectories. I then added and subtracted the corresponding standard deviations to the intercepts of the growth factors to provide the model implied growth trajectories as a function of high versus low standing on the initial status factors. Finally, I plotted these means to probe the nature of the relation among the latent growth factors. These trajectories are shown in Fig. 1.7.

Figure 1.7 reflects that, although all developmental growth trajectories are positive, there are differences in the *magnitudes* of the trajectories for both constructs. Positive growth in adolescent alcohol use was *accelerating* over time as a function of peer alcohol use; that is, higher initial levels of peer use were associated with steeper rates of positive growth in adolescent use. In comparison, positive growth in peer alcohol use was *decelerating* over time as a function of adolescent alcohol use; that is, higher initial levels of adolescent alcohol use were associated with less steep rates of positive growth in peer use. Thus, the initial status of both peer and adolescent alcohol use was predictive of later changes in the other construct, but the *magnitude* of the rate of positive change varied within each construct.

Summary of Findings

The series of latent curve models allowed us to explore all six of our research questions described at the beginning of the chapter. Growth in adolescent alcohol use was characterized by a positive linear trajectory over the three time periods, and there were large individual differences in both the starting point of the trajectory and the rate of change over time (*Q1*). Individual differences in developmental trajectories of adolescent alcohol use were systematically related to age, gender, and parental alcoholism diagnosis (*Q2*). Growth in peer alcohol use was also characterized by a positive linear trajectory with large individual differences in both the starting point of the trajectory and the rate of change over time (*Q3*). Individual differences in developmental trajectories of peer alcohol use were systematically related to age and parental alcoholism, but not gender (*Q4*). Individual differences in developmental trajectories of adolescent alcohol use were closely related to those of peer alcohol use. Results indicated that the intercepts of the two growth processes were strongly and positively correlated with one another, as were the rates of change of the two growth processes, thus reflecting that the characteristics of growth in adolescent alcohol use were similar to the characteristics of growth in peer alcohol use (*Q5*). Finally, higher initial levels of adolescent alcohol use were predictive of smaller positive changes in peer alcohol use, and higher initial levels of peer alcohol use were predictive of larger positive changes in adolescent alcohol use (*Q6*).

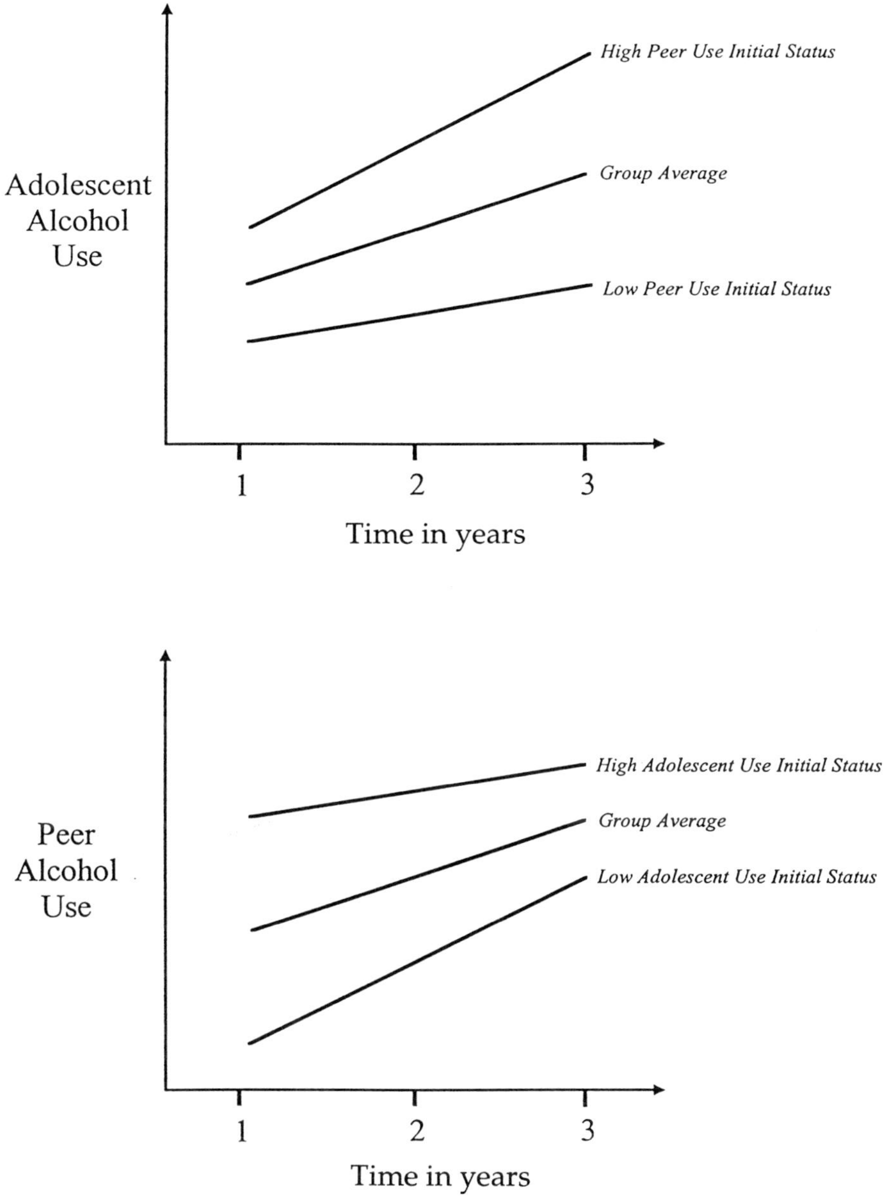

Fig. 1.7. Model implied growth trajectories for adolescent alcohol use as a function of initial status on peer alcohol use, and model implied growth trajectories for peer alcohol use as a function of initial status on adolescent alcohol use.
Note: *High* and *Low* defined as one standard deviation above or below the mean on the initial status growth factor.

Taken together, these results are most consistent with the *bidirectional peer influences* model of adolescent alcohol use, at least for adolescents similar in characteristics to those studied here.

Extensions of the Latent Curve Model

There are many interesting ways in which the latent curve models can be extended to test a variety of types of research hypotheses often encountered in studies of substance use and abuse. For example, whereas I used manifest variables for the repeated measures of alcohol use over time, the model can easily be extended to incorporate multiple indicator latent factors within each time period to estimate and remove the effects of measurement error (McArdle, 1988). The latent curve model can provide formal tests of both mediation and moderation across a variety of research designs (McArdle, 1988, 1989, 1991). The latent curve approach is particularly well suited for evaluating the efficacy of prevention and treatment programs targeted at altering normative developmental trajectories (e.g., deflecting substance use trajectories during adolescence; Curran & Muthén, in press; Muthén & Curran, 1997). Given an adequate number of repeated measures, curvilinear trajectories can be examined (McArdle & Epstein, 1987) as well as alternative functional forms of growth (Browne & Du Toit, 1993). These nonlinear growth models play a particularly important role when studying development in substance use during times of acceleration or deceleration when growth might not follow a linear trajectory. Finally, the latent curve model can be used to provide important estimates of statistical power and effect size to be used in the planning of future longitudinal studies of development and change (Muthén & Curran, 1997). In sum, latent curve analysis is a highly flexible technique that can be adapted to test many different types of questions about individual differences in change over time and is particularly well suited for the empirical study of the development of substance use across the life span.

Limitations of the Latent Curve Model

The previous analyses have highlighted a number of distinct advantages of the latent curve model for studying individual differences in change over time. There are, of course, a number of limitations as well. First, the sample used for the previous analyses was characterized by a 97% retention rate across the three time periods making missing data a nonissue here. In general, this is not typically the case, especially in longitudinal studies of substance use and high risk behavior, and issues of missing data must be closely considered in analyses of this kind (Seltzer & Raudenbush, 1997). Although recent advances have been made in the incorporation of missing data structures within the structural equation modeling framework (e.g., Graham, Hofer, & MacKinnon, 1996; McArdle, 1994; Muthén, 1993; Muthén, Kaplan, & Hollis, 1987), this still poses a limitation in latent curve analysis. The latent curve model as typically estimated makes the further assumption that all subjects are assessed at the same time periods (so-called *time-structured* data; Willett & Sayer, 1994), and this can be a limitation in many types of substance use research settings. However, this limitation is one of current software implementation, and individually varying assessments can be appropriately modeled in a latent variable framework using certain software packages (e.g., Mx; Neale, 1994). Further, normal theory maximum likelihood (ML) estimation was used for all of the latent curve models here. ML estimation assumes that the observed data follow a multivariate normal distribution (Browne, 1984), and violations of this assumption are associated with both inflated test statistics (Curran, West, & Finch, 1996; Hu, Bentler, & Kano, 1992) and attenuated standard errors (Muthén & Kaplan, 1985, 1992). This is a significant issue in the empirical study of substance use given the severely non-normal distributions commonly encountered in practice. Alternative methods of estimation such as two-staged least-squares (Bollen, 1996) and pseudo maximum likelihood (Arminger & Schoenberg, 1985) that are less affected by non-normality are becoming increasingly available to the applied researcher, but further work is needed to better understand these estimators in applied settings. These limitations should be closely examined when considering whether latent curve analysis is an appropriate statistical model to be used in a given research application.

CONCLUSIONS

There is rarely a situation in which there exists one single statistical model that is the only "correct" choice for evaluating a given research hypothesis. There are a variety of powerful methods currently available to the applied researcher, each of which is characterized by certain advantages and certain disadvantages. The optimal method depends on the specific data characteristics and research hypotheses of the particular study at hand. The applied researcher will do well to strive to maintain flexibility in the selection of an analytic technique that best meets the needs of the particular research hypothesis. One never wants to be in the position of having a single statistical model in desperate search of a question. The old adage "When you only own a hammer, everything begins to look like a nail" rings especially true here.

It is also extremely important to stress that the selection of an appropriate statistical model is *not* tantamount to the selection of a particular software package. That is, the key analytic decision is not whether to use EQS or LISREL or HLM or PROC MIXED to evaluate the proposed research hypothesis. Instead, the statistical model that best corresponds to the theoretical model must first be selected, and only then will the choice of software logically follow. It is ultimately the responsibility of the individual researcher to ascertain the particular advantages and disadvantages of each of the available analytic techniques, and to make a fully informed decision about the selection of the statistical technique that is best suited to evaluate the given research hypothesis.

I have dedicated a large portion of this chapter to a discussion of matching the statistical model to the theoretical model. However, I also believe that the relation between theory and statistics can be a symbiotic one. That is, the statistical model is not merely a tool with which to empirically evaluate the theoretically derived research hypothesis. Although this is a critical role filled by the statistical model, it can also be used to prompt us to more clearly articulate our developmental theories and associated research hypotheses. Given the ever increasing complexity of the available statistical modeling techniques, there is a corresponding need to pay much greater attention to the specific details of the influences and mechanisms that we postulate in our theories. In the past, predictions about development could often be broadly discussed in general terms of "change," and there was not a pressing need to explicitly articulate precisely what *type* of change the theory was inferring. However, given that the selection of an appropriate statistical model depends directly on the type of change that we hypothesize to exist, we are in turn obligated to clearly enunciate these theoretical details to optimally guide our statistical analyses. The subsequent result of this increased focus of attention on the selection of an appropriate statistical model is the strengthening of both our developmental theories and our corresponding research hypotheses.

Appendix A

```
/title
autoregressive crosslagged panel model for adolescent alcohol use and
peer alcohol use with age, gender and parental alcoholism as predictors
/specifications
cases=363; variables=9; method=ML; matrix=cov; analysis=cov;
/labels
v1=age; v2=gen; v3=coa; v4=alc1; v5=alc2; v6=alc3;
v7=peer1; v8=peer2; v9=peer3;
/equations
v4= *v1 + *v2 + *v3 + e4;
v5= *v4 + *v7 + e5;
v6= *v5 + *v8 + e6;
v7= *v1 + *v2 + *v3 + e7;
v8= *v7 + *v4 + e8 + *v1;
v9 = *v8 + *v5 + e9 + *v2;
/variances
v1 to v3=*; e4 to e9=*;
/covariances
v3,v2=.003*; v3,v1=-.1*; v2,v1=-.04*;
e7,e4=.020*; e8,e5=.773*; e9,e6=.133*;
/matrix
1.968
-0.022  0.250
-0.104  0.001  0.246
1.238   0.019  0.164  7.906
1.750   0.110  0.331  7.634  15.828
1.532   0.239  0.478  6.726  12.917  23.013
1.179  -0.081  0.078  3.246   4.021   3.853  3.047
1.118  -0.022  0.107  2.691   4.749   4.555  2.077  3.321
0.822  -0.132  0.159  2.147   4.055   5.849  1.691  2.126  4.039
/end
```

Appendix B

```
/title
two factor unconditional latent curve model for
three repeated measures of adolescent alcohol use
/specifications
cases=363; variables=9; method=ml; matrix=cov; analysis=moment;
/labels
v1=age; v2=gen; v3=coa; v4=alc1; v5=alc2; v6=alc3;
v7=peer1; v8=peer2; v9=peer3;
```

```
f1=alcint; f2=alcslp;
/equations
v4= 1.0f1 + 0f2 + e4;
v5= 1.0f1 + 1f2 + e5;
v6= 1.0f1 + 2f2 + e6;
f1= *v999 + d1;
f2= *v999 + d2;
/variances
e4 to e6=*;
d1 to d2=*;
/covariances
d2,d1=*;
/means
12.9146 0.5152 0.5647 1.3636 2.1185 3.1846 1.3289 1.7543 2.4749
/matrix
1.968
-0.022 0.250
-0.104 0.001 0.246
1.238 0.019 0.164 7.906
1.750 0.110 0.331 7.634 15.828
1.532 0.239 0.478 6.726 12.917 23.013
1.179 -0.081 0.078 3.246 4.021 3.853 3.047
1.118 -0.022 0.107 2.691 4.749 4.555 2.077 3.321
0.822 -0.132 0.159 2.147 4.055 5.849 1.691 2.126 4.039
/end
```

Appendix C

```
/title
two factor conditional latent curve model for adolescent alcohol use
regressed on age, gender, and parent alcoholism diagnosis
/specifications
cases=363; variables=9; method=ml; matrix=cov; analysis=moment;
/labels
```

```
v1=age; v2=gen; v3=coa; v4=alc1; v5=alc2; v6=alc3;
v7=peer1; v8=peer2; v9=peer3;
f1=alcint; f2=alcslp;
/equations
v1= *v999 + e1;
v2= *v999 + e2;
v3= *v999 + e3;
v4= 1.0f1 + 0f2 + e4;
v5= 1.0f1 + 1f2 + e5;
v6= 1.0f1 + 2f2 + e6;
f1= *v999 + *v1 + *v2 + *v3 + d1;
f2= *v999 + *v1 + *v2 + *v3 + d2;
/variances
e1 to e6=*; d1 to d2=*;
/covariances
e3,e2=*; e3,e1=*; e2,e1=*; d2,d1=*;
/means
12.9146 0.5152 0.5647 1.3636 2.1185 3.1846 1.3289 1.7543 2.4749
/matrix
1.968
-0.022 0.250
-0.104 0.001 0.246
1.238 0.019 0.164 7.906
1.750 0.110 0.331 7.634 15.828
1.532 0.239 0.478 6.726 12.917 23.013
1.179 -0.081 0.078 3.246 4.021 3.853 3.047
1.118 -0.022 0.107 2.691 4.749 4.555 2.077 3.321
0.822 -0.132 0.159 2.147 4.055 5.849 1.691 2.126 4.039
/end0
```

Appendix D

```
/title
multivariate growth model between adolescent and peer alcohol use
regressed on age, gender, and parent alcoholism diagnosis
/specifications
cases=363; variables=9; method=ml; matrix=cov; analysis=moment;
/labels
```

```
v1=age; v2=gen; v3=coa; v4=alc1; v5=alc2; v6=alc3;
v7=peer1; v8=peer2; v9=peer3;
f1=alcint; f2=alcslp; f3=peerint; f4=peerslp;
/equations
v1= 12.9*v999 + e1;
v2= .515*v999 + e2;
v3= .554*v999 + e3;
v4= 1.0f1 + 0f2 + e4;
v5= 1.0f1 + 1f2 + e5;
v6= 1.0f1 + 2f2 + e6;
v7= 1.0f3 + 0f4 + e7;
v8= 1.0f3 + 1f4 + e8;
v9= 1.0f3 + 2f4 + e9;
f1= -7*v999 + .5*v1 + .5*v2 + .5*v3 + d1;
f2= -.14*v999 + *v2 + *v3 + *f3 + d2;
f3= -6.6*v999 + .5*v1 + .5*v2 + .5*v3 + d3;
f4= .5*v999 + *f1 + d4;
/variances
e1 to e9=*; d1 to d4=*;
/covariances
e3,e2=0*; e3,e1=0*; e2,e1=0*; e7,e4=.5*; e8,e5=.5*; e9,e6=.5*;
d2,d1=*; d4,d3=*; d4,d2=*; d3,d1=*;
/means
12.9146 0.5152 0.5647 1.3636 2.1185 3.1846 1.3289 1.7543 2.4749
/matrix
1.968
-0.022 0.250
-0.104 0.001 0.246
1.238 0.019 0.164 7.906
1.750 0.110 0.331 7.634 15.828
1.532 0.239 0.478 6.726 12.917 23.013
1.179 -0.081 0.078 3.246 4.021 3.853 3.047
1.118 -0.022 0.107 2.691 4.749 4.555 2.077 3.321
0.822 -0.132 0.159 2.147 4.055 5.849 1.691 2.126 4.039
/end
```

NOTES

Correspondence concerning this chapter should be sent to Patrick J. Curran, Department of Psychology, University of North Carolina-Chapel Hill, CB# 3270, Davie Hall, Chapel Hill, NC, 27599-3270.

ACKNOWLEDGMENTS

The Adolescent and Family Development Project was supported by Grant DA05227 from the National Institute on Drug Abuse awarded to Laurie Chassin and Manuel Barrera.

I would like to thank Laurie Chassin, Craig Colder, Andrea Hussong, Robert MacCallum, Bengt Muthén, Jen Rose, and Eric Stice for their valuable input.

REFERENCES

Aiken, L .S., & West, G. W. (1991). *Multiple regression: Testing and interpreting interactions.* Newbury Park, CA: Sage.

Arminger, G., & Schoenberg, R. (1989). Pseudo maximum likelihood estimation and a test for misspecification in mean and covariance structure models. *Psychometrika, 54,* 409–425.

Bauman, K. E., & Ennett, S. T. (1994). Peer influence on adolescent drug use. *American Psychologist, 49,* 820–822.

Bentler, P. M. (1980). Multivariate analysis with latent variables: Causal modeling. *Annual Review of Psychology, 31,* 419–456.

Bentler, P. M. (1983). Some contributions to efficient statistics for structural models: Specification and estimation of moment structures. *Psychometrika, 48,* 493–517.

Bentler, P. M. (1997). *EQS: Structural equations program manual.* Los Angeles: BMDP Statistical Software.

Bollen, K. A. (1996). An alternative two stage least squares (2SLS) estimator for latent variable equations. *Psychometrika, 61,* 109–121.

Brook, J. S., Brook, D. W., Gordon, A. S., Whiteman, M., & Cohen, A. (1990). The psychosocial etiology of adolescent drug use: A family interactional approach. *Genetic, Social, and General Psychology Monographs, 116,* 111–267.

Browne, M. W. (1984). Asymptotic distribution free methods in the analysis of covariance structures. *British Journal of Mathematical and Statistical Psychology, 37,* 127–141.

Browne, M. W., & Du Toit, S. H. C. (1993). Models for learning data. In L. Collins & J. Horn (Eds.), *Best methods for the analysis of change: Recent advances, unanswered questions, future directions* (pp. 47–68). Washington, DC: American Psychological Association.

Bryk, A. S., & Raudenbush, S. W. (1987). Application of hierarchical linear models to assessing change. *Psychological Bulletin, 101,* 147–158.

Bryk, A. S., & Raudenbush, S. W. (1992). *Hierarchical linear models: Applications and data analysis methods.* Newbury Park, CA: Sage.

Burstein, L. (1980). The analysis of multi-level data in educational research and evaluation. *Review of Research in Education, 8,* 158–233.

Chassin, L., Barrera, M., Bech, K., & Kossak-Fuller, J. (1992). Recruiting a community sample of adolescent children of alcoholics: A comparison of three subject sources. *Journal of Studies on Alcohol, 53,* 316–320.

Chassin, L., Curran, P. J., Hussong, A. M., & Colder, C. R. (1996). The relation of parent alcoholism to adolescent substance use: A longitudinal follow-up study. *Journal of Abnormal Psychology, 105,* 70–80.

Chassin, L., Presson, C. C., Sherman, S. J., & Curran, P. J. (1992). Social psychological factors in adolescent substance use and abuse. In F. P. Medway & T. P. Cafferty (Eds.), *Theories of school psychology* (pp. 397–424). Hillsdale: Lawrence Erlbaum Associates.

Chassin, L., Rogosch, R., & Barrera, M. (1991). Substance use and symptomatology among adolescent children of alcoholics. *Journal of Abnormal Psychology, 100,* 449–463.

Cronbach, L. J. (1976). Research on classrooms and schools: Formulation of questions, design, and analysis. *The Stanford Evaluation Consortium*, Occasional Papers, July, 1976.

Curran, P. J. (1997). The bridging of quantitative methodology and applied developmental research. In P. J. Curran (Chair), *Comparing three modern approaches to longitudinal data analysis: An examination of a single developmental sample*. Symposium conducted at the 1997 meeting of the Society for Research on Child Development, Washington, DC.

Curran, P. J., & Muthén, B. O. (in press). The application of latent curve analysis to testing developmental theories in intervention research. *American Journal of Community Psychology*.

Curran, P .J., Muthén, B. O., & Harford, T. C. (1998). The influence of changes in marital status on developmental trajectories of alcohol use in young adults. *Journal of Studies on Alcohol, 59*, 647–658.

Curran, P. J., Stice, E., & Chassin, L. (1997). The relation between adolescent and peer alcohol use: A longitudinal random coefficients model. *Journal of Consulting and Clinical Psychology, 65*, 130–140.

Curran, P. J., West, S. G., & Finch, J. (1996). The robustness of test statistics to non-normality and specification error in confirmatory factor analysis. *Psychological Methods, 1*, 16–29.

Diggle, P. J., Liang, K. Y., & Zeger, S. L. (1994). *Analysis of longitudinal data*. Oxford: Clarendon Press.

Duncan, T. E., Duncan, S. C., Alpert, A., & Hops, H. (1997). Latent variable modeling of longitudinal and multilevel substance use data. *Multivariate Behavioral Research, 32*, 275–318.

Dwyer, J. H. (1983). *Statistical models for the social and behavioral sciences*, New York: Oxford.

Farrell, A. D. (1994). Structural equation modeling with longitudinal data: Strategies for examining group differences and reciprocal relationships. *Journal of Consulting and Clinical Psychology, 62*, 477–487.

Farrell, A., & Danish, S. (1993). Peer drug associations and emotional restraint: Causes or consequences of adolescents' drug use? *Journal of Consulting and Clinical Psychology, 43*, 522–527.

Fisher, L. A., & Bauman, K. E. (1988). Influence and selection in the friend–adolescent relationship: Findings from studies of adolescent smoking and drinking. *Journal of Applied Social Psychology, 18*, 289–314.

Goldstein, H. I. (1987). *Multilevel models in educational and social research*. London: Oxford University Press.

Graham, J. W., Hofer, S. M., & MacKinnon, D. P. (1996). Maximizing the usefulness of data obtained with planned missing value patterns: An application of maximum likelihood procedures. *Multivariate Behavioral Research, 31*, 197–218.

Hawkins, J. D., Catalano, R. F., & Miller, J. Y. (1992). Risk and protective factors for alcohol and other drug problems in adolescence and early adulthood: Implications for substance abuse prevention. *Psychological Bulletin, 112*, 64–105.

Hu, L., Bentler, P. M., & Kano, Y. (1992). Can test statistics in covariance structure analysis be trusted? *Psychological Bulletin, 112*, 351–362.

Jöreskog, K. G. (1971a). Simultaneous factor analysis in several populations. *Psychometrika, 36*, 409–426.

Jöreskog, K. G. (1971b). Statistical analysis of sets of congeneric tests. *Psychometrika, 36*, 109–133.

Jöreskog, K. G., & Sörbom, D. (1979). *Advances in factor analysis and structural equation models*. Cambridge, MA: Abt Books.

Kandel, D. B. (1985). On processes of peer influences in adolescent drug use: A developmental perspective. *Advances in Alcohol and Substance Abuse, 4*, 139–163.

Kaplan, D. (1990). Evaluating and modifying covariance structure models: A review and recommendation. *Multivariate Behavioral Research, 25*, 137–155.

Laird, N. M., & Ware, H. (1982). Random-effects models for longitudinal data. *Biometrics, 38*, 963–974.

MacCallum, R. (1986). Specification searches in covariance structure modeling. *Psychological Bulletin, 100*, 107–120.

MacCallum, R. C., Kim, C., Malarkey, W., & Kiecolt-Glaser, J. (1997). Studying multivariate change using multilevel models and latent curve models. *Multivariate Behavioral Research, 32*, 215–253.

Maisto, S. A., & Carey, K. B. (1985). Origins of alcohol abuse in children and adolescents. In B. B. Lahey & A. E. Kazdin (Eds.), *Advances in clinical child psychology, volume 8*. New York: Plenum.

McArdle, J. J. (1988). Dynamic but structural equation modeling of repeated measures data. In J. R. Nesselroade & R. B. Cattell (Eds.), *Handbook of multivariate experimental psychology, second edition*. New York: Plenum.

McArdle, J .J. (1989). Structural modeling experiments using multiple growth functions. In P. Ackerman, R. Kanfer, & R. Cudeck (Eds.), *Learning and individual differences: Abilities, motivation and methodology* (pp. 71–117). Hillsdale, NJ: Lawrence Erlbaum Associates.

McArdle, J. J. (1991). Structural models of developmental theory in psychology. In P. Van Geert & L. P. Mos (Eds.), *Annals of theoretical psychology, volume VII.*, (pp. 139–160). New York: Plenum.

McArdle, J. J. (1994). Structural factor analysis experiments with incomplete data. *Multivariate Behavioral Research, 29* (4), 409–454.

McArdle, J .J,. & Epstein, D. (1987). Latent growth curves within developmental structural equation models. *Child Development, 58*, 110–133.

Meredith, W., & Tisak, J. (1984). "Tuckerizing" curves. Paper presented at the annual meeting of the Psychometric Society, Santa Barbara, CA.

Meredith, W., & Tisak, J. (1990). Latent curve analysis, *Psychometrika, 55*, 107–122.

Muthén, B. O. (1991). Analysis of longitudinal data using latent variable models with varying parameters. In L. M. Collins & J. L. Horn (Eds.), *Best methods for the analysis of change: Recent advances, unanswered questions, future directions*. Washington, DC: American Psychological Association.

Muthén, B. O. (1993). Latent variable modeling of growth with missing data and multilevel data. In C. M. Cuadras & C. R. Rao (Eds.), *Multivariate analysis: Future directions 2* (pp. 199–210). Amsterdam: North Holland.

Muthén, B. O., & Curran, P. J. (1997). General longitudinal modeling of individual differences in experimental designs: A latent variable framework for analysis and power estimation. *Psychological Methods, 4*, 371–402.

Muthén, B. O., & Kaplan, D. (1985). A comparison of some methodologies for the factor analysis of non-normal Likert variables. *British Journal of Mathematical and Statistical Psychology, 38*, 171–189.

Muthén, B. O., & Kaplan, D. (1992). A comparison of some methodologies for the factor analysis of non-normal Likert variables: A note on the size of the model. *British Journal of Mathematical and Statistical Psychology, 45*, 19–30.

Muthén, B. O., Kaplan, D., & Hollis, M. (1987). On structural equation modeling with data that are not missing completely at random. *Psychometrika, 52*, 431–462.

Muthén, B. O., & Muthén, L. (in press). The development of heavy drinking and alcohol related problems from ages 18 to 37 in a U.S. National Sample. *Journal of Studies on Alcohol Use.*

Neale, M. C. (1994). *Mx: Statistical modeling (2nd ed.)*. Box 710 MCV, Richmond, VA 23298: Department of Psychiatry.

Oetting, E. R., & Beauvais, F. (1986). Peer cluster theory: Drugs and the adolescent. *Journal of Counseling and Development, 65*, 17–22.

Oetting, E. R., & Beauvais, F. (1987). Peer cluster theory, socialization characteristics and adolescent drug use: A path analysis. *Journal of Counseling Psychology, 34*, 205–213.

Petraitis, J., Flay, B. R., & Miller, T. Q. (1995). Reviewing theories of adolescent substance use: Organizing pieces of the puzzle. *Psychological Bulletin, 117*, 67–86.

Rao, C. R. (1958). Some statistical methods for comparison of growth curves. *Biometrika, 51*, 83–90.

Rogosa, D. R. (1988). Myths about longitudinal research. In K. W. Schaie, R. T. Campbell, W. Meredith, & S. C. Rawlings (Eds.), *Methodological Issues in Aging Research*. New York: Springer.

Rogosa, D. R. (1995). Myths and methods: "Myths about longitudinal research" plus supplemental questions. In J. Gottman (Ed.), *The Analysis of Change*, (pp. 2–65). Mahwah, NJ: Lawrence Erlbaum Associates.

Rogosa, D. R. & Willett, J. B. (1985). Understanding correlates of change by modeling individual differences in growth. *Psychometrika, 50*, 203–228.

Schulenberg, J., O'Malley, P. M., Bachman, J. G., Wadsworth, K. N., & Johnston, L. (1996). Getting drunk and growing up: Trajectories of frequent binge drinking during the transition to young adulthood. *Journal of Studies on Alcohol, 57*, 289–304.

Seltzer, M., & Raudenbush, S. (1997). The application of hierarchical linear models in developmental research. In P. J. Curran (Chair), *Comparing three modern approaches to longitudinal data analysis: An examination of a single developmental sample*. Symposium conducted at the 1997 meeting of the Society for Research on Child Development, Washington, DC.

Sher, K. J. (1991). *Children of alcoholics: A critical appraisal of theory and research*. Chicago, IL: University of Chicago Press.

Stoolmiller, M. (1994). Antisocial behavior, delinquent peer association, and unsupervised wandering for boys: Growth and change from childhood to early adolescence. *Multivariate Behavioral Research, 29*, 263–288.

Swaim, R. C., Oetting, E. R., Edwards, R. W., & Beauvais, F. (1989). Links from emotional distress to adolescent drug use: A path model. *Journal of Consulting and Clinical Psychology, 57*, 227–231.

Tucker, L. R. (1958). Determination of parameters of a functional relation by factor analysis. *Psychometrika, 23*, 19–23.

Willett, J. B. (1988). Measuring change: The difference score and beyond. In H. J. Walberg & G. D. Haertel (Eds.), *The international encyclopedia of educational evaluation*. Oxford, England: Pergamon Press.

Willett, J. B., & Sayer, A. G. (1994). Using covariance structure analysis to detect correlates and predictors of change. *Psychological Bulletin, 116*, 363–381.

2

Longitudinal Data on Families: Growth Modeling Alternatives

Siek-Toon Khoo
Arizona State University

Bengt Muthén
University of California, Los Angeles

Family data are often collected to study influences that occur within the context of home environments. Longitudinal family data are important for investigating and understanding adolescent development. The purpose of this chapter is to demonstrate a number of possible alternative methods for analyzing longitudinal family data with multiple siblings per household. Longitudinal family data is an example of hierarchical data with repeated observations over time nested within individuals and individuals nested within families. The repeated measurements offer opportunities for longitudinal modeling to study individual differences in the development. Two approaches to analyzing family data are discussed. The first is the multilevel approach that treats the siblings within families as statistically equivalent. The longitudinal family data are analyzed in a three-level model. The second approach is the multivariate approach that analyzes the longitudinal family data in a multivariate longitudinal model, taking the family as the unit of analysis and allowing multiple growth processes per family. Latent variable modeling is used for the second approach. It offers a flexible multivariate framework that is richer and presents more opportunities for the study of developmental growth and interrelationships of developments between family members. To illustrate the methodology, we use the subset of heavy drinking data spanning 15 years for 1,614 two-sibling families from the National Longitudinal Survey of Youth (NLSY). Individual and family background variables are included in the analyses as predictor variables.

Longitudinal data are useful for studying individual differences in development. Family data with multiple siblings are often collected for genetic studies or for studies that investigate influences that occur within the context of home environments. Data obtained from observing families over time are important for investigating and understanding adolescent development. Here, the interest is in modeling familial resemblance, interactions within families and influences of the family environment and family characteristics. The purpose of this chapter is to demonstrate a number of possible alternative methods for analyzing longitudinal family data with multiple siblings per household.

When individual data are collected from surveys that sample organizational units or naturally occurring clusters rather than individuals, the individuals are nested and observed within the larger units. This type of data has a hierarchical structure. For example, we may have students observed within classrooms, family members within households, or residents within neighborhoods. Observations and measurements may be made at each of the different levels of the hierarchy. In the case of sibling data, there are variables describing the individual siblings and there are variables describing the family. Since the individuals within the same larger unit share values on the higher level variables, observations on individuals within the larger units or clusters may be correlated. The nonindependence of these observations, if ignored, may give standard errors of parameter estimates that are underestimated if there is a substantial correlation between the observations. Aggregating and averaging the individual variables to the higher level and doing the analysis at the higher level would be throwing away valuable individual-level information. More importantly, ignoring the hierarchical structure may mean missing the opportunities to model the effects occurring at each level and to explicitly formulate and test hypotheses pertaining to each level.

In developmental research where repeated observations are made over time on a group of individuals, the multiple observations per individual can also be viewed as nested within persons. The repeated measurements offer opportunities for longitudinal modeling to study developmental trajectories and for describing individual differences in development over time. In longitudinal family data, we have repeated observations nested within individuals who, in turn, are nested within families. This structure constitutes three levels of hierarchical data.

In analyzing longitudinal family data with multiple siblings, there are several areas of interest. The first concerns the interest in describing individual development and change over time and the relationships between the developmental trajectories and the individual characteristics and experiences. The second concerns the study of how shared family environment and characteristics may influence the developmental trajectories. The third concerns the study of interrelationships of developments between siblings and sibling influences. For example, it is likely that the developmental process of an older sibling may affect that of a younger sibling. The fourth concerns the study of how family variables interact with individual characteristics in the developmental process. For example, ethnicity may interact with sibling order in its influence on a developmental process, with the older sibling being expected to take on more responsibilities in a certain culture. In order to answer these research questions efficiently, there is a need to take advantage of the newer methods and software now available.

Two approaches to analyzing family data are discussed in this chapter. The first is the multilevel approach that considers the siblings to be nested within families and treats the siblings within a family as statistically equivalent. This means that they behave the same statistically, sharing the same means and variances on the same set of variables. This approach models the longitudinal family data in a three-level hierarchical model with a separate model at each level. The second is the multivariate latent growth approach that takes the family as the unit of analysis. This approach considers the siblings to be different by sibling order and allows multiple growth processes per family in a multivariate context. Each sibling in the family has his or her own set of variables that can have different means and different variances. This approach takes advantage of the small cluster sizes of families. The within-family dependence is modeled within the family unit in the multivariate model.

The Multilevel Approach

Recent advances in the analysis of hierarchical data draw on statistical techniques that have been given several different terms, for example, *random effects models*, *mixed effects models*, and *hierarchical linear models* (Bock, 1989; Bryk & Raudenbush, 1987, 1992; Goldstein, 1986, 1995; Laird & Ware, 1982; Strenio, Weisberg, & Bryk, 1983). Random effects modeling offers integrated and flexible approaches for the analysis of hierarchical data in hierarchical linear models or multilevel models. The analysis of longitudinal family data using multilevel modeling would call for a three-level hierarchical model. These models can be estimated using readily available multilevel analysis software such as HLM (Bryk, Raudenbush, & Congdon, 1996), MLn (Woodhouse, Rabash, Goldstein, Yang, & Plewis, 1996) and SAS PROC MIXED (SAS Institute, Inc.,1995).

In the multilevel framework, a growth model is specified as a two-level model where the Level-1 units are the repeated measurements that are nested within the Level-2 units of individuals. The Level-1 model gives the within-subject model that represents individual growth trajectories. The Level-2 model gives the between-subject model that represents the variation of growth across individuals. When individuals are also nested in families, a Level-3 model can be formulated and tested to study variation across families and family-level influence. These multilevel models are discussed later.

The Multivariate Approach

Family data can also be analyzed taking the family as the unit of analysis in multivariate models without throwing away individual information. The family data consist of data from each member of the family (allowing responses from each member to be represented by separate variables) and also shared family information. Cross-sectional family data have often been analyzed this way, especially in genetic analyses, in which researchers seek to model family resemblance. Neale, Walters, Eaves, Maes, and Kendler (1994) analyzed fears using multivariate models based on data from the twin pair and their parents. Pike, McGuire, Hetherington, Reiss, and Plomin (1996) studied family environment and adolescent depressive symptoms and antisocial behavior in a multivariate model using data from a sample of two-parent families with same-sex sibling pairs. These multivariate models are usually implemented in the latent variable framework using structural equation modeling software. Recent advances in latent growth curve modeling make it possible to extend this method to longitudinal family data.

Use of random effects models for growth modeling has been extended and applied in the context of latent variable modeling (McArdle & Epstein, 1987; Meredith & Tisak, 1984, 1990; Muthén, 1991, 1993, 1996; Muthén & Khoo, 1998; Willett & Sayer, 1994). This development lends powerful tools to the study of developmental changes (see also the chapter by Curran in this volume). These latent growth curve models can be implemented using structural equation modeling software in a multivariate model instead of in a multilevel model. The combination of latent growth curve modeling and structural equation modeling offers great flexibility in dealing with multivariate outcomes, multiple processes, measurement models in the covariates, and modeling of mediational effects (Muthén & Curran, 1997).

Taking family as the unit of analysis, longitudinal family data can be analyzed using multivariate models with multiple growth trajectories, each corresponding to a sibling in the family. This type of analysis can be carried out in either the multilevel framework or the multivariate latent variable framework. In the multilevel framework, the multilevel context is only applied to the modeling of growth. At level 1 is the within-family measurement model with multiple sets of repeated measurements for each sibling (multivariate) and at level 2 is the between-family model in a two-level model. This technique is illustrated in Raudenbush, Brennan, and Barnett (1995) in a multivariate hierarchical model for studying psychological change within married couples. This was specified in a two-level model with two growth functions by using indicator variables for male and female. The data structure is seen as *repeated observations nested within families.* This technique combines growth modeling with cross-sectional methods for the study of matched pairs. A more natural and flexible framework for this type of analysis is the multivariate latent variable framework, mentioned in the last paragraph, that combines growth modeling with the flexibility of structural equation modeling in multivariate latent growth models.

In the multivariate latent variable framework, the growth model with multiple trajectories per family can be specified using the multiple sets of repeated multivariate outcomes. The trajectory of each sibling is modeled using his or her repeated observations over time. This is the measurement part of the model. The trajectories are then related to the individual and family predictor variables. This is the structural part of the model. The latent variable framework offers flexibility for studying developmental relationships in the structural part of the model, for example, in the modeling of mediational effects and the adding of measurement models in the predictor variables. Later, we describe the second approach in several multivariate latent growth models with multiple processes to demonstrate possible ways of investigating developmental relationships between family members.

Besides the multilevel approach and the multivariate approach described, there are other approaches that can also be used for the analysis of longitudinal family data. For example, there is the multilevel latent growth approach (Muthén, 1997); which models the hierarchical nature of the data in a multilevel covariance framework. This model is very similar to the three-level multilevel approach. There are also the McArdle (1988) factor-of-curves and curve-of-factors methods that model the family curve. These are also multivariate latent growth models. Duncan and Duncan (1996) demonstrated these methods with substance use data.

The chapter by Muthén in this volume analyzes heavy drinking data from the National Longitudinal Survey of Youth (NLSY). The analyses on longitudinal family data in this chapter uses a subset of the same heavy drinking data from the NLSY. For illustrative purpose of this chapter, only the two-sibling families are included. The aim is to study the relationship in the development of heavy drinking between the older sibling and the younger sibling. This method can be generalized to families with more than two siblings. The subset of data used and the data structure are described in the next section.

DATA AND GROWTH FUNCTIONS

The NLSY is a nationally representative household survey of young adults living in the United States in 1979. The sample includes every age-eligible household member. The two-sibling families from the NLSY data set are used to illustrate the models in this chapter. Table 2.1 shows the sibling ages in the 1,614 two-sibling families.

The ages shown are their ages in the year 1982. There are eight birth-year cohorts from 1957 to 1964. The age difference of most sibling pairs is about 2 years. Only about 16% of the families have siblings who are more than 3 years in age difference. There are longitudinal alcohol measures available covering ages 18 to 37. This chapter analyzes the frequency of heavy drinking variable covering ages 18 to 32. This variable is obtained from answers to the questionnaire item "How often have you had 6 or more drinks on one occasion during the last 30 days?" There are 7 response categories, from 0 to 6 (the categories are given in

TABLE 2.1

Sibling Ages in 1,614 Two-Sibling Families (Frequency Tables)

Frequency of Siblings by Age

	Older sibling			*Younger sibling*	
AGE1	*Frequency*	*Percent*	*AGE2*	*Frequency*	*Percent*
18	6	0.4	18	364	22.6
19	88	5.5	19	395	24.5
20	225	13.9	20	287	17.8
21	274	17.0	21	204	12.6
22	277	17.2	22	169	10.5
23	267	16.5	23	109	6.8
24	241	14.9	24	56	3.5
25	236	14.6	25	30	1.9

Frequency of Sibling-pairs by Age

	Age 2								
Age 1	*18*	*19*	*20*	*21*	*22*	*23*	*24*	*25*	*Total*
18	6	0	0	0	0	0	0	0	6
19	80	8	0	0	0	0	0	0	88
20	122	96	7	0	0	0	0	0	225
21	66	122	82	4	0	0	0	0	274
22	47	80	90	52	8	0	0	0	277
23	23	41	58	74	54	17	0	0	267
24	14	30	35	46	61	37	18	0	241
25	6	18	15	28	46	55	38	30	236
Total	364	395	287	204	169	109	56	30	1614

Table 2.2). For the illustrative purposes of this chapter, the outcome is treated as a continuous variable.

Some background variables describing individual characteristics and family characteristics are included in the analyses to demonstrate how these variables can be included in the models to study their influences on the outcome at individual level and at family level. The individual variables included are gender, early start (early onset of drinking), and dropping out of high school. The family variables included are ethnicity variables and family history of alcoholism variables. The variables are considered family-level variables if the siblings logically share the same value.

The early start variable, which represents early onset of drinking, is obtained from answers to the question "How old were you when you first started drinking, for example, having two or more drinks a week?" The early start variable is coded as 1 if the starting age is 14 or younger, and coded as 0 otherwise. The family history variables are obtained from answers to the question "Have any of your relatives listed on this card been alcoholics or problem drinkers at any time in their lives?" Three family history variables are constructed from the answers, based on whether alcoholism occurs only among first-order relatives (FH1), second or third-order relatives (FH23) or both first-order and second/third-order relatives (FH123).

The outcome variable and the background variables are summarized in Table 2.2.

TABLE 2.2

Descriptions of Variables

Heavy drinking	(HD, 0–6)	never (0), once (1), 2 or 3 times (2), 4 or 5 times (3), 6 or 7 times (4), 8 or 9 times (5), 10 or more times (6)
Gender	(Male,1/0)	male (1), female (0)
Early start	(ES,1/0)	1 if onset on or before 14, 0 otherwise
High school dropout	(HSdrp,1/0)	1 if not having completed high school by age 22
Ethnicity	(Black,1/0)	Black (1), non-Black (0)
	(Hisp,1/0)	Hispanic (1), non-Hispanic (0)
Family history	(FH123,1/0)	1 if family history among 1st and 2nd/3rd-degree relatives
	(FH1,1/0)	1 if family history among 1st-degree relatives only
	(FH23,1/0)	1 if family history among 2nd/3rd-degree relatives only

The observed sample means of heavy drinking frequencies for both the older and the younger siblings over time are shown in Fig. 2.1.

It can be seen that the means follow a nonsymmetric trajectory that increases at a faster rate than it decreases. This means that a standard quadratic growth function will not do well in terms of model fit. The growth analyses of the same heavy drinking outcome in the Muthén chapter estimated some of the time steps to fit a very flexible functional form. In this chapter, a logarithmic transformation of the time scale using fixed time-steps is carried out to obtain the necessary growth function. The transformed time scale λ_t is obtained by the following transformation:

$$\lambda_t = 2 * [\log_e(Age_t - 16) - \log_e(21-16)],\ t = 1,..,15 \tag{2.1}$$

where Age_t is the age of an individual at a particular time of measurement *t*. Preliminary growth analysis using this transformation had shown a good model fit to the estimated mean trajectories for both the older

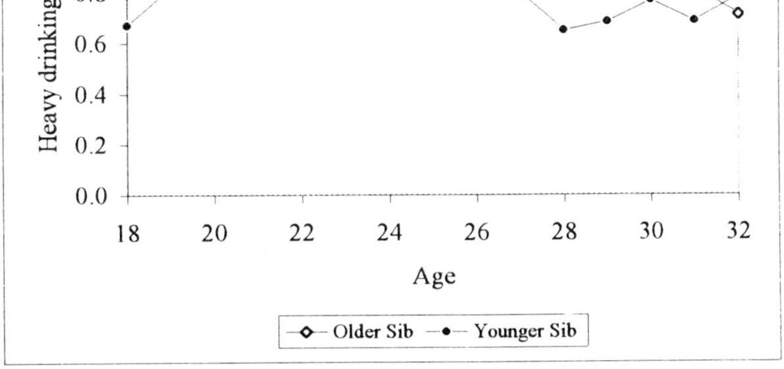

FIG. 2.1. Heavy drinking frequency.

and the younger siblings. The same logarithmic transformation of the time scale is used in both the multilevel modeling next, and the multivariate longitudinal modeling in later sections.

The growth function for the heavy drinking outcome variable in this chapter is specified as a quadratic function of λ_t. Since $\lambda_t = 0$ when $Age_t = 21$, the intercept of the growth function is defined at age 21. The growth function is explained further in the next sections.

Taking age as a basis for the time scale instead of using the measurement occasion means that there is a missing data structure due to the different age cohorts being interviewed at different ages. Issues concerning the missing data structure due to the multiple-cohort design are dealt with in the same way as the analyses in the Muthén chapter in this volume. The missing data are treated as missing completely at random, that is assuming that the different age cohorts are from the same population and there are no cohort effects. This same assumption is made for both the multilevel modeling and the multivariate longitudinal modeling.

ANALYSIS USING THE MULTILEVEL MODELING APPROACH

The subset of longitudinal family data from NLSY consists of heavy drinking data spanning 15 years for 1,614 two-sibling families. One approach for analyzing this type of hierarchical longitudinal data is the multilevel approach. Bryk and Raudenbush (1987, 1992) among others have shown how a multilevel model involving growth can be formulated.

The NLSY data with repeated measurements nested within siblings who are in turn, nested within families can be modeled using a three-level model. Individual growth trajectories or the measurement model is specified at level 1. The observed outcome at each time point is specified as a function of a set of growth parameters over time with measurement error, assuming systematic growth trajectories. At level 2, the growth parameters that characterize individual growth trajectories are specified to vary across the population of siblings. Individual differences in growth can be studied by introducing predictor variables of individual characteristics at this between-individual level. At level 3, the model specified at the family level expresses variation across a population of families. Family-level predictor variables can be included in this

Level-3 model to describe variation across families. Two multilevel models are presented here: Model A and Model B. Model A is a three-level unconditional model without covariates. Model B includes individual covariates at level 2 and family covariates at level 3. A growth model with a linear growth rate and a quadratic growth rate is specified at level 1:

$$y_{tij} = \pi_{0ij} + \pi_{1ij}a_{tij} + \pi_{2ij}a_{tij}^{2} + e_{tij},\ t = 1, 2, 3,... \tag{2.2}$$

where y_{tij} is the heavy drinking frequency for individual i in family j at time t. a_{tij} is the time-related variable at time t for individual i in family j. This corresponds to the λ_t described previously for individual i in family j. The π's are the growth parameters that vary across individuals. Note that a_{tij} is 0 when λ_t is 0 and this occurs at age 21. Therefore, π_{0ij} is defined as the heavy drinking status at 21 years of age. The coefficient π_{1ij} is the linear growth rate and π_{2ij} is the quadratic growth rate, which gives the acceleration or the rate of increase of the linear growth rate. The term e_{tij} is the random measurement error term for individual i in family j at time t, which is assumed normally distributed with a mean of 0 and common variance σ^2.

In Model A, each of the growth parameters from level 1 is specified as varying randomly across individuals from the mean values at level 2:

$$\begin{aligned} \pi_{0ij} &= \beta_{00j} + r_{0ij},\ r_{0ij} \sim N(0, \tau_{\pi 00}) \\ \pi_{1ij} &= \beta_{10j} + r_{1ij},\ r_{1ij} \sim N(0, \tau_{\pi 11}) \\ \pi_{2ij} &= \beta_{20j} + r_{2ij},\ r_{2ij} \sim N(0, \tau_{\pi 22}) \end{aligned} \tag{2.3a}$$

where β_{00j}, β_{10j}, and β_{20j} represent the growth parameter means of family j and r_{0ij}, r_{1ij}, and r_{2ij} are the random components of the growth parameters representing the deviations of individual i from the family mean values. These random components are assumed normally distributed with variances $\tau_{\pi 00}$, $\tau_{\pi 11}$, and $\tau_{\pi 22}$, respectively and covariances $\tau_{\pi 01}$, $\tau_{\pi 02}$, and $\tau_{\pi 12}$.

In Model B, individual covariates are added to the Level-2 model:

$$\begin{aligned} \pi_{0ij} &= \beta_{00j} + \beta_{010}(Male)_{ij} + \beta_{020}(ES)_{ij} + \beta_{030}(HSdrp)_{ij} + r_{0ij} \\ \pi_{1ij} &= \beta_{10j} + \beta_{110}(Male)_{ij} + \beta_{120}(ES)_{ij} + \beta_{130}(HSdrp)_{ij} + r_{1ij} \\ \pi_{2ij} &= \beta_{20j} + \beta_{210}(Male)_{ij} + \beta_{220}(ES)_{ij} + \beta_{230}(HSdrp)_{ij} + r_{2ij} \end{aligned} \tag{2.3b}$$

Here π_{0ij}, π_{1ij}, and π_{2ij} are modeled as functions of individual covariates, Male, ES, and HSdrp. β_{00j}, β_{10j}, and β_{20j} are the intercepts for family j in the respective growth parameters and β_{010}, β_{020},.., β_{230} are the regression coefficients. These regression coefficients are modeled as fixed effects, that is, they are not allowed to vary across families. Here, the random components r_{0ij}, r_{1ij}, and r_{2ij} have the same distributional assumptions and represent the deviations of individual i from their model predicted values.

In Model A, the family growth parameters are specified to vary across a population of families at level 3:

$$\begin{aligned} \beta_{00j} &= \gamma_{000} + u_{00j},\ u_{00j} \sim N(0, \tau_{\beta 00}) \\ \beta_{10j} &= \gamma_{100} + u_{10j},\ u_{10j} \sim N(0, \tau_{\beta 11}) \\ \beta_{20j} &= \gamma_{200} + u_{20j},\ u_{20j} \sim N(0, \tau_{\beta 22}) \end{aligned} \tag{2.4a}$$

where γ_{000}, γ_{100}, and γ_{200} represent the grand means in this unconditional model and u_{00j}, u_{10j}, and u_{20j} are the random parts representing the deviations of family j from the overall means. u_{00j}, u_{10j}, and u_{20j} are assumed normally distributed with variances $\tau_{\beta 00}$, $\tau_{\beta 11}$, and $\tau_{\beta 22}$, and their respective covariances.

In Model B, family-level covariates are introduced to the Level-3 model to describe variation across the population of families. A variable is considered to be a family-level variable if siblings within a family logically share the same value for the variable. In this analysis, ethnicity (Black and Hispanic) and family history variables (FH123, FH1, and FH23) are considered family variables. These predictor variable are added to the Level-3 model to study their influences on the family growth parameter intercepts:

$$\begin{aligned} \beta_{00j} &= \gamma_{000} + \gamma_{001}(\textit{Black})_j + \gamma_{002}(\textit{Hispanic})_j + \gamma_{003}(\textit{FH}123)_j + \gamma_{004}(\textit{FH}1)_j + \gamma_{005}(\textit{FH}23)_j + u_{00j} \\ \beta_{10j} &= \gamma_{100} + \gamma_{101}(\textit{Black})_j + \gamma_{102}(\textit{Hispanic})_j + \gamma_{103}(\textit{FH}123)_j + \gamma_{104}(\textit{FH}1)_j + \gamma_{105}(\textit{FH}23)_j + u_{10j} \\ \beta_{20j} &= \gamma_{200} + \gamma_{201}(\textit{Black})_j + \gamma_{202}(\textit{Hispanic})_j + \gamma_{203}(\textit{FH}123)_j + \gamma_{204}(\textit{FH}1)_j + \gamma_{205}(\textit{FH}23)_j + u_{20j} \end{aligned} \tag{2.4b}$$

In this Level-3 model β_{00j}, β_{10j}, and β_{20j} are modeled as functions of family covariates, Black, Hispanic, FH123, FH1, and FH23. γ_{000}, γ_{100}, and γ_{200} are the intercepts and γ_{001}, γ_{002},.., γ_{205} are the regression coefficients. Here, the random components, u_{00j}, u_{10j}, and u_{20j} have the same distributional assumptions and represent the deviations of family j from their model predicted values. Note that if predictor variables are included at level 2, then β_{00j}, β_{10j}, and β_{20j} are intercept terms that represent conditional means after adjusted for the Level-2 predictor variables.

Results of the Multilevel Modeling

Model A (characterized by Equations 2.2, 2.3a, 2.4a) and Model B (characterized by Equations 2.2, 2.3b, 2.4b) are estimated using the multilevel software HLM (Bryk, Raudenbush, & Congdon, 1996). Table 2.3 shows the results for Model A.

The fixed-effect estimates give the overall means for the three growth parameters. The variances of the growth parameters are partitioned among individuals into the within-family components and the between-family components. These are given by the Level-2 variance components and the Level-3 variance components in Table 2.3.

The estimated mean status of heavy drinking frequency at 21 years of age is 1.060 and has a significant variation across individuals (variance = 1.112, χ^2= 5,134.2, df = 1,614, $p < 0.001$). The linear growth rate mean of -0.011 is not significantly different from zero (se = 0.018, $p = 0.526$), but there is significant variation (variance = 0.194, χ^2 = 3,693.7, df = 1,614, $p < 0.001$) across individuals, showing that there are individuals whose growth rate is still increasing at 21, and there are also those whose rates are already on the decline at 21 with an average of about zero. The quadratic rate has a negative estimated mean of -0.127, which is significant (se = 0.014, $p < 0.001$). This shows that there is generally a deceleration in the growth rate of heavy drinking frequency. Heavy drinking frequency tends to increase, but with a slowing rate, and decreases after reaching a peak. The quadratic rate also has a significant variation (variance = 0.069, χ^2 = 3,390.4, df = 1,614, $p < 0.001$) showing that the deceleration in the growth rate varies across individuals. In sum, these results show that there are a lot of individual differences in the drinking trajectories.

As shown by the χ^2 statistics in Table 2.3, there are significant between-family variances for all three growth parameters that are given by the Level-3 variance components. These are considerable compared to the total variations (individuals' within-family components plus between-family components). The between-family variations are 25.9%, 31.7%, and 23.3% of the total variations for Status at age 21, linear rate, and quadratic rate, respectively.

Table 2.4 shows the results for Model B.

Looking at the fixed effects, FH123 have significant influence (γ_{003} = 0.400, $p < 0.001$) on heavy drinking status at 21. This shows that individuals with family history in both first degree and the second/third degree relatives tend to drink more by age 21. Both Blacks (comparing

TABLE 2.3

Three-Level Hierarchical Model for Two-Sibling Families (Unconditional Model)

Number of two-sibling families = 1,614				
Fixed Effect	*Mean*	*SE*	*t-ratio*	*P-value*
Status at 21 years of age	1.060	0.029	36.878	0.000
Linear Growth Rate	-0.011	0.018	-0.634	0.526
Quadratic Growth Rate	-0.127	0.014	-9.083	0.000
Random Effect	*Variance Components*	*df*	χ^2	*p-value*
Level 1				
Measurement errors	1.130			
Level 2				
Individual Status at 21	1.112	1614	5134.2	0.000
Individual Linear Growth Rate	0.194	1614	3693.7	0.000
Individual Quadratic Growth Rate	0.069	1614	3390.4	0.000
Level 3				
Family mean Status at 21	0.389	1613	2127.3	0.000
Family mean Linear Growth Rate	0.093	1613	1908.7	0.000
Family mean Quadratic Growth Rate	0.021	1613	1771.3	0.004
Percentage of Variance Between Families				
Status at 21 years of age	25.9%			
Linear Growth Rate	31.7%			
Quadratic Growth Rate	23.3%			
Correlations Among the Random Effects				
Level 2				
Status at 21 years of age	1.000			
Linear Growth Rate	-0.050	1.000		
Quadratic Growth Rate	-0.937	0.362	1.000	
Level 3				
Status at 21 years of age	1.000			
Linear Growth Rate	-0.590	1.000		
Quadratic Growth Rate	-0.877	0.86	1.000	
Deviance = 54,768.8				
Number of estimated parameters = 16				

TABLE 2.4

Three-Level Hierarchical Model for Two-Sibling Families (With Covariates)

Number of two-sibling families = 1,614

Fixed Effect	*Coefficients*	*SE*	*t-ratio*	*p-value*
Status at 21 years of age, π_{0ij}				
Intercept, γ_{000}	0.644	0.050	12.768	0.000
Black, γ_{001}	-0.575	0.064	-9.041	0.000
Hisp, γ_{002}	-0.279	0.076	-3.668	0.000
FH123, γ_{003}	0.400	0.086	4.631	0.000
FH1, γ_{004}	0.022	0.079	0.280	0.780
FH23, γ_{005}	0.048	0.068	0.701	0.483
Male, β_{010}	0.893	0.048	18.598	0.000
ES, β_{020}	0.421	0.077	5.486	0.000
HSdrp, β_{030}	0.179	0.063	2.846	0.005
Linear Growth Rate, π_{1ij}				
Intercept, γ_{100}	-0.052	0.036	-1.566	0.117
Black, γ_{101}	0.196	0.042	4.704	0.000
Hisp, γ_{102}	0.124	0.049	2.548	0.001
FH123, γ_{103}	-0.138	0.056	-2.447	0.015
FH1, γ_{104}	-0.021	0.052	-0.410	0.682
FH23, γ_{105}	-0.110	0.044	-2.480	0.013
Male, β_{110}	0.031	0.033	0.938	0.349
ES, β_{120}	-0.114	0.051	-2.219	0.026
HSdrp, β_{130}	0.124	0.043	2.918	0.004
Quadratic Growth Rate, π_{2ij}				
Intercept, γ_{200}	-0.151	0.026	-5.715	0.000
Black, γ_{201}	0.171	0.033	5.156	0.000
Hisp, γ_{202}	0.140	0.040	3.540	0.001
FH123, γ_{203}	-0.018	0.045	-0.407	0.684
FH1, γ_{204}	0.055	0.042	1.307	0.191
FH23, γ_{205}	0.017	0.036	0.466	0.641
Male, β_{210}	-0.096	0.027	-3.562	0.001
ES, β_{220}	0.043	0.043	0.983	0.326
HSdrp, β_{230}	-0.017	0.035	-0.505	0.613

Table 2.4 continued

Table 2.4 continued

Random Effect	*Variance Components*	*df*	χ^2	*p-value*
Level 1				
Measurement errors, e_{tij}	1.135			
Level 2				
Status at 21, r_{0ij}	0.829	1611	4706.2	0.000
Linear Growth Rate, r_{1ij}	0.174	1611	3675.5	0.000
Quadratic Growth Rate, r_{2ij}	0.064	1611	3358.2	0.000
Level 3				
Status at 21, u_{00j}	0.334	1608	2146.9	0.000
Linear Growth Rate, u_{10j}	0.718	1608	1883.9	0.000
Quadratic Growth Rate, u_{20j}	0.017	1608	1758.3	0.005

Deviance = 54,063.3

Number of estimated parameters = 40

to non-Blacks) and Hispanics (comparing to non-Hispanics) have significantly lower drinking frequencies at 21 (γ_{001} = -0.575, $p < 0.001$; γ_{002} = -0.279, $p < 0.001$) after controlling for the other variables. All the individual covariates, Male, ES, and HSdrp, have positive influence on the drinking status at 21 (β_{010} = 0.893, $p < 0.001$; β_{020} = 0.421, $p < 0.001$; β_{030} = 0.179, $p = 0.005$). This indicates that males, early starters and high school dropouts tend to drink more at age 21.

For the regression of the linear rate at age 21, the intercept term is not significant (γ_{100} = -0.052, $p = 0.117$) and could have been set to 0 in the model. Both ethnicity groups, Blacks and Hispanics, have significantly higher linear rates at 21 (γ_{101} = 0.196, $p = 0.000$; γ_{102} = 0.124, $p = 0.001$). This may be because they started later and were still on the upward trend because these two group also had lower drinking frequencies at 21. Two of the family history variables (FH123, FH23) have significant negative influence on the linear rate (γ_{103} = -0.138, $p = 0.015$; γ_{105} = -0.110, $p = 0.013$). Early Start (ES) is also negatively related to the linear rate (β_{120} = -0.114, $p = 0.026$). This shows that individuals with family history and who started drinking early might have reached their peak and were probably slowing down by 21. HSdrp is positively related to the linear rate (β_{130} = 0.124, $p = 0.004$) showing that high school dropouts were still at an upward trend in their drinking pattern at age 21.

Only ethnicity variables and gender show significant effects on the quadratic growth rate. Both Blacks and Hispanics are positively related (γ_{201} =0.171, $p < 0.001$; γ_{202} = 0.140, $p = 0.001$) to the quadratic growth rate, showing the linear rate is decreasing slower. Being male is negatively related to the quadratic growth rate, showing that growth rate is decreasing faster (β_{210} = -0.096, p = 0.001). Taking the three growth parameters together, the results show that Blacks and Hispanics were not drinking as much at 21 but were on an upward trend that decreased slower, whereas males were drinking a lot at 21, but had reached their peak and were on a faster downward trend at 21.

Table 2.4 also shows that there are still considerable unexplained variances in the three growth parameters for both the within-family model at Level 2 and the between-families model at Level 3. There are probably more key covariates that can be included in the model to explain both the within-family variations across individuals and the between-family variations.

Multilevel models are very useful for partitioning the variations into the within-family variance components and the between-family variance components and for formulating and testing hypothesis at each level separately.

ANALYSIS USING THE LATENT VARIABLE MODELING APPROACH

The growth model with a linear growth rate and a quadratic growth rate specified by Equation 2.2 can be reformulated to fit into the general latent variable framework (Jöreskog & Sorböm, 1979). In the latent growth formulation, the repeated measurements are strung out as a multivariate vector, and we have a regular single-level latent variable model where all the flexibility of structural equation modeling can be applied. Figure 2.2 shows such a model.

The outcome Y_{ti} of individual i at time t is expressed as a quadratic function of the time-related variable λ_t,

$$Y_{ti} = 1.S_i + \lambda_t\, Lrate_i + \lambda_t^{\,2}\, Qrate_i + \varepsilon_{ti},\ t = 1,..,15 \qquad (2.5)$$

with the three latent factors S_i, $Lrate_i$, and $Qrate_i$ representing the three growth parameters, S_i being the random intercept, $Lrate_i$ the linear

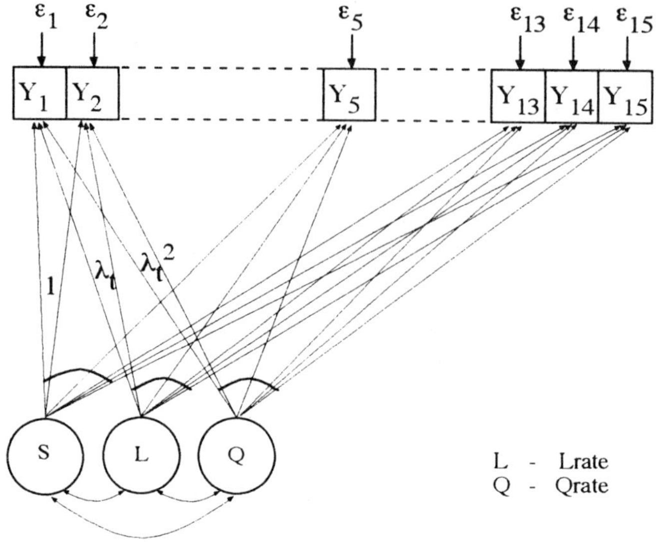

FIG. 2.2 A longitudinal growth model of heavy drinking.

growth rate, and $Qrate_i$ the quadratic growth rate or the acceleration. ε_{ti} is the measurement error at time t for individual i. In order to achieve the latent variable formulation, λ_t is the fixed factor loading of Y_{ti} on L_i. The factor loading of Y_{ti} on Q_i is λ_t^2.

Longitudinal sibling data can be analyzed using multivariate models with multiple growth trajectories, one for each sibling. This is possible because of the small number of siblings per family. The unit of analysis is the family, and the growth factors of the siblings can be interrelated by correlating the factors or by adding structural paths as in any standard structural equation model. In this way, the nonindependence between siblings within family is modeled within the structural model. These multivariate growth models can be estimated using any structural equation modeling (SEM) software that handles mean structures. That covers practically all the commercially available SEM software, for example, EQS (Bentler, 1985), LISCOMP(Muthén, 1987), LISREL (Jöreskog & Sörbom, 1989), and MPlus (Muthén & Muthén, 1998).

This multivariate approach is illustrated through Model C and Model D. The main aim of these examples is not in the theory-building but in pointing out possibilities. Both models consist of two growth processes, one for the older sibling and one for the younger sibling. These models can be generalized to any reasonable number of siblings in a family. The analysis is carried out without covariates in Model C. Individual variables, gender (Male), early start (ES), and high school dropout (HSdrp) and family variables, ethnicity (Black, Hispanics), and family drinking history (FH123, FH1, FH23) are included in Model D.

The growth trajectories are characterized by quadratic growth functions with three growth factors each (*S21*, *LRate*, *QRate*), where *S21* is the individual status factor at 21 years of age, *LRate* is the individual linear growth rate factor, and *QRate* is the individual quadratic growth rate factor. The factors are subscripted with *O* for the older sibling and with *Y* for the younger sibling. The model of longitudinal heavy drinking for family *j* is given by two simultaneous growth functions:

$$O_{jt} = S21_{oj} + \lambda_t \, Lrate_{oj} + \lambda_t^2 \, Qrate_{oj} + \varepsilon_{jot} \tag{2.6}$$

for the older sibling, and

$$Y_{jt} = S21_{Yj} + \lambda_t \, Lrate_{Yj} + \lambda_t^2 \, Qrate_{Yj} + \varepsilon_{jYt} \tag{2.7}$$

for the younger sibling,

where O_{jt} and Y_{jt} are heavy drinking frequency at time $t = (18,19,..,32)$ for the older sibling and the younger sibling, respectively. The λ_ts are the time-related variables; ε_{jOt} and ε_{jYt} are the measurement errors. If the equivalent parameters are equated across siblings, and if we assume that the measurement error variances are common across time and across siblings, then the pair of growth functions above would reduce to Equation (2.2) for the HLM model at level 1.

Model C (Fig. 2.3) is the basic family growth model with the two quadratic growth trajectories, one for each sibling.

All the measurement error variances are allowed to be different across sibling groups and across time. The factor variances are also allowed to be different across sibling groups. In this general preliminary model, the growth factor means are allowed to be different across siblings. The equality of each parameter across siblings can be tested by reanalyzing the same model but with the parameter in question equated across siblings. The two models are said to be nested. The equality hypothesis can then be tested using the chi-square difference test with 1 degree of freedom of the two nested models. The growth factors are

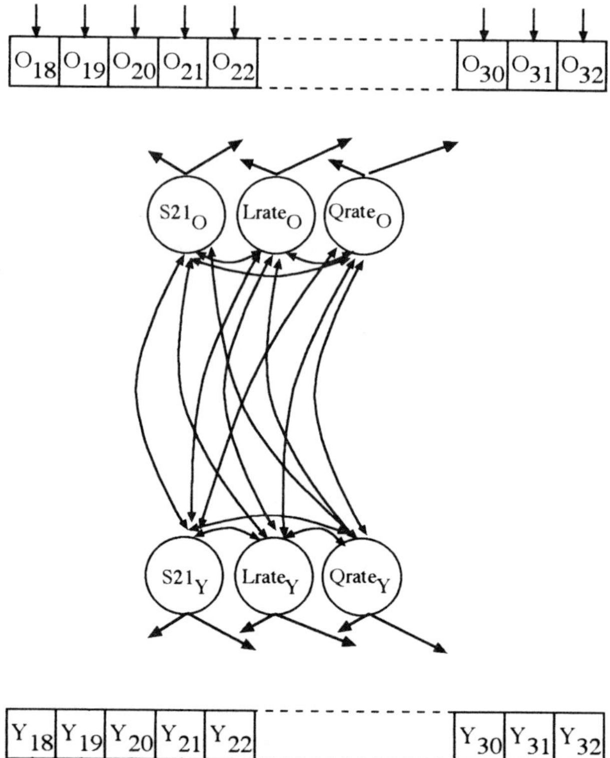

FIG. 2.3. A Longitudinal Growth Model of Heavy Drinking for Two-Sibling Families - Model C.

correlated among one another to investigate the strength of associations among the two sets of growth factors.

In Model D (see Fig. 2.4), the growth model remains the same as in Model C but with covariates introduced into the model.

Each sibling's individual covariates are allowed to influence his or her own growth trajectory by having paths from the covariates to the three growth parameters. The family-level covariates are allowed to influence both trajectories by including paths from these covariates to each of the six growth parameters. Because it is not possible to include all the predictors that will explain the associations among the growth parameters, the residual unexplained variances are allowed to correlate freely. In this model, the coefficients of family covariates and individual covariates are allowed to be different across siblings. The equality of each of these coefficients across the two sibling groups can be tested by

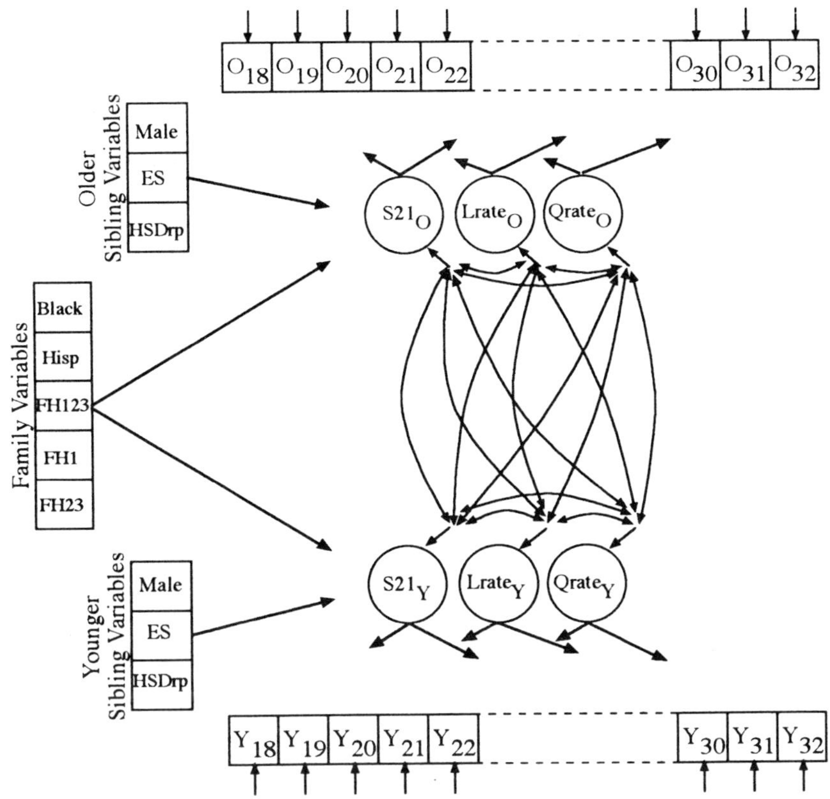

FIG. 2.4. A Longitudinal Growth Model of Heavy Drinking for Two-Sibling Families - Model D.

using a chi-square difference test of two nested models as before. It would be interesting to investigate whether a family variable has the same impact on the drinking patterns of the older siblings and the younger siblings. The influence of a family variable may interact with sibling order. For example, a family environment with alcoholic parents may affect an older child and a younger child differently.

The growth factors are allowed to be freely correlated in Model D as in Model C. Model D can be modified to include direct paths among the growth factors. If we hypothesize that the trajectories of the older siblings influence those of the younger siblings, directional paths can be introduced from $S21_O$ to $S21_Y$, $LRate_Y$ and $QRate_Y$, from $LRate_O$ to

$LRate_Y$ and $QRate_Y$ and from $QRate_O$ to $QRate_Y$. Such a possible revised model is shown in Figure 2.4R.

The two multivariate models are estimated using Mplus, a newly developed modeling software (Muthén & Muthén, 1998) that handles the multiple-cohort data structure in the NLSY data.

Results of the Multivariate Modeling

Table 2.5 shows the results of Model C.

This model does not have any predictor variables. Here, we are interested in examining the goodness of fit for the growth model and

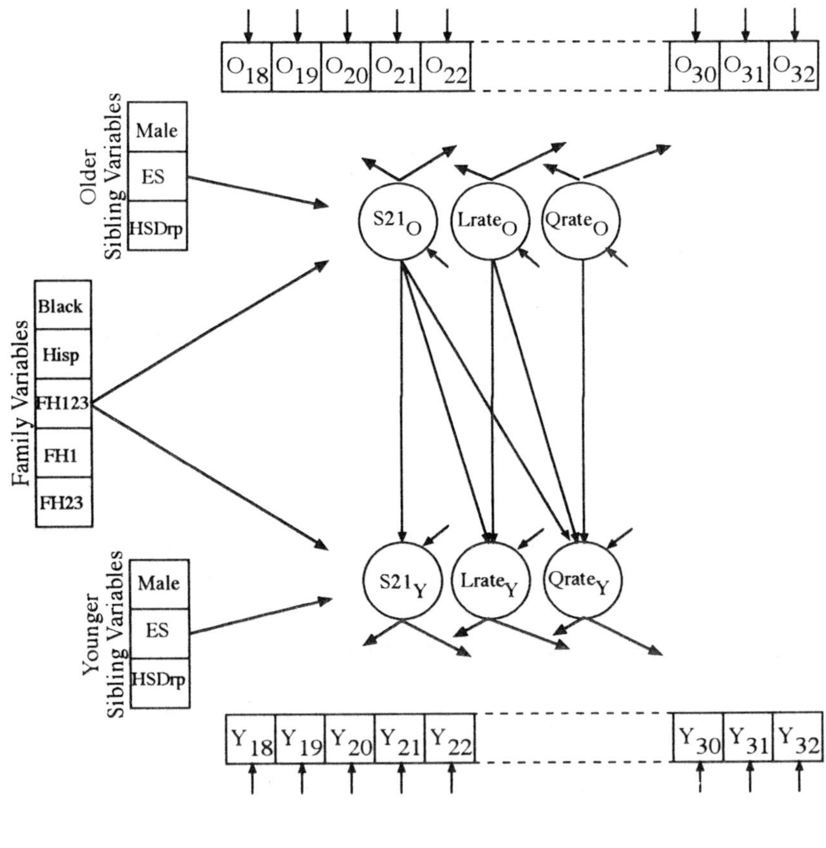

FIG. 2.4R. A Longitudinal Growth Model of Heavy Drinking for Two-Sibling Families - Revised Model D.

TABLE 2.5

Multivariate Longitudinal Model for Two-Sibling Families (Without Covariates)

Number of two-sibling families =1,614

$\chi^2 = 762.434$ (df = 438, $p = 0.000$); RMSEA = 0.0214, *p(RMSEA < 0.05) =1*

	Older Sibling			*Younger Sibling*		
	Estimate	*SE*	*t-ratio*	*Estimate*	*SE*	*t-ratio*
Means						
Status at 21 years of age	1.069	0.038	27.811	1.018	0.034	30.049
Linear Growth Rate	0.006	0.039	0.153	0.029	0.016	1.812
Quadratic Growth Rate	-0.066	0.018	-3.774	-0.088	0.010	-8.780
Variances						
Status at 21 years of age	1.345	0.081	16.629	1.400	0.066	21.274
Linear Growth Rate	0.075	0.045	1.665	0.101	0.013	7.544
Quadratic Growth Rate	0.058	0.15	3.943	0.035	0.007	4.998
Covariances & Correlations	*Older Sibling*			*Younger Sibling*		
Older Sibling	*OS21*	*OL*	*OQ*	*YS21*	*YL*	*YQ*
Status at 21 years of age OS21	1.345					
	(1.000)					
Linear Growth Rate OL	0.065	0.075				
	(0.205)	*(1.000)*				
Quadratic Growth Rate OQ	-0.193	-0.002	0.058			
	(-0.691)	*(-0.030)*	*(1.000)*			

Table 2.5 continued

Table 2.5 continued

Covariances & Correlations		*Older Sibling*			*Younger Sibling*	
Younger Sibling						
Status at 21 years of age YS21	0.376	-0.173	0.025	1.400		
	(0.274)	*(-0.534)*	*(0.088)*	*(1.000)*		
Linear Growth Rate YL	-0.043	0.040	-0.002	0.290	0.101	
	(-0.117)	*(0.460)*	*(-0.026)*	*(0.771)*	*(1.000)*	
Quadratic Growth Rate YQ	-0.031	0.019	0.001	-0.15	-0.003	0.035
	(-0.143)	*(0.371)*	*(0.022)*	*(-0.678)*	*(-0.050)*	*(1.000)*

to compare in general the two sets of growth parameters in terms of their means and variances. We will also look at the strength of associations among the growth parameters.

Figure 2.5 shows the mean trajectories of the siblings over time as estimated in Model C.

This model has a chi-square value of 762.43 with 438 degrees of freedom (p = 0.000, N = 1,614). A test of close fit using the root mean squared error of approximation (RMSEA; Browne & Cudeck, 1993) shows an acceptable model fit (RMSEA = 0.0214, p(RMSEA < 0.05) = 1). Looking at the drinking status at age 21, the older siblings average slightly higher than the younger siblings (1.069, 1.018). The linear-rate means appear to be not significantly different from 0 (at 0.05 level) for both siblings showing that on average, the heavy drinking patterns tend to peak around 21 years of age for both siblings. The linear growth rate variance at age 21 is significant for the younger siblings but not for the older siblings showing that although most of the older siblings peaked at 21, there are considerable variations for the younger siblings. It is interesting to compare this result to the findings for Model A where there is only one set of growth parameters for both siblings.

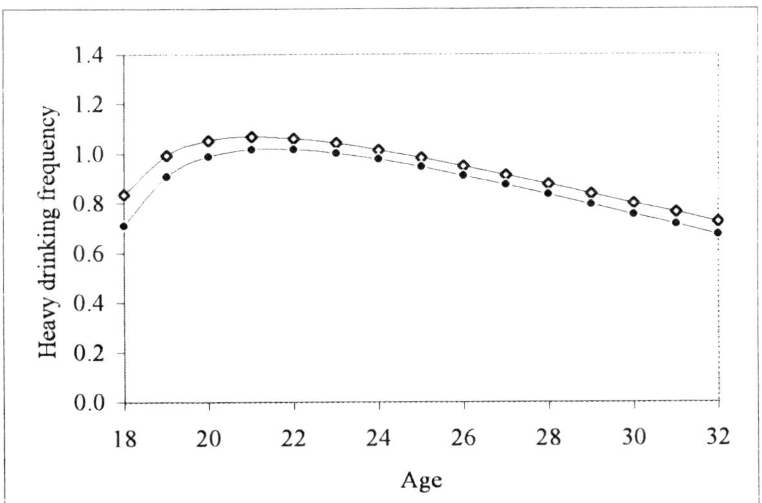

FIG. 2.5. Estimated Mean Trajectories

The difference in variation in the linear rate between the two sibling groups would not have been detected in Model A. The quadratic-rate means are negative and significant for both siblings (-0.066, -0.088). The variances of all the growth factors are significant except for the linear growth rate of the older siblings, showing considerable individual differences in status at age 21 and growth over time. The variance–co-variances of the two sets of growth factors are also shown in Table 2.5. The corresponding correlations are given in parentheses underneath the covariances. For the within-siblings factor covariances, there is a significant negative association between the status at 21 and the quadratic rate that is consistent across both siblings (r = -0.691, -0.678). This shows that individuals who peak higher would tend to decrease at a faster rate in terms of heavy drinking. Looking at factor covariances across siblings, there are significant positive associations between the status at age 21 (r = 0.274), and a significant positive association between the linear rates at age 21 (r = 0.460), showing that there are some similarities in the drinking patterns of the siblings. There is also a significant negative association between the older siblings' linear growth rate at 21 and the younger siblings' drinking status at 21 (r = -0.534) and a significant negative association between the older siblings' status at 21 and the younger siblings' quadratic growth rate. Family-level predictor variables are introduced in Model D in an attempt to model these associations.

An individual's trajectory or growth pattern is established by taking all three growth factors together. For example, a high status at 21 with near-zero linear rate and a negative quadratic rate characterize a trajectory that is peaking at age 21 (near zero rate) and on a downward trend (zero rate decreasing to become negative rate). A low status at 21 with a positive linear rate and a negative quadratic rate characterize a trajectory low at 21 that is still on the up trend but is slowing down.

Table 2.6 shows the results of Model D.

With individual-level and family-level predictor variables in the model, we are interested in the influences of these predictors on the trajectories of the two siblings. Model D has a chi-square value of 1,077.85 with 720 degrees of freedom (p = 0.000, N = 1,614), a test of close fit shows a RMSEA (e) of 0.00175 and p(e < = 0.05) = 1, which also gives acceptable model fit. Table 2.6 mainly shows the results from the added structural part of the model. The estimates are the regression coefficients in regressing the growth factors/parameters on the individ-

TABLE 2.6

Multivariate Longitudinal Model for Two-Sibling Families (With Covariates) $N = 1,614$

$\chi^2 = 1,077.85$(df = 720, $p = 0.000$);	RMSEA = 0.0175, p(RMSEA < 0.05) = 1					
	Older Sibling			*Younger Sibling*		
	Estimate	*SE*	*t-ratio*	*Estimate*	*SE*	*t-ratio*
Status 21 Regression						
Intercept	0.687	0.071	9.703	0.612	0.060	10.181
Individual variables						
Gender (Male=1, Female=0)	0.777	0.072	10.733	0.908	0.061	14.882
Early Start	0.550	0.130	4.232	0.367	0.092	3.979
High School-Dropout	0.065	0.091	0.713	0.198	0.078	2.540
Family variables						
Black	-0.532	0.088	-6.067	-0.621	0.075	-8.314
Hispanic	-0.239	0.106	-2.267	-0.253	0.090	-2.826
FH123	0.375	0.120	3.127	0.260	0.102	2.544
FH1	0.163	0.105	1.550	0.110	0.090	1.226
FH23	0.020	0.098	0.203	0.116	0.083	1.389
Linear Growth Rate Regression						
Intercept	-0.274	0.081	-3.363	-0.019	0.030	-0.628
Individual variables						
Gender (Male=1, Female=0)	0.181	0.084	2.142	0.050	0.031	1.609

Table 2.6 continued

Table 2.6 continued

	Older Sibling			*Younger Sibling*		
	Estimate	*SE*	*t-ratio*	*Estimate*	*SE*	*t-ratio*
Early Start	-0.007	0.151	-0.045	-0.133	0.047	-2.836
High School-Dropout	0.306	0.106	2.900	0.096	0.040	2.418
Family variables						
Black	0.312	0.100	3.104	0.080	0.037	2.140
Hispanic	-0.016	0.121	-0.135	0.071	0.045	1.574
FH123	0.187	0.137	1.362	0.019	0.051	0.367
FH1	0.307	0.120	2.555	-0.036	0.045	-0.811
FH23	-0.077	0.112	-0.689	0.009	0.042	0.215
Quadratic Growth Rate Regression						
Intercept	0.036	0.037	0.995	-0.095	0.019	-4.981
Individual variables						
Gender (Male=1, Female=0)	-0.106	0.038	-2.793	-0.071	0.020	-3.591
Early Start	-0.057	0.068	-0.843	0.035	0.030	1.181
High School-Dropout	-0.115	0.048	-2.420	-0.005	0.025	-0.196
Family variables						
Black	-0.017	0.045	-0.383	0.103	0.024	4.346
Hispanic	0.081	0.054	1.492	0.069	0.028	2.419
FH123	-0.130	0.062	-2.117	-0.023	0.032	-0.721
FH1	-0.124	0.054	-2.306	0.024	0.028	0.860
FH23	0.024	0.050	0.477	-0.003	0.050	-0.130

ual covariates and the family covariates. In this preliminary model, the coefficients of family covariates are allowed to be different across siblings to investigate differential impact. Equality of the coefficients across siblings can be tested by comparing nested models but is not carried out in this illustrative example.

Looking at the coefficients in Table 2.6, it is interesting to see that there are many coefficients that appear similar across siblings but also several coefficients that appear to be quite different across siblings. For the regressions on the drinking status at 21, the coefficients appear to be similar across siblings except for the coefficients of HSdrp (0.065,0.198), which is significant for the younger siblings and not for the older siblings. This result shows that a younger sibling who is a high-school dropout has the tendency to be drinking more at age 21 compared to a non-high school dropout, but not so for an older sibling. Both being male and starting drinking early are related to a significantly higher drinking status at 21, this is the same for both siblings. With the family variables, Blacks (compared to non-Blacks) and Hispanics (compared to non-Hispanics) have a significantly lower drinking status at 21 for both siblings. In terms of family history, both siblings drink more at age 21 if they have family drinking history among the first-degree relatives as well as the second- or third-degree relatives.

For the linear growth rate regression, there appears to be differential impact for early starters (ES). This is negatively significant for the younger sibling only, which shows that younger siblings who started drinking early tend to have lower linear rate at 21. This may mean that they peak earlier, and by 21, they are either at the peak or already on a downward path. It is interesting to note that only 8.4% of the older siblings are early starters, whereas 12.6% of the younger siblings are early starters. There also appears to be greater gender differences for the older siblings than for the younger siblings (0.181, 0.05). This coefficient (Male effect) is positive and significant for the older siblings and not for the younger siblings. For the family history variables, only FH1 has significant influence and only for the older siblings, showing that their heavy drinking frequency is at a higher growth rate. This seems to indicate that the older sibling is affected more by family environment with alcoholic parents or other first-degree relatives. High school dropouts also have a higher growth rate at 21 and for both siblings. The Black effect has a positive influence on the linear growth rate at 21 for both siblings.

The Male effect has significant negative coefficients on the quadratic growth rate for both siblings showing decelerations. Taking gender effects on all three growth parameters together, it shows that older male siblings tend to drink more than female older siblings at age 21; they also tend to be on a steeper trajectory that is decelerating faster compared to the older female siblings. The younger male siblings also tend to drink more than the females at age 21. There is no significant difference in the rate of increase in the drinking frequency, but the males tend to decelerate faster from a higher drinking status at 21 compared to younger female siblings.

Both Blacks and Hispanics have significant positive coefficients on the quadratic growth rate, probably showing a slower decline for the younger siblings only. Both high school dropouts and having family history involving first-degree relatives seem to result in negative coefficients on the quadratic growth rate, showing a slowing ascent that is leveling out faster or a faster decline but only for the older siblings. For older siblings with family history, this may be a result of having a higher growth rate at 21 than those without family history to start with.

To illustrate some of these differential effects of the predictor variables on the heavy drinking patterns of the older siblings and the younger siblings, we contrast the effects by showing some estimated mean trajectories in Fig. 2.6a to Fig. 2.6d. In Fig. 2.6a, we show the estimated average heavy drinking trajectories (from age 20 to age 26) of the older siblings and the younger siblings who are males from Black families with family history of alcoholism among the first-degree relatives (FH1 or FH123). A combined trajectory for both siblings is also shown; this is obtained from the multilevel analysis (Model B).

Figure 2.6b shows the same set of trajectories for Black males without family history.

Here, the trajectories approximately coincide showing practically no difference between the older sibling group and the younger sibling group. In contrast, Fig. 2.6a shows that family history of alcoholism has impact on both the older siblings and the younger siblings, but the older siblings are more severely affected. The mean trajectory of the younger siblings shows a flat trajectory that is elevated by about 0.2 points, yet the mean trajectory of the older siblings shows an upward trend that results in an increase of about 0.4 points when peaking at age 24, when it starts to decline slowly.

Figures 2.6c and 2.6d show the same set of trajectories for White males.

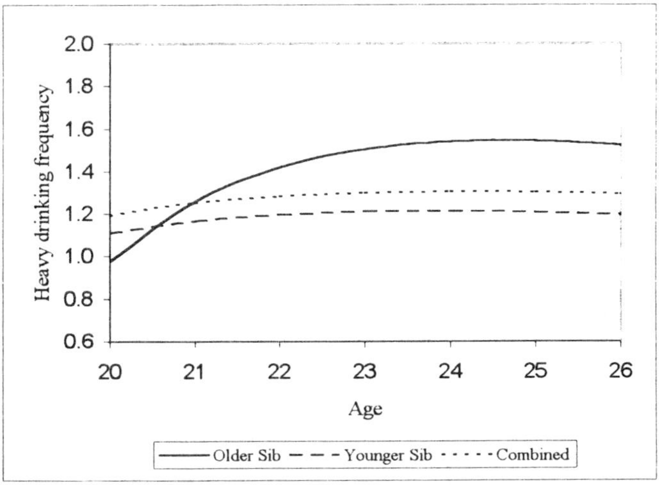

FIG. 2.6a. Black Male with Family History (estimated mean trajectories)

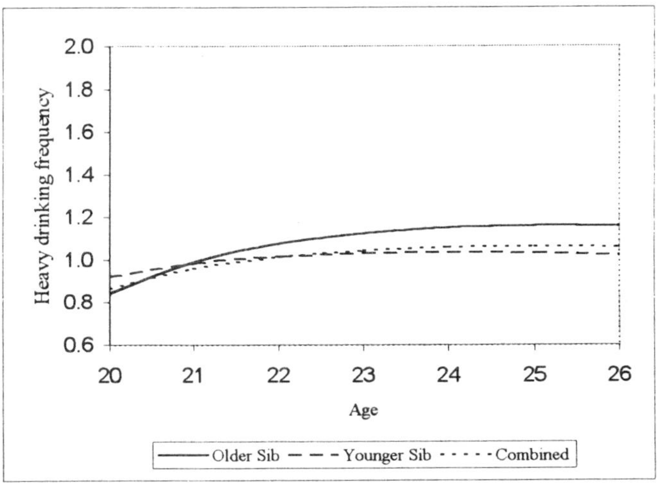

FIG. 2.6b. Black Male without Family History (estimated mean trajectories)

The White males have, in general, a higher heavy drinking frequency level than the Black males. Figure 2.6d shows the trajectories for White males without family history. Here again, the trajectories approximately coincide showing no difference between the older sibling group and the

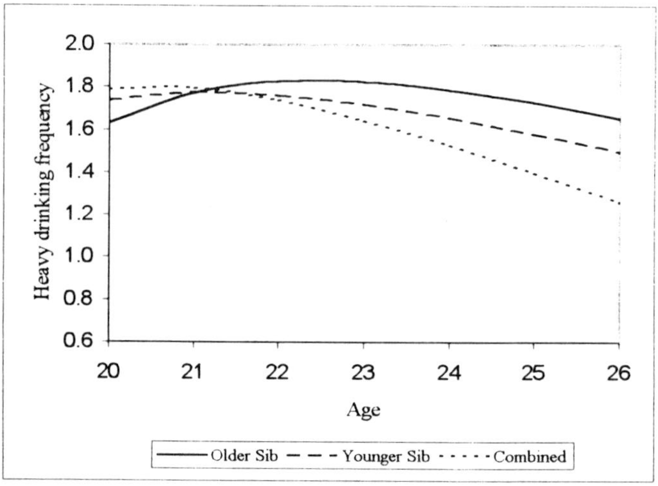

FIG. 2.6c. White Male with Family History (estimated mean trajectories)

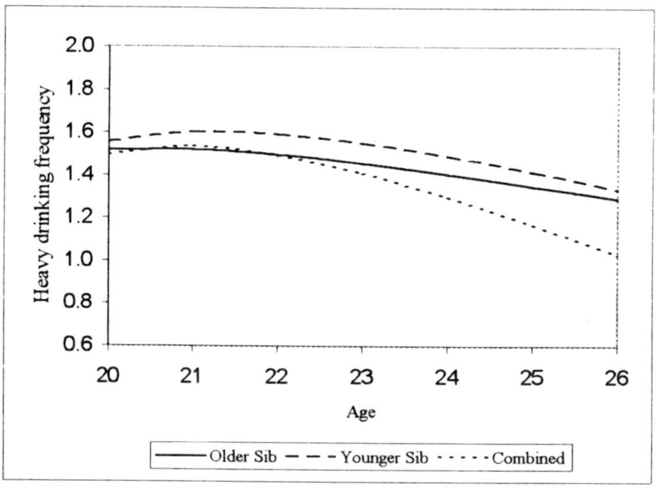

FIG. 2.6d. White Male without Family History (estimated mean trajectories)

younger sibling group. Figure 2.6c shows a pattern that is quite consistent with the results shown in Figure 2.6a for Black males with family history of alcoholism. The older siblings from the White families are also affected more by family history of alcoholism compared to the younger siblings. The mean trajectory of the younger siblings shows a trajectory that is elevated by about 0.2 points and the mean trajectory of the older siblings shows an upward trend that results in an increase of about 0.4 points when peaking at age 22 and then starts to decline slowly.

These preliminary results have generated some hypotheses to be tested. For example, which of the coefficients are equal across siblings and which are not equal? There may also be some challenging new research questions for the substantive researchers. For example, for those coefficients that are not equal and show differential effects across siblings, would there be plausible explanations in the current developmental theories as to why the effects are different?

CONCLUSIONS

Two approaches to analyzing family data were discussed. The first is the multilevel approach that takes the hierarchical data structure into account and deals efficiently with the three-level hierarchical data. The multilevel modeling allows the introduction of individual covariates and family covariates into the appropriate levels, thus encouraging clear conceptualization of relationships at each level. In the modeling of individual development and change over time, the hierarchical linear modeling framework is very flexible about the time structure because observations are viewed as nested within individuals and not as a fixed set of measurements across individuals. The multilevel modeling also allows for the partitioning of variances into the across-siblings, within-family components and the between-family components.

In this approach, we can formulate and test hypotheses explicitly at the individual level and at the family level separately. Another advantage of this approach is that if the families are again nested in larger organizational units, for instance, in neighborhoods, then the modeling can further be extended to another higher level to study neighborhood effects on families. Models can be built to answer questions effectively in the first and second areas of interest in modeling longitudinal family data mentioned in the introduction. The three-level model will not model sibling relationships and sibling influences mentioned as the third area of interest. However, interaction effects between family variables and individual characteristics including sibling order mentioned in the fourth area of interest can be studied in some limited ways by introducing interaction terms at Level 2 as in traditional linear regression models.

The second approach is the multivariate approach that analyzes the longitudinal family data in a multivariate latent growth model. This approach takes the family as the unit of analysis and allows the siblings to have different growth processes. Having multiple growth processes in the model, one for each sibling, allows separate modeling of the developmental growth of each of the siblings. It takes into account the nonindependence of the siblings within family by modeling the dependence within the covariance structure but this approach does not explicitly partition the variance into the within-family component and the between-family component. Models can be set up to answer questions from all four areas of interests in analyzing longitudinal family data mentioned earlier. It is useful when studying small numbers of siblings per family, for example, studying first-born and second-born; oldest child, middle child, and youngest child; or twins. This method allows for the testing of differential effects of family variables on the development of siblings due to sibling order by comparing the goodness of fit of explicitly specified nested models.

The multivariate approach also has the advantage of the full utility of the structural equation modeling framework. We can have multiple indicators for the outcome variables as well as for the predictor variables. The model can include direct effects as well as indirect effects through mediators. Combined with multiple-group analysis, this approach presents more opportunities for the study of similarities and differences in the developmental growth of the siblings. One example is to have the two-sibling families divided into a same-gender siblings group and an opposite-gender siblings group. Multiple-group analyses can then be carried out to see if the developmental relationships between siblings and influences of family variables are the same across the two groups. It would also be interesting to investigate if families with family history of alcoholic problems are the same as those without in terms of siblings' developmental relationships and effects of other family predictor variables.

The multilevel approach and the multivariate approach described are useful in different research settings and objectives and with different data structures. The multivariate latent growth approach is rich and full of potential to make it possible to answer many research questions in the developmental area. The intent of this chapter is to stimulate analysis using this framework. The illustrative examples serve as an introduction to this approach.

ACKNOWLEDGMENT

This research was supported by Grant 1 R21 AA10948-01A1 from NIAAA.

REFERENCES

Bentler, P. M. (1985). Theory and implementation of EQS: A structural equations program. Los Angeles: BMDP Statistical Software.

Bock, R. D. (1989). *Multilevel analysis of educational data*. San Diego, CA: Academic Press.

Browne, M. W., & Cudeck, R. (1993). Alternative ways of assessing model fit. In K. A. Bollen & J. S. Long (Eds.), *Testing structural equation models* (pp. 136–162). Newbury Park, CA: Sage.

Bryk, A. S., & Raudenbush, S. W. (1987). Application of hierarchical linear models to assessing change. *Psychological Bulletin, 101*, 147–158.

Bryk, A. S., & Raudenbush, S. W. (1992). *Hierarchical linear models: Applications and data analysis methods*. Newbury Park, CA: Sage Publications.

Bryk, A. S., Raudenbush, S. W., & Congdon, R. T. (1996). *HLM: Hierarchical Linear and Nonlinear Modeling with the HLM/2L and HLM/3L Programs*. Chicago, IL: Scientific Software International, Inc.

Duncan, S. C., & Duncan, T. E. (1996). A multivariate latent growth curve analysis of adolescent substance use. *Structural Equation Modeling, 3*, 323–347.

Goldstein, H. I. (1986). Multilevel mixed linear model analysis using iterative generalized least squares. *Biometrics, 73*, 43–56.

Goldstein, H. I. (1995). *Multilevel Statistical Models*. London: Edward Arnold.

Jöreskog, K. G., & Sörbom, D. (1979). *Advances in factor analysis and structural equation models*. Cambridge, MA: Abt Books.

Jöreskog, K. G., & Sörbom, D. (1989). LISREL 7 user's reference guide. Mooresville, IN: Scientific Software.

Laird, N. M., & Ware, J. H. (1982). Random-effects models for longitudinal data. *Biometrics, 38*, 963–974.

McArdle, J. J., & Epstein, D. (1987). Latent growth curves within developmental structural equation models. *Child Development, 58*, 110–133.

McArdle, J. J. (1988). Dynamic but structural equation modeling of repeated measures data. In R. B. Cattel & J. Nesselroade (Eds.), *Handbook of Multivariate Experimental Psychology* (2nd ed.; pp. 561–614). New York: Plenum.

Meredith, W., & Tisak, J. (1984). *"Tuckerizing" curves*. Paper presented at the annual meeting of the Psychometric Society, Santa Barbara, California.

Meredith, W., & Tisak, J. (1990). Latent curve analysis. *Psychometrica, 55*, 107–122.

Muthén, B. O. (1987). LISCOMP: Analysis of linear structural equations with a comprehensive measurement model. Mooresville, IN: Scientific Software.

Muthén, B. O. (1991). Analysis of longitudinal data using latent variable models with varying parameters. In L. Collins & J. Horn (Eds.), *Best methods for the analysis of change: Recent advances, unanswered questions, future directions* (pp. 1–17). Washington DC: American Psychological Association.

Muthén, B. O. (1993). Latent variable modeling of growth with missing data and multilevel data. In C. R. Rao & C. M. Cuadras (Eds.), *Multivariate analysis: Future directions 2* (pp. 199–210). Amsterdam: North-Holland.

Muthén, B. O. (1996). *Longitudinal studies of achievement growth using latent variable modeling.* Technical report # 412, National Center for Research on Evaluation, Standards, and Student Testing (CRESST), University of California, Los Angeles.

Muthén, B. O. (1997). Latent variable modeling of longitudinal data and multilevel data. In A. E. Raftery (Ed.), *Sociological methodology* (pp. 453–480). Washington, DC: Blackwell.

Muthén, B.O., & Curran, P. J. (1997). General longitudinal modeling of individual differences in experimental designs: A latent variable framework for analysis and power estimation. *Psychological Methods, 2,* 371–402.

Muthén, B.O., & Khoo, S. T. (1998). Longitudinal studies of achievement growth using latent variable modeling. *Learning and Individual Differences* V10 N2, 73–101.

Muthén, L. K., & Muthén, B. O. (1998). Mplus User's Guide: The Comprehensive Modeling Program for Applied Researchers. Los Angeles: Muthén & Muthén.

Neale, M. C., Walters, E. E., Eaves, L. J., Maes, H. H., & Kendler, K. S. (1994). Multivariate genetic analysis of twin-family data on fears: Mx models. *Behavior Genetics, 24,* 119–139.

Pike, A., McGuire, S., Hetherington, E. M., Reiss, D., & Plomin, R. (1996). Family environment and adolescent depressive symptoms and antisocial behavior: A multivariate genetic analysis. *Developmental Psychology, 32,* 590–603.

Raudenbush, S. W., Brennan, R. T., & Barnett, R. C. (1995). A multivariate hierarchical model for studying psychological change within married couples. *Journal of Family Psychology, 9,* 161–174.

SAS Institute (1995). SAS/STAT Software:Changes and Enhancements, Release 6.11, Cary, NC: Author.

Strenio, J. L. F., Weisberg, H. I., & Bryk, A. S. (1983). Empirical Bayes estimation of individual growth curve parameters and their relationship to covariates. *Biometrics, 39,* 71–86

Willett, J. B., & Sayer, A. G. (1994). Using covariance structure analysis to detect correlates and predictors of individual change over time. *Psychological Bulletin, 116,* 363–381.

Woodhouse, G., Rabash, J., Goldstein, H., Yang, M., & Plewis, I. (1996). MLn, version 1.0a. London: Multilevel Models Project, Institute of Education.

3

The Natural History of Smoking: A Pattern–Mixture Random-Effects Regression Model

Donald Hedeker, PhD
University of Illinois at Chicago

Jennifer S. Rose, PhD
Indiana University

This chapter describes and illustrates use of random-effects regression models (RRM) to examine the natural history of smoking from adolescence to adulthood. For longitudinal data analysis, RRMs are useful because they allow for the presence of missing data, time-varying or invariant covariates, and subjects measured at different time points. Thus, a key advantage of RRM is that it can accommodate unbalanced longitudinal data, where a sample of subjects are not all measured at each and every time point. Also, variants of RRM have been developed to model dichotomous and ordinal outcomes, that are common in substance use research. A pattern–mixture approach (Little, 1995) can also be accommodated within RRM to further handle and describe the influence of missing data in longitudinal studies. For this approach, subjects are first divided into groups depending on their missing-data pattern, and then variables based on these groups are used as model covariates. Researchers are then able to examine the effect of missing-data patterns on the outcome(s) of interest. In this chapter, we illustrate these methods using an example from a study examining smoking status from early adolescence to young adulthood.

Longitudinal studies play a prominent role in investigations of substance use. In these studies, the same individuals are measured repeatedly on a number of variables over a series of time points. For example, individuals may be assessed in terms of their substance use across repeated weeks, months, or years. The aim of these studies is often to examine whether individuals change in their substance use across time, and whether other variables can produce a change in an individual's substance use across time (e.g., treatment group). Although data from longitudinal substance use studies are potentially very useful, investigators do not always make the best use of these data in their statistical analysis. One reason for this is that missing data invariably occur in longitudinal studies, and investigators may not know how to analyze such incomplete data. For this reason, individuals with incomplete data across time are sometimes removed from the analysis. This complete-data approach can provide biased results to the degree that individuals with complete data are not representative of the total population sampled for the study. Clearly, in substance-use research, individuals with incomplete data may be very different from those with complete data across time. Also, even if complete-data individuals are representative of the larger population, there is a loss in statistical power for hypothesis testing if the total data set is not used in the analysis.

One approach for analysis of incomplete longitudinal data is random-effects regression models (RRM). For RRM, individuals contribute as many repeated observations to the analysis for which they have data. Furthermore, because time can be treated as a continuous variable in RRM, individuals do not have to be measured at the same time points. This is a useful feature for longitudinal studies that have follow-up times that are not uniform across all subjects. Covariates can be included in the model and can be either time-varying or invariant. Thus, changes in the repeated outcomes may be due to both stable characteristics of the subject (e.g., their gender or race) as well as characteristics that change across time (e.g., life-events). Finally, although most traditional approaches to longitudinal data analysis estimate average change (across time) in a population, RRM can also estimate change for each subject when there is sufficient data.

Variants of RRM have been developed under a variety of names including: random-effects models (Laird & Ware, 1982); variance component models (Dempster, Rubin, & Tsutakawa, 1981); hierarchical linear models (Bryk & Raudenbush, 1987); multilevel models (Goldstein, 1995); two-stage models (Bock, 1989); random coefficient models (DeLeeuw & Kreft, 1986); mixed models (Longford, 1987); empirical

Bayes models (Strenio, Weisberg, & Bryk, 1983); unbalanced repeated-measures models (Jennrich & Schluchter, 1986); and, random regression models (Bock, 1983a and 1983b). Generalizations of RRM have also been developed for the case of dichotomous response data (Gibbons & Bock, 1987; Goldstein, 1991; Stiratelli, Laird, & Ware, 1984) and ordinal response data (Ezzet & Whitehead, 1991; Hedeker & Gibbons, 1994; Jansen, 1990;), thus allowing a general framework for analysis of both continuous and categorical outcome variables. Recent articles that describe RRM usage for longitudinal substance use and/or mental health data include Gibbons et al. (1993) and Hedeker and Mermelstein (1996). Several book-length texts (Bryk & Raudenbush, 1992; Diggle, Liang, & Zeger, 1994;Goldstein, 1995; Jones, 1993; Longford, 1993) further describe use of RRM.

For longitudinal studies with missing data across time, a further development that can be implemented in RRM is the *pattern–mixture* approach for incomplete longitudinal data described by Little (1993, 1994, 1995). In this approach, subjects are divided into groups based on their missing-data pattern across time. These groups can then be used, for example, to examine the effect of the missing-data pattern on the outcome(s) of interest. Also, interactions of these groups with other important model terms (i.e. group and group by time interaction) can be included to examine whether changes across time, for example, depend on a subject's missing-data pattern. Overall estimates can be obtained by averaging over the missing-data patterns. Hedeker and Gibbons (1997) illustrate use of this approach applied to psychiatric clinical trials data.

In this chapter we first describe use of RRM applied to substance use data and then indicate how the pattern–mixture approach can be implemented. In addition to illustrating how the missing-data patterns are used as grouping variables in the analysis, we will also describe how overall estimates are obtained by averaging over patterns ("mixing" the patterns). Our aim is to illustrate an approach for analysis of incomplete longitudinal data that is both practical and informative. Although we will apply the pattern–mixture approach within RRM, it can also be used with other longitudinal models that allow for incomplete data across time (e.g., structural equation models or GEE-based models). For use within structural equation models, interested readers should consult Muthén, Kaplan, and Hollis (1987) or McArdle and Hamagami (1991).

RANDOM-EFFECTS REGRESSION MODELS (RRM)

To introduce RRM, consider a simple linear regression model for the measurement y of individual i (i = 1,2, ... ,N subjects) on occasion j (j = 1,2, ... ,n_i occasions):

$$y_{ij} = \beta_0 + \beta_1 t_{ij} + \varepsilon_{ij} \tag{3.1}$$

This model represents the regression of the outcome variable y on the independent variable time (denoted t). The subscripts indicate whose observation it is (subscript i) and the relative timing of the observation (the subscript j). The actual timing is represented by the independent variable t which may represent time in weeks, months, and so on. Both y and t carry the i and j subscripts, and so they are allowed to vary both by individuals and occasions. In a linear regression model, such as Equation 3.1, the errors ε_{ij} are assumed to be normally and *independently* distributed in the population with zero mean and common variance σ^2. This assumption of independence is generally unreasonable for longitudinal data. Because outcomes y are observed repeatedly from the same individuals, it is unlikely that the model errors of a given individual will be independent. It is much more likely to assume that errors within an individual are correlated. In RRM, individual-specific effects are added to the model to account for this data dependency, as in

$$y_{ij} = \beta_0 + \beta_1 t_{ij} + \upsilon_{0i} + \varepsilon_{ij}, \tag{3.2}$$

Where the additional term υ_{0i} indicates the influence of individual i on his/her repeated observations. Specifically, β_0 is the overall population intercept, υ_{0i} is the intercept deviation for subject i, and β_1 is the overall population slope. Thus, individuals deviate from the regression of y on t in a parallel manner in this model.

Because individuals in a sample are usually thought to be representative of a larger population of individuals, the individual-specific effects υ_{0i} are treated as random effects. This population distribution is usually assumed to be a normal distribution with zero mean and variance σ^2_υ. With the random effects υ_{0i} in model 3.2, the errors ε_{ij} are now assumed to be normally and *conditionally independently* distributed in the population with zero mean and common variance σ^2. That is, the errors are independent conditional on the random individual-specific effects υ_{0i}. Because the errors now have an influence due to individuals removed from them, this conditional independence assumption is much more reasonable than the ordinary independence assumption associated with Equation 3.1.

In the model of Equation 3.2, individuals deviate from the regression of y on t in a parallel manner. A slightly more sophisticated model allows both the intercept and trend due to time to vary by individuals:

$$y_{ij} = \beta_0 + \beta_1 t_{ij} + \upsilon_{0i} + \upsilon_{1i} t_{ij} + \varepsilon_{ij}, \quad (3.3)$$

where, β_0 is the overall population intercept, β_1 is the overall population slope, υ_{0i} is the intercept deviation for subject i, υ_{1i} is the slope deviation for subject i, and ε_{ij} is an independent error term distributed normally with mean 0 and variance σ^2. The errors are independent conditional on both υ_{0i} and υ_{1i}. With two random individual-specific effects, the population distribution of intercept and slope deviations is assumed to be a bivariate normal $N(0, \Sigma_u)$, where Σ_υ is a 2 × 2 variance–covariance matrix. This model can be thought of as a personal trend or change model because it represents the measurements of y as a function of time, both at the individual (υ_{0i} and υ_{1i}) and population (β_0 and β_1) levels. The intercept parameters indicate the starting point, and the slope parameters indicate the degree of change over time points. The population intercept and slope parameters represent the trend for the population, whereas the individual parameters express how the individual deviates from the population trend.

In this model, occasions range from $j = 1$ to n_i, with each person measured at n_i time points. Because n carries the subject subscript i, this indicates an important feature of the model, namely, that each subject may vary in terms of the number of measured occasions. The underlying assumption of the model is that the data that are available for a given individual are representative of how that individual deviates from the population trend across the time frame of the study. Later, more detail is given about the assumption of the model with regard to missingness.

If an individual deviates from the population trend in only a random manner, the individual parameters (i.e., the intercept υ_{0i} and slope υ_{1i}) will be approximately equal to zero for that person. In this case, that person's trend across time is well represented by the population trend parameters: the intercept β_0 and slope β_1. More likely, though, is that each individual will deviate in a personal (i.e., systematic) manner from the population trend, and so the individual parameters are necessary in the model to characterize these personal trends. The variability in the individual parameters is assessed by the intercept variance $\sigma^2_{\upsilon 0}$, slope variance $\sigma^2_{\upsilon 1}$, and the covariance of the intercept and the slope $\sigma_{\upsilon 0 \upsilon 1}$. If each individual's deviation from the population trend is only due to random error, then the individual intercepts υ_{0i} and slopes υ_{1i} equal 0 for all subjects, and these variance terms will also equal zero. Alternatively, as each individual's deviation from the population trend is non-random (i.e., υ_{0i} and/or υ_{1i} are nonzero) these variance terms increase from zero.

Often a researcher is interested in assessing the influence of covariates, such as treatment group, on the responses across time. For this, covariates that either do not change over time (time invariant) or that vary across measured occasions (time varying) can be added to the model:

$$y_{ij} = \beta_0 + \beta_1 t_{ij} + \beta_2 x_i + \beta_3 x_{ij} + \upsilon_{0i} + \upsilon_{1i} t_{ij} + \varepsilon_{ij}, \tag{3.4}$$

Here, β_2 is the coefficient for the time-invariant covariate x_i, and β_3 is the coefficient for the time-varying covariate x_{ij}. Interactions between the covariates can be included in the same way as interactions are included into an ordinary multiple regression model. For example, x_i might represent the treatment group that a subject is assigned to (for the course of the study), and x_{ij} might be the treatment by time interaction that is obtained as the product of x_i by t_{ij}.

RRM FOR LONGITUDINAL CATEGORICAL DATA

Generalizations of RRM for categorical outcomes have typically used either a probit or logistic regression model for the categorical outcomes, and various methods for incorporating and estimating the influence of the random effects. A recent review article (Pendergast et al., 1996) presents and describes many of these approaches. In general, parameter estimation is computationally more intensive for these models than for models of continuous outcome data, and so, often only a single random subject effect is assumed. In this chapter, we describe RRM for categorical data with a single random effect; for examples with multiple random effects see Hedeker and Gibbons (1994) or Gibbons and Hedeker (1994).

For dichotomous outcomes, denoted Y_{ij} (with values, say, of 0 or 1), a logistic regression model can be used to relate the log odds of the outcome in terms of explanatory variables:

$$\log\left[\frac{P(Y_{ij}=1\mid x_{ij})}{1-P(Y_{ij}=1\mid x_{ij})}\right]=\beta_0+\beta_1 t_{ij}+\beta_2 x_i+\beta_3 x_{ij}\,, \tag{3.5}$$

where, throughout the article, $\boldsymbol{x}_{ij}$ will denote the vector of covariates (e.g. t_{ij}, $\boldsymbol{x}_i$, $\boldsymbol{x}_{ij}$). The numerator is the probability of a 1 response, and the denominator 1 - $P(Y_{ij} = 1)$ equals the probability of a 0 response. The ratio of these probabilities is the odds of a 1 response, and the log of this ratio is the log odds of a 1 response (sometimes called the logit of P). Notice that the log odds is equal to 0 when the probability of a 1 response equals .5 (i.e., equal odds of a response in category 0 and category 1), is negative when the probability is less than .5 (i.e., odds favoring a response in category 0), and is positive when the probability is greater than .5 (i.e., odds favoring a response in category 1).

Again, because longitudinal data are not independent, use of this ordinary logistic regression model would be inappropriate. As in the linear regression case, we can augment the logistic

$$\log\left[P(Y_{ij}=1\mid \boldsymbol{x}_{ij},\upsilon_i)\ _{\flat}\ 1-P(Y_{ij}=1\mid \boldsymbol{x}_{ij},\upsilon_i)\right]=\beta_0+\beta_1 t_{ij}+\beta_2 \boldsymbol{x}_i+\beta_3 \boldsymbol{x}_{ij}+\upsilon_i\,. \tag{3.6}$$

The individual-varying effect, υ_i represents an individual's influence on the log odds of a response for that individual. As in the model for continuous outcomes, these subject-varying effects are assumed to be normally distributed in the population of subjects with mean 0 and variance σ^2_υ. The subject effects υ_i reflect the influence that subject i has on his/her repeated observations. To the degree that subjects exert little influence on their responses, over and above the influence of the other model terms, values of υ_i, and correspondingly σ^2_υ,will not deviate from zero. However, as subjects exert influence on their repeated observations, values of υ_i will deviate from zero, resulting in a population variance σ^2_υ that is greater than zero.

PATTERN-MIXTURE MODELS

When missing data are observed across time, Laird (1988) has shown that RRM provide valid statistical tests if the missing data are ignorable.

By ignorable, it is meant that the missingness can depend on observed covariates *and* previous values of the dependent variable from the subjects with missing data. Thus, if missingness is related to previous performance, in addition to other observable subject characteristics (i.e., covariates), then RRM provides valid statistical tests for the model parameters. However, by itself, RRM does not provide valid tests if the missing data are nonignorable. Nonignorable missing data result when the missingness depends on the value of the dependent variable that would have been observed if the observation was not missing. For example, in smoking research, missing smoking status observations are often recoded to *smoking* under the belief that the observation is missing because the individual is smoking (i.e., the missingness is nonignorable). Besides recoding missing to smoking, and therefore assuming that missingness is always indicative of smoking, more statistical methods for dealing with nonignorable missingness are becoming available. Little and Schenker (1995) describe a variety of techniques for dealing with nonignorable missingness; here, we describe one of these techniques: the pattern–mixture approach.

In a series of articles, Little (1993, 1994, 1995) described pattern–mixture models for analysis of missing data, and has provided a general statistical treatment of these models.Using a longitudinal psychiatric data set, Hedeker and Gibbons (1997) described and illustrated the use of pattern–mixture RRM for continuous outcomes. Here, we describe use of a pattern–mixture RRM for dichotomous substance use outcomes. As illustrated, the pattern–mixture approach is appealing because it provides useful results and is relatively easy to implement.

To apply the pattern–mixture approach, subjects are classified into groups based on their missing-data pattern across time. For example, if subjects are measured at three time points, then eight (2^3) missing-data patterns are possible. One of these patterns (missing at all three time points), however, provides no data on the outcome variable, and although this pattern can be included in the general pattern–mixture approach by making various missing-data assumptions (see Little, 1993), it is not included here. To include the missing-data pattern information in a statistical model like RRM, the remaining seven patterns can be represented by six dummy-coded variables, for example the codings D1 to D6 given in Table 3.1.

These six dummy-codes represent deviations from the complete-data pattern (OOO); D1 compares pattern MOO to completers, D2 compares

TABLE 3.1

Examples of Dummy-Variable Codings for Missing-Data Patterns: A Three Time Point Study

	Dummy-Codes by Patterns of Missing Data											
	General						*Monotone*		*Last Wave*		*In-complete*	*Not at Final*
Pattern	*D1*	*D2*	*D3*	*D4*	*D5*	*D6*	*M1*	*M2*	*L1*	*L2*	*I1*	*F1*
OOO	0	0	0	0	0	0	0	0	0	0	0	0
MOO	1	0	0	0	0	0			0	0	1	0
OMO	0	1	0	0	0	0			0	0	1	0
MMO	0	0	1	0	0	0			0	0	1	0
OOM	0	0	0	1	0	0	0	1	0	1	1	1
MOM	0	0	0	0	1	0			0	1	1	1
OMM	0	0	0	0	0	1	1	0	1	0	1	1

pattern OMO to completers, etc. Depending on substantive considerations, alternative coding schemes that provide other comparisons among the seven pattern groups can be used (see Darlington, 1990, pp. 232–241; or Cohen & Cohen, 1983, chapter 5). These variables are then entered into the model as main effects and as interactions with other model variables. Thus, the pattern–mixture approach allows one to examine: (a) the degree to which the outcome variable differs across missing-data patterns (i.e., a main effect of the missing-data pattern dummy-coded variables), and (b) the degree to which the missing-data pattern moderates the influence of other model terms (i.e., interactions with missing-data pattern). From the full model, submodels can be obtained for each of the missing-data pattern groups, and overall averaged estimates (i.e., averaging over the missing-data patterns) derived for the model parameters.

In some cases, it may not be feasible, theoretically interesting, or interpretable to model differences between all potential missing-data pattern groups. Also, some of the patterns, either by design or by chance, may not be realized in the sample. For example, in some studies, once a subject is missing at a given wave they are missing at all later waves. In this case, with three waves, three monotone missing-data patterns (OMM, OOM, and OOO) would result. Then, two dummy-coded variables M1 and M2 given in Table 3.1 could be used to represent differences between each of the two dropout groups and the group of subjects observed at all time points. An investigator might want to combine some missing-data patterns to increase interpretability. For example, groups might be formed based on the last available measurement wave. Table 3.1 lists dummy-codes L1 and L2 for this purpose: L1 contrasts individuals who were not measured after the first time point with those who were measured at the last time point, and L2 contrasts those subjects not measured after the second time point with those who were measured at the last timepoint. Other recodings that may be used include a simple grouping of complete data versus incomplete data, as given by contrast I1 in Table 3.1, or missing at the final timepoint versus available at the final timepoint, as given by contrast F1 in Table 3.1. From an analytic perspective, missing-data pattern is simply an additional between-subjects factor. In determining which set of contrasts to use, similar considerations arise as when dealing with the coding of other between-subjects factors (e.g., race or marital status). Thus, substantive considerations and data sparseness are important factors in determining how many missing-data patterns to consider and which set of contrasts to use.

EXAMPLE

We first illustrate RRM applied to longitudinal substance use data, and then augment the model to accommodate the pattern–mixture approach. The data for this example are taken from an ongoing cohort–sequential study of the natural history of cigarette smoking from adolescence to adulthood (Chassin, Presson, Sherman, Corty, & Olshavsky, 1984; Chassin, Pressman, Sherman, & Edwards, 1990; Chassin, Presson, Rose, & Sherman, 1996). The current example expands upon the results reported by Chassin et al. (1996), which investigated the natural history of cigarette smoking from adolescence to adulthood in a community sample.

The following description is taken from Chassin et al. (1996). Between 1980 and 1983, all consenting 6th through 12th graders in a midwestern county school system completed up to four annual, classroom administered surveys (a total sample of 8,556 students were assessed at least once). Follow-up surveys were conducted in1987 (young adulthood) and 1993 (adulthood) with retention rates of 73% of the original adolescent sample at each follow-up. For this study, we used the five oldest cohorts to ensure that sample participants had graduated from high school by 1987, and thus could be considered young adults ($N = 3{,}897$). At all three time points, participants were defined as current regular smokers if they reported smoking weekly or more often. For the measure of adolescent smoking status, we used reports of 11th- or 12th-grade smoking status so that this measurement could best represent participants' "final" adolescent smoking status. A very small number of infrequent smokers (less often than weekly) were eliminated from the analyses. Participants were primarily Caucasian (96%) and gender was split evenly. In 1993, 72% reported having had some post-high school education. The mean age at the adolescent, young adult, and adult time points were 17, 23, and 29 years, respectively; Table 3.2 lists the frequency distribution of age for each of the three time points.[1]

For data analysis, Chassin et al. (1996) used RRM to identify age-related trajectories in smoking across time (i.e., adolescence, young adulthood, and adulthood). The outcome of interest was a dichotomous variable measuring regular (at least weekly) smoking at the three time points. The observed prevalences of smoking were .20 (adolescence), .27 (young adulthood), and .26 (adulthood). RRM results indicated that smoking increased significantly between adolescence and young adulthood, but remained stable from young adulthood to adulthood (i.e., a nonsignificant decrease between these 2 time points). The Chassin et al. (1996) study included participants with missing data in the RRM analyses (i.e., individuals with smoking data on at least one of the three time points were included), but did not specifically explore the influence of different patterns of missingness. The current example extends the previous analyses by using a pattern–mixture approach to explore the effects of patterns of missing data on smoking trajectories from adolescent to adulthood.

In this example, time is a categorical variable that takes on three values (adolescence, young adult, and adult). Subjects may have different numbers of observations, but the measurement time points are the same for all subjects. With three time points, there are two degrees of freedom available to represent changes across time in our model. Thus, in performing the analysis, one must consider how to represent change attributable to time in the model. One choice would be to perform a

[1]©(1996) by the American Psychological Association. Adapted with permission.

TABLE 3.2

Frequency Distributions of Age Across the Three Study Time Points

Adolescence		*Young Adulthood*		*Adulthood*	
Age	*%*	*Age*	*%*	*Age*	*%*
10	0.05	17	0.02	23	0.05
11	0.02	18	0.08	24	0.02
12	0.08	19	0.02	25	0.05
13	0.80	20	3.80	26	0.90
14	3.30	21	17.40	27	14.00
15	10.90	22	22.90	28	23.20
16	27.70	23	24.00	29	24.30
17	37.20	24	18.80	30	20.70
18	18.70	25	10.30	31	13.10
19	1.10	26	1.70	32	3.20
20	0.20	27	0.08	33	0.20
21	0.05	28	0.05	34	0.02

trend analysis and use polynomials to examine whether there are linear and/or quadratic trends in smoking across time. Alternatively, one of the time points could be designated as the reference time period (e.g., adolescence) and tests for differences between this time period and each of the other two time periods (i.e., young adult and adult) could be performed. Because the levels of time are ordinal, Helmert contrasts provide a useful choice (Bock, 1975). Here, with three levels, the first Helmert contrast compares adolescence (time point 1) to the average of young adult and adult time periods (average of time points 2 and 3). The second Helmert contrast compares the young adult (time point 2) to the adult (time point 3) time period. Thus, in simple terms, the first contrast is between adolescence and adulthood, and the second is a comparison within adulthood. Another, somewhat similar, choice would be to use sequential contrasts and compare (a) adolescence to young adult and (b) young adult to adult. In general, the choice of contrasts should be specified prior to the analysis and should be determined by the research question(s) of interest. As noted in Chassin et al. (1996), specific questions of interest in this study were (a) whether there was an increase in smoking following adolescence, and (b) whether smoking began to decline between young adulthood and later adulthood. These two questions are directly examined by the two successive Helmert contrasts.

In order to examine the possibility of a cohort effect in these data, we created a variable indicating age in 1993. Individuals were separated into two age cohort groups: those 28 or younger in 1993 versus those older than 28 in 1993. Of the 3,613 individuals, 1,250 (34.6%) were 28 or younger, and 2,363 (65.4%) were older than 28. Table 3.3 lists the missing-data patterns for these two cohorts.

As can be seen from Table 3.3, there is an approximate linear relationship between the amount of missing data and proportion smoking. Clearly, subjects with complete data at all time points have the lowest proportion of smoking. Also, there appears to be more missing data for the younger cohort. Classifying missing-data patterns as either completion (i.e., pattern OOO) or noncompletion (i.e., all other patterns) gives completion rates of 48% (596 of 1,250) and 55% (1,303 of 2,363) for the younger and older cohorts, respectively. This difference in data completion differs significantly between the two cohorts ($\chi^2 = 18.3$, df = 1, $p < .001$), indicating that missingness was more pronounced for the younger cohort.[2]

Though the association of missingness and cohort is interesting, a more vital question is whether missingness is related to the outcome variable, smoking status. To the degree that missingness and smoking are related, it is important to control for missingness when assessing the cohort effect on smoking. It is also important to examine potential interactive effects of missingness (with time and cohort-related influences) on smoking status. For example, the interaction of missingness and cohort will indicate whether the cohort effect on smoking differs between missing data patterns.

[2]Participants in the younger cohort were less likely to have been in 11th or12th grade during adolescent data collection. Specifically, 61.8% and 84.5% were in either 11th or 12th grade during adolescent data collection for the young and old cohorts, respectively, the remaining 38.2% and 15.4% were measured somewhere between grades 7 to 10 for this assessment. Consequently, adolescent smoking status was ambiguous for those participants measured in grades 7 to 10 because we could not be sure that they did or did not take up smoking before leaving high school. These participants were coded as missing on the measure of adolescent smoking status, resulting in greater missingness among younger cohort participants.

TABLE 3.3

Missing-Data Pattern Frequencies and Proportion Smoking[1] by Cohort (N = 3,613)

	Cohort: Age in 1993					
	≤28		*> 28*		*Total*	
Pattern	*N*	$\hat{p}$	*N*	$\hat{p}$	*N*	$\hat{p}$
OOO	596	.210	1303	.194	1899	.199
MOO	329	.304	203	.397	532	.339
OMO	83	.361	223	.305	306	.320
MMO	83	.530	69	.449	152	.493
OOM	91	.253	179	.263	270	.259
MOM	66	.439	44	.477	110	.455
OMM	2	.500	342	.295	344	.297
Total	1,250	.256	2,363	.234	3,613	.242

[1]These proportions, denoted $\hat{p}$, are estimated across time points.

Separate Analysis of Completers and All Available Cases

Let us first consider a model for changes in smoking status across time as a function of cohort and the Helmert time-related contrasts:

$$\log\left[\frac{P(Y_{ij}=1\mid x_{ij},\upsilon_i)}{1-P(Y_{ij}=1\mid x_{ij},\upsilon_i)}\right] = \beta_0+\beta_1 T1_j+\beta_1 T2_j+\beta_3 C_i + \beta_4(C_i\times T1_j)+\beta_5(C_i\times T2_j)+\upsilon_i\ . \quad (3.7)$$

Because $Y = 1$ designates smoking and $Y = 0$ designates abstinence, this is a model for the log-odds of smoking. The explanatory variables and their codings are as follows: T1 represents the Helmert contrast comparing adolescent versus the average of the young adult and adult time points (coded as -1, .5, .5, for the three time points, respectively), T2 represents the second Helmert contrast comparing young adult versus adult time points (coded as 0, -1, 1, for the three time points, respectively), and C represents Cohort (0 for ≤ 28 and 1 for >28). Due to the coding of cohort, β_1 and β_2 represent the time effects (i.e., T1 and T2) for the younger individuals, and $\beta_1 + \beta_4$ and $\beta_2 + \beta_5$ represent the time effects for the older individuals. Thus, β_4 and β_5 represent differences in the two time effects between age cohort groups. Specifically, if there is no difference in smoking between cohorts when comparing the adolescent versus the two adult time points, β_4 will equal 0. Similarly, if the difference in smoking between the two adult time points is the same for the two cohorts, then β_5 will equal 0.

This logistic random-effects regression model was first fit including only completers (i.e., only the 1,899 subjects with data at all time points) and then re-estimated using all 3,613 subjects. Parameter estimates are presented in the first two sets of columns in Table 3.4.

Conclusions based on these two analyses agree in terms of indicating that there is a significant increase in smoking between adolescence and the two later time points ($p < .001$ for both analyses). Calculating odds ratios (OR) based on the all-subjects analysis, subjects are 2.08 (= exp[1.5 × .489] and 1.59 (= exp[1.5 × (.489 - .181)]) times as likely to be smoking at the adult time points, relative to adolescence, for the young and old cohorts, respectively.[3] The multiplication factor of 1.5 is necessary to account for the Helmert contrast coding that differed by 1.5 units between adolescent (-1) and adult time points (both .5).

In terms of the comparison between young adult and adult time points (i.e., T2), the completer analysis, but not the all-subjects analysis, indicates a significant decline in smoking between these two time points for the young cohort ($p < .036$); the odds of smoking is 1.64 times as likely at the young adult, relative to adult time point (OR for adult relative to young adult smoking = exp[2 × (-.247)] = .61; OR for young adult relative to adult smoking = 1/.61 = 1.64). A few marginally significant results are also obtained and differ somewhat between the two analyses. The completer analysis suggests that the increase in smoking between adolescence and adult time points (i.e., the T1 contrast) is not as great for the old cohort ($p < .074$; OR = exp[1.5 × (-.245)] = .69), whereas the analysis using all data instead suggests an overall decrease in smoking for the old cohort ($p < .078$; OR = exp[-.304] = .74).

The p-values in the table correspond to the so-called *Wald test* (Wald, 1943), based on the ratio of the parameter estimate to its standard error. These test statistics (i.e., z = ratio of the parameter estimate to its standard error) are compared to a standard normal frequency table to test the null hypothesis that a given parameter equals 0. These z-statistics

[3]All odds ratios reported in this article are subject-specific odds ratios, that is, they are adjusted for the level of the random subject effect (see appendix).

TABLE 3.4

Smoking Status Across Time (N = 3,613) Logistic RRM Parameter Estimates (est.), Standard Errors (se), and p-values

	Completers (N = 1,899)			*All Subjects (N = 3,613)*			*Pattern–Mixture (N = 3,613)*		
	est.	*se*	*p <*	*est.*	*se*	*p <*	*est.*	*se*	*p <*
Intercept β_0	-3.406	.244	.001	-2.528	.150	.001	-3.343	.232	.001
T1 β_1	.519	.109	.001	.489	.094	.001	.513	.109	.001
T2 β_2	-.247	.118	.036	-.111	.081	.171	-.243	.118	.038
Cohort β_3	-.220	.258	.394	-.304	.172	.078	-.222	.252	.378
Cohort by T1 β_4	-.245	.137	.074	-.181	.117	.122	-.242	.137	.077
Cohort by T2 β_5	.172	.147	.243	.030	.110	.785	.169	.147	.250
Miss β_6							1.633	.292	.001
Miss by T1 β_7							-.355	.229	.121
Miss by T2 β_8							.286	.163	.080
Miss by Cohort β_9							.258	.353	.465
Miss by Cohort by T1 β_{10}							.685	.276	.013
Miss by Cohort by T2 β_{11}							-.369	.232	.112
Subject sd σ_υ	3.776	.184		3.667	.146		3.697	.159	
-2 log L	4451.4			7771.9			7648.2		

T1 and T2 represent two Helmert contrasts for time (coded as -1, .5, .5; 0, -1, 1).

Cohort coded as 0 for age ≤ 28 or 1 for age > 28.

Miss coded as 0 for complete-data subjects or 1 for incomplete-data subjects.

p-values not given for variance estimates (see Bryk and Raudenbush, 1992, p. 55)

can also be squared, in which case the resulting test statistic is distributed as chi-square on 1 degree of freedom. In either case, the p-values are identical. There are some concerns in using the standard errors in constructing a hypothesis test for the random-effect variance term (i.e., the subject standard deviation σ_υ), particularly when the variance is near-zero and the number of subjects is small (Bryk & Raudenbush, 1992). As a result, we do not provide p-values for the subject standard deviations listed in Table 3.4.

Figures 3.1 and 3.2, plot the observed and estimated group proportions across the three time points corresponding to these two analyses. Later, we describe how these estimated group proportions were obtained. For now, we note that, as the figures illustrate, the models fit the observed data reasonably well.

A Comparison of Completers Versus Incompleters

As stated earlier, we are also interested in examining whether missingness is related to smoking, and whether missingness moderates the effect of cohort on smoking. For this, we define a variable Miss with two values: 0 if the person was measured at all time points (completers), and 1 if the person was not measured at all time points (incompleters). Dividing subjects into these two groups is a simple characterization of the missing-data patterns, however it provides a direct way of assessing whether subjects who were assessed at all time points differ from those who were not. Notice that based on the data presented in Table 3.3, there does seem to be a lower level of smoking for completers relative to incompleters. For these data, this dichotomization of the missing-data patterns appears reasonable, although a more sophisticated analysis could be used to further distinguish the incompleter missing-data patterns.

Figure 3.3 depicts the observed proportions across time for incompleters. Notice that the sample size used to determine these observed proportions varies across time points. Contrasting Fig. 3.3 to Fig. 3.1 (the figure for completers) clearly illustrates a higher level of smoking for incompleters relative to completers. Regarding cohort, Figs. 3.1 and 3.3 differ in their interpretation. For completers, Fig. 3.1 suggests a higher smoking level at the two adult time points for the younger cohort (relative to the older cohort), whereas, for incompleters, Fig. 3.3 suggests the opposite.

To more formally examine the effect of missing-data pattern on smoking, we augment the model given by eqn. 3.7 in the following way:

$$\begin{aligned}\log\left[\frac{P(Y_{ij}=1\mid \boldsymbol{x}_{ij},\upsilon_i)}{1-P(Y_{ij}=1\mid \boldsymbol{x}_{ij},\upsilon_i)}\right] &= \beta_0+\beta_1 T1_j+\beta_2 T2_j+\beta_3 C_i \\ &+\beta_4(C_i\times T1_j)+\beta_5(C_i\times T2_j) \\ &+\beta_6 Miss_i+\beta_7(Miss_i\times T1_j)+\beta_8(Miss_i\times T2_j) \\ &+\beta_9(Miss_i\times C_i)+\beta_{10}(Miss_i\times C_i\times T1_j) \\ &+\beta_{11}(Miss_i\times C_i\times T2_j)+\upsilon_i\end{aligned} \quad (3.8)$$

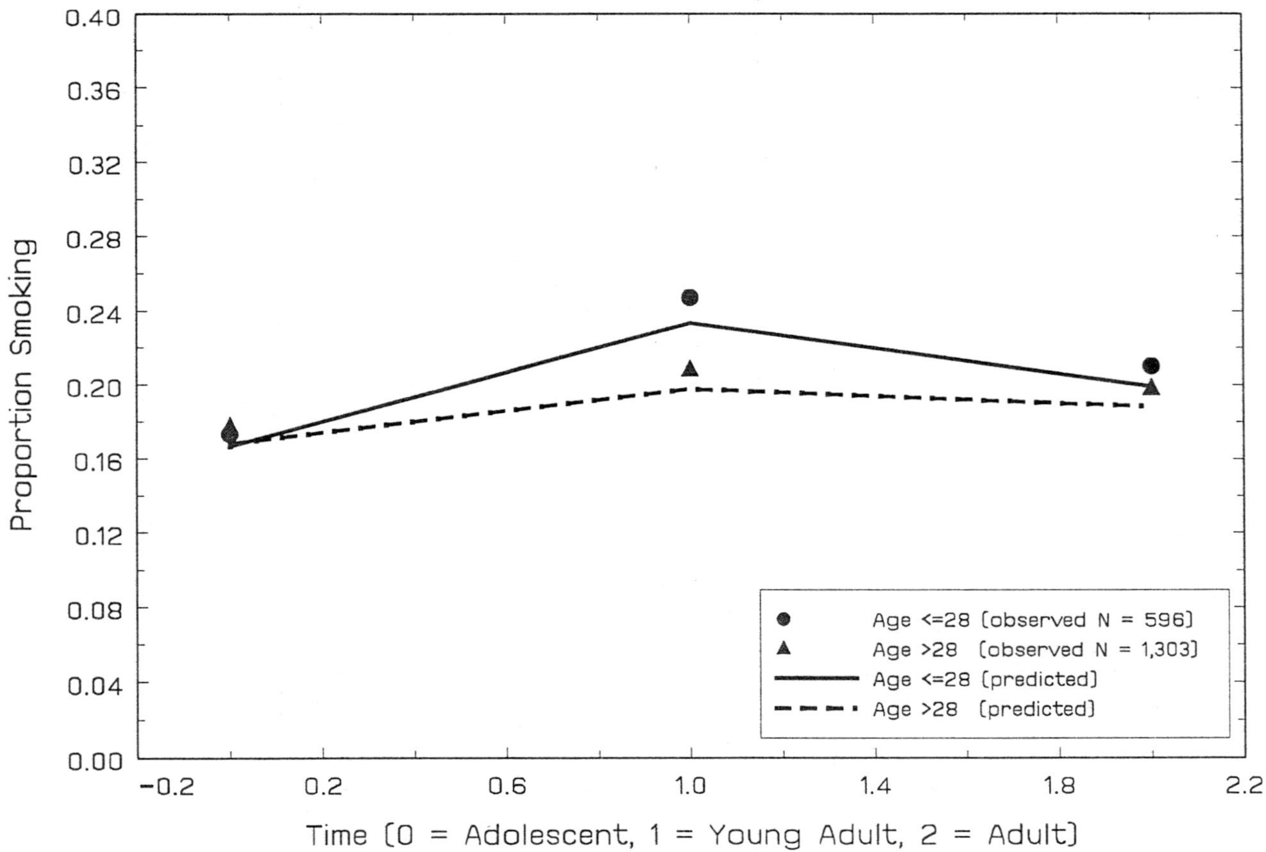

FIG. 3.1. Proportion smoking across time by cohort: Complete data

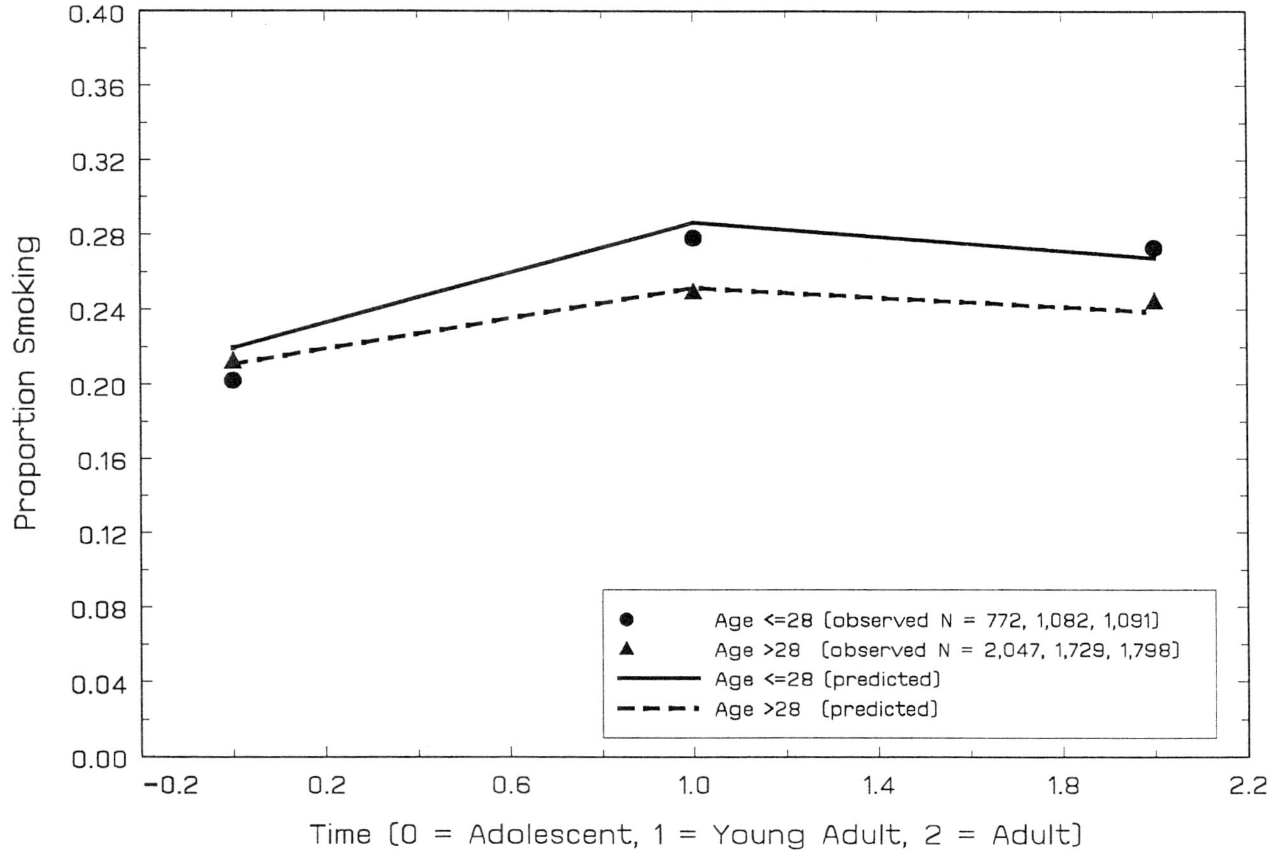

FIG. 3.2. Proportion smoking across time by cohort: Available data

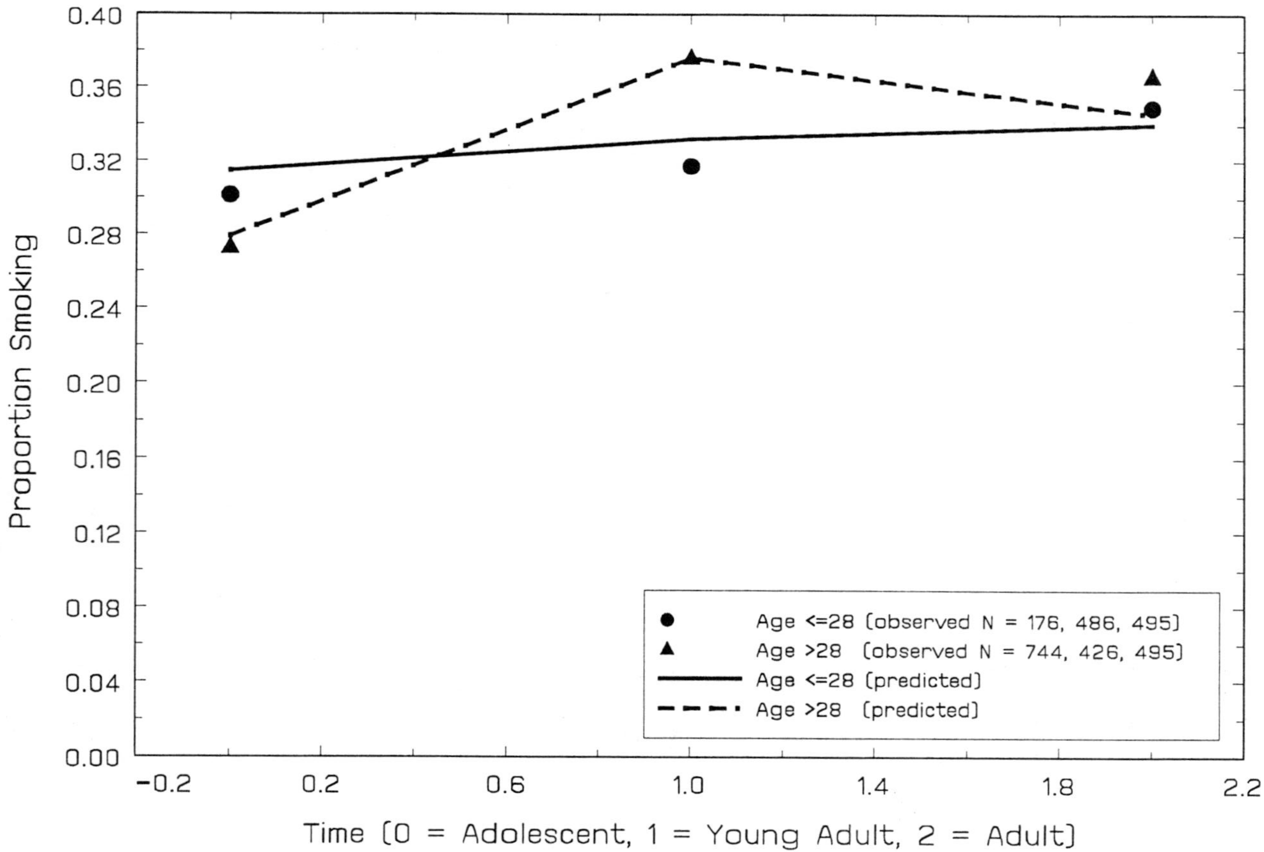

FIG. 3.3. Proportion smoking across time by cohort: Incomplete data

Because of the interactions with time in the model and the use of the (centered) Helmert contrasts for time, β_3 and β_6 represent main effects (i.e., averaged across time) of Cohort and Miss, respectively, and β_9 is the averaged two-way interaction of these two variables. With the coding of Cohort and Miss as described, the regression coefficients β_1 and β_2 represent the time effects for younger completers (both Cohort and Miss equal 0). Parameters β_4 and β_5 compare the time-related effects between old and young completers, whereas β_7 and β_8 compare the time-related effects between young completers and young incompleters. Finally, β_{10} and β_{11} represent the interaction of Cohort by Miss on the time-related effects. The results of this analysis are listed in the last set of columns in Table 3.4 (i.e., labeled pattern–mixture).

The likelihood ratio test for the joint significance of the Miss-related model terms (β_6 through β_{11}) yields $\chi^2 = 7{,}771.9 - 7{,}648.2 = 123.7$, which on six degrees of freedom is highly significant ($p < .001$). This test confirms that missingness and interactions with missingness are significantly related to smoking, over and above the influences of Cohort, Time, and Cohort by Time. In terms of the significance of the individual regression coefficients, from Table 3.4 we see that subjects with incomplete data have a significantly higher smoking level than completers ($p < .001$). Expressed as an odds ratio, an incomplete-data subject is more than five times as likely as a completer to be smoking (OR = exp[1.633] = 5.12). Also, a significant 3-way interaction term is observed for Miss by Cohort by T1 ($p < .013$). Comparing Figs. 3.1 and 3.3 helps to explain this interaction. Basically, the adult versus adolescence increase in smoking (averaging the young adult and adult time points) operates in an opposite manner for completers and incompleters: For completers it is more pronounced for younger subjects, whereas for incompleters it is more pronounced for older subjects. There is also a marginally significant Miss by T2 interaction ($p < .080$) suggesting that the significant drop in smoking from young adult to adult time points that is observed in young completers ($p < .038$) does not occur for young incompleters.

Notice that the results for β_0 through β_5 are almost exactly the same for the pattern–mixture analysis and the analysis done only on completers. Because of our coding of Miss (0 = no and 1= yes), these parameters have exactly the same meaning in these two models. Namely, the intercept, Cohort, Time, and Cohort by Time parameters for completers. The estimates are not exactly the same because the model variance terms differ. In Model 1, σ_υ represents the subject standard deviation for completers only, whereas in Model 3 this same parameter represents subject variation for completers and incompleters. This coding of Miss also allows us to address an important substantive point: does the inclusion of subjects with incomplete data matter? If their inclusion does not make any difference, then from a statistical perspective, $\beta_6 = \beta_7 = \beta_8 = \beta_9 = \beta_{10} = \beta_{11} = 0$. As mentioned before, the likelihood-ratio test of this null hypothesis was easily rejected. Since statistical significance is influenced by sample size, it is also important to consider the practical significance of separating complete and incomplete data subjects. In this regard, comparison of Figs. 3.1 and 3.3 strongly conveys the importance of distinguishing between those with complete and incomplete data.

Thus, missingness does matter for these data and our analysis has uncovered two interesting and statistically significant results. First and foremost, there is a large significant increase in smoking that is observed for subjects with incomplete data, relative to complete data subjects. Second, the cohort difference across time varies depending on missingness. Our first result supports the often-reported finding of increased baseline use of tobacco, alcohol, and other drugs for study dropouts compared to completers in adolescent substance use studies (Tebes, Snow, & Arthur, 1992). Our second result is more difficult to interpret, however, it does indicate that the cohort by time interaction is not robust across missing data patterns.

Estimating Trend Lines

Results from the analyses, which are based on modeling the log-odds of smoking, can be used to obtain estimated probabilities of smoking for various subgroups. In particular, we can use the model estimates to derive the estimated trend lines for the four subgroups that are depicted in Figs. 3.1 and 3.3. To do this, some additional calculations using the parameter estimates presented in Table 3.4 are necessary. First, note that Equation 3.8 can be algebraically reexpressed to provide the probability of smoking in terms of the model parameters (e.g., see Kleinbaum, 1994, pp. 17–18):

$$P(Y_{ij} = 1 \mid x_{ij}, \upsilon_i) = \frac{1}{1 + \exp(-z_{ij})} \quad , \tag{3.9}$$

where

$$\begin{aligned} z_{ij} &= x'_{ij}\beta + \upsilon_i \\ &= \beta_0 + \beta_1 T1_j + \beta_2 T2_j + \beta_3 C_i + \beta_4 (C_i \times T1_j) + \beta_5 (C_i \times T2_j) \\ &\quad + \beta_6 Miss_i + \beta_7 (Miss_i \times T1_j) + \beta_8 (Miss_i \times T2_j) \\ &\quad + \beta_9 (Miss_i \times C_i) + \beta_{10} (Miss_i \times C_i \times T1_j) \\ &\quad + \beta_{11} (Miss_i \times C_i \times T2_j) + \upsilon_i \end{aligned} \tag{3.10}$$

Using the parameter estimates and Equation 3.9 we can derive estimated probabilities for the four subgroups (based on the codings of C, Miss, T1, and T2). For this, estimated values of z_{ij} for the four subgroups are obtained as:

Young Completers ($C = 0$ and $Miss = 0$)

$$\begin{aligned} \hat{z}_{ij} &= \hat{\beta}_0 + \hat{\beta}_1 T1_j + \hat{\beta}_2 T2_j + \upsilon_i \\ &= -3.343 + .513 T1_j - .243 T2_j + \upsilon_i \end{aligned} \tag{3.11}$$

Old Completers ($C = 1$ and $Miss = 0$)

$$\begin{aligned} \hat{z}_{ij} &= (\hat{\beta}_0 + \hat{\beta}_3) + (\hat{\beta}_1 + \hat{\beta}_4) T1_j + (\hat{\beta}_2 + \hat{\beta}_5) T2_j + \upsilon_i \\ &= (-3.343 - .222) + (.513 - .242) T1_j + (-.243 + .169) T2_j + \upsilon_i \\ &= -3.565 + .271 T1_j - .074 T2_j + \upsilon_i \end{aligned} \tag{3.12}$$

Young Incompleters ($C = 0$ and $Miss = 1$)

$$\begin{aligned} \hat{z}_{ij} &= (\hat{\beta}_0 + \hat{\beta}_6) + (\hat{\beta}_1 + \hat{\beta}_7) T1_j + (\hat{\beta}_2 + \hat{\beta}_8) T2_j + \upsilon_i \\ &= (-3.343 + 1.633) + (.513 - .355) T1_j + (-.243 + .286) T2_j + \upsilon_i \\ &= -1.710 + .158 T1_j + .043 T2_j + \upsilon_i \end{aligned} \tag{3.13}$$

Old Incompleters ($C = 1$ and $Miss = 1$)

$$\begin{aligned} \hat{z}_{ij} &= (\hat{\beta}_0 + \hat{\beta}_3 + \hat{\beta}_6 + \hat{\beta}_9) + (\hat{\beta}_1 + \hat{\beta}_4 + \hat{\beta}_7 + \hat{\beta}_{10}) T1_j + (\hat{\beta}_2 + \hat{\beta}_5 + \hat{\beta}_8 + \hat{\beta}_{11}) T2_j + \upsilon_i \\ &= (-3.343 + -.222 + 1.633 + .258) + (.513 - .242 - .355 + .685) T1_j + (-.243 + .169 + .286 - .369) T2j + \upsilon_i \\ &= -1.674 + .601 T1j - .157 T2j + \upsilon i \end{aligned} \tag{3.14}$$

To apply these equations, we still need to indicate values for the Helmert time contrasts $T1$ and $T2$, and also a value for the random subject effect υ_i. For the Helmert contrasts, $T1 = -1, .5, .5$ and $T2 = 0, -1, 1$ for the three time points, respectively. For the random subject effects, based on our analysis, these came from a normal distribution with mean 0 and standard deviation estimated as 3.697. Thus, we could use specific values or sampled values from this normal distribution to yield probabilities that are specific to the particular subject effect υ_i used. For this reason, estimates from the random-effects model are sometimes termed subject-specific estimates (Bryk, Raudenbush, & Congdon, 1996; Neuhaus, Kalbfleisch, & Hauck, 1991). To reiterate, using eqn. 3.9 yields the subject-specific, or conditional, probability of the outcome for a specific value of the random effect υ_i. The observed proportions, alternatively, are marginal estimates that indicate the estimated probability of the outcome for a group of subjects, not a specific subject. For nonlinear models like the logistic model used here, marginal and subject-specific estimates are not on the same scale, and so cannot be directly compared. However, marginal estimates can be obtained from the subject-specific estimates, but additional calculations are required (see Appendix). Essentially, the subject-specific estimates need to be averaged across the population of subjects to obtain marginal estimates. The trendlines depicted in the figures can then be obtained from these marginal estimates.

Averaging Across Patterns

Thus far, the analysis has indicated that the Cohort and Time effects vary by missing-data pattern (i.e., whether a subject has complete data or not). Trying to understand the reasons that effects vary by missing-data pattern is often of primary importance. For example, in the pattern–mixture analysis in Hedeker and Gibbons (1997) of schizophrenic patients' severity scores across time, the significant drug by time by dropout interaction indicated that placebo dropouts experienced little change in severity levels across time, whereas active drug dropouts experienced dramatic improvement across time. In this case, the reasons for dropout might have been that placebo subjects who did not improve were removed from the study and given active drug treatment, whereas active drug patients who dropped out left the study because of their dramatic improvement. In the present example, the meaning of the significant Miss by Cohort by T1 and marginally significant Miss by T2 interactions is less apparent. Clearly, when important interactions with the missing-data patterns exist, interpretation of the interactions can be enlightening.

It is also useful, in many cases, to obtain overall population estimates averaging over the missing-data patterns. Little (1993 and 1995) and Hogan and Laird (1997) describe this additional step of the pattern–mixture approach. In the present case, based on the final model, we can obtain estimates for the six fixed effects (Intercept, T1, T2, Cohort, Cohort by T1, and Cohort by T2) separately for completers

$$\hat{\beta}^{(c)} = [-3.343\ .513\ -.243\ -.222\ -.242\ .169]' \tag{3.15}$$

and incompleters

$$\begin{aligned}\hat{\beta}^{(ic)} &= \hat{\beta}^{(c)} + [1.633\ -.355\ .286\ .258\ .685\ -.369]' \\ &= [-1.710\ .158\ .043\ .036\ .443\ -.200]' \end{aligned} \tag{3.16}$$

Averaged estimates for these four parameters (denoted $\hat{\beta}$) are then equal to:

$$\hat{\beta} = \pi^{(c)}\hat{\beta}^{(c)} + \pi^{(ic)}\hat{\beta}^{(ic)} \tag{3.17}$$

where $\pi^{(c)}$ and $\pi^{(ic)}$ represent the population weights for completers and incompleters, respectively. Although these weights are not usually known, they can be estimated by the sample proportions (1,899/3,613 and 1,714/3,613 for completers and incompleters, respectively). This yields:

$$\hat{\beta} = [-2.568\ .344\ -.108\ -.099\ .083\ -.006]' \tag{3.18}$$

as the averaged overall estimates. To obtain corresponding estimates of the standard errors, the delta method as described in Hogan and Laird (1997) can be used:

$$\hat{V}(\hat{\beta}_h) = (\hat{\pi}^{(c)})^2\,\hat{V}(\hat{\beta}_h^{(c)}) + (\hat{\pi}^{(ic)})^2\,\hat{V}(\hat{\beta}_h^{(ic)}) + \frac{\hat{\pi}^{(c)}\hat{\pi}^{(ic)}}{N}\,(\hat{\beta}_h^{(c)} - \hat{\beta}_h^{(ic)})^2 \tag{3.19}$$

where $h = 1, \ldots$ denotes the six fixed effects, $N = 3{,}613$ is the total number of subjects, and $\hat{V}(\hat{\beta}_h)$ denotes the estimate of the variance of β_h (i.e., the square of its estimated standard error). The last term in the sum is the contribution to the variance that is added because the proportion of completers (and incompleters) is estimated in the sample. The estimated standard errors for these six overall terms are then .163, .112, .082, .178, .135, and .115. Calculating Wald statistics (i.e., estimate divided by its standard error) for these six parameters yield the following *p*-values: .001, .002, .189, .575, .539, and .961.

It is interesting to compare these pattern-mixture averaged estimates and standard errors to those obtained from the RRM analysis ignoring missing-data patterns (i.e., the RRM analysis of all 3,613 subjects presented in the second set of columns in Table 3.4). Comparing these two indicates that the estimates for the young subjects ($\hat{\beta}_0$, $\hat{\beta}_1$, and $\hat{\beta}_2$) are very similar. However, the estimates for the Cohort effect ($\hat{\beta}_3$) and Cohort by Time interaction effects ($\hat{\beta}_4$ and $\hat{\beta}_5$) are not so similar between these approaches. Most notably, although a marginally significant Cohort effect ($\hat{\beta}_3 = -.304$, $p < .078$) is indicated by ordinary RRM analysis, the pattern-mixture averaged result is highly nonsignificant for this parameter ($\hat{\beta}_3 = -.108$, $p < .575$). Essentially, in computing the averaged estimates across patterns the pattern–mixture approach gives more relative weight to the subgroup of subjects with missing data (whose Cohort estimate $\hat{\beta}_3 = .036$) than does the ordinary RRM, which weights each subject's data by its degree of precision. All other things equal, the degree of precision is largely determined by the amount of data. Thus, in ordinary RRM, because incompleters provide less data (than completers) their data receives less influence than in the pattern–mixture approach. The bottom line is that misleading results can be obtained by ignoring the missing-data patterns when they have influence on the outcome variable or when they moderate the effects of other model variables.

DISCUSSION

For incomplete longitudinal data, RRM provide a useful approach because subjects are not assumed to be measured at the same number of time points. As a result, subjects who are missing at a given interview wave are not excluded from the analysis. The assumption of the model is that the data that are available for a given subject are representative of that subject's deviation from the (average) trend lines based on the model covariates. Also, RRM can include both time-varying and time-invariant covariates and are available for continuous and categorical outcomes.

In this chapter, we have illustrated how to augment these random-effects models by including variables defined by a subject's pattern of missing data. The pattern–mixture approach provides assessment of degree to which (the influence of) model terms vary by the missing-data patterns, and provides a way of obtaining estimates averaged over the missing-data patterns. In our example, we considered a very simple characterization of the pattern–mixture approach, namely, was the subject measured at all study time points or not. Though simple, it showed that subjects within complete data had higher levels of smoking than those measured at all time points. Also, there was evidence that the cohort by time interaction varied between completers and incompleters. Thus, the addition of missing data pattern led to substantive findings that were overlooked in our ordinary RRM analysis. In particular, our finding of increased smoking for incompleters from the pattern–mixture analysis supports previous research from adolescent substance use studies (Tebes, Snow, & Arthur, 1992).

In dealing with incomplete longitudinal data, the complexity of the statistical model used may depend on the amount of missing data. For example, if a study has only a few subjects with missing data across time, it may not matter whether the subjects with missing data are ignored or included as a group in the analysis. The statistical power for detecting effects (i.e., main effects and interactions) due to missing-data patterns is lower, all other things equal, as the numbers of subjects in those patterns are reduced. Although it is difficult to give general guidelines for the treatment of missing-data patterns in a given analysis (i.e., number of missing-data patterns, or grouping together of missing-data patterns), the considerations are similar to those encountered when dealing with other between-subjects grouping variables.

Another consideration in applying the pattern–mixture approach is the number of interaction terms to form with the missing data patterns. In the present example, we had a fairly small number of model covariates and only one variable to represent missing data patterns, so that including interactions with missingness for all covariates was not problematic. However, if many covariates are included in the model (e.g., gender, education level beyond high school, ethnicity) then the number of possible interactions can become quite large. In this case, a stratified analysis by missing-data pattern can be helpful in identifying covariates that have substantially different effects across missing-data patterns. Including missing-data pattern interactions for this subset of covariates would then provide a more parsimonious pattern–mixture model.

Methods for appropriately handling missing data have been increasingly developed in the last decade or so, allowing researchers the ability to more effectively deal with this potential threat to validity. Besides the pattern–mixture approach presented here, "selection models" have also been proposed to handle missing data in longitudinal studies (Diggle & Kenward, 1994; Heckman, 1976; Heyting, Tolboom, & Essers, 1992; Leigh, Ward, & Fries, 1993). In selection models, whether or not a subject drops out is modeled in terms of variables obtained prior to the dropout, often the variables measured at baseline. Because dropout is dichotomous, a logistic or probit regression model is often used for this, yielding a predicted dropout probability or propensity for each subject. These dropout propensity scores can then be used in the longitudinal data model (e.g., a RRM) as a covariate to adjust for the potential influence of dropout. So, instead of using the missing-data pattern as a model covariate in the longitudinal data analysis (i.e., the pattern mixture approach), the selection model approach uses the predicted probability of dropout as a covariate. By modeling dropout, selection models provide information about the predictors of study dropout, however, an advantage of pattern–mixture models is that they can be used even when no such predictors are available. Currently, there is much debate in the statistical literature about these two approaches; further discussion on some of the differences between pattern–mixture and selection models can be found in Glynn, Laird, and Rubin (1986), Little (1993, 1994, 1995), and in the discussion of Diggle and Kenward (1994).

In sum, it is important that researchers consider the reasons for missing data in their own data sets, and to choose among statistical methods with these reasons in mind. For example, if it can be assumed that missing observations on the dependent variable y are predicted well by the available y measurements for those individuals with missing data, then an ordinary RRM is reasonable (because the missingness is ignorable). Alternatively, if the missingness cannot be assumed to be ignorable, but there are good predictors of missingness, then a selection model approach may be used to obtain results that are adjusted for the predicted probability or propensity to be missing. As illustrated in this chapter, even when there are no good predictors of missingness, the pattern–mixture approach can be used to examine and adjust for the influence of missing-data pattern. Of course, any method of "dealing" with missing data makes some assumptions about the missing data. For instance, in our example, ordinary RRM assumes that Fig. 3.2 is representative of the population trends in smoking across time for the two age cohorts, whereas pattern–mixture RRM assumes that Figs. 3.1 and 3.3 are representative of these trends for complete and incomplete-data subjects, respectively. The bottom line is that researchers must choose a statistical model or models to represent both the observed and unobserved data.

CONCLUSIONS

As demonstrated, RRM provides a useful way of analyzing longitudinal substance use data. Specifically, RRM allows for the presence of missing data, irregularly spaced measurements across time, time-varying and invariant covariates, accommodation of individual-specific deviations from the average time trend, and estimation of the population variance associated with these individual effects. Methods and software exist for analysis of continuous, dichotomous, and ordinal outcomes. Additionally, when missing data are present for the outcome variable, the pattern–mixture approach provides a way of assessing the influence of missing-data pattern on the outcome variable and on the effects in the model. Hopefully, this chapter has demonstrated that the combined approach of pattern–mixture RRM represents an important and useful tool for analyzing incomplete longitudinal data.

Computer Programs

Software for RRM is increasingly available, especially for normal-theory models (MLn, Rasbash, Yang, Woodhouse, & Goldstein, 1995; HLM, Bryk, Raudenbush, & Congdon, 1996; VARCL, Longford, 1986; the BMDP 5V procedure, Schluchter, 1988; the SAS procedure MIXED; MIXREG, Hedeker & Gibbons, 1996b).A detailed comparison of some of these programs is included in Kreft, DeLeeuw, and van der Leeden (1994). For categorical outcomes, software is available for dichotomous (EGRET, Statistics and Epidemiology Research Corporation, 1991) and dichotomous/ordinal (MIXOR, Hedeker & Gibbons, 1996a) outcomes. Also, MLn, HLM, and VARCL provide facilities for analysis of dichotomous outcomes. MIXOR was used for the analyses presented in this article. To obtain the pattern–mixture averaged results based on the MIXOR estimates, a SAS IML program was written. Both MIXREG and MIXOR can be obtained free of charge through the internet (http://www.uic.edu˜hedeker/mix.html). Readers interested in the specifications that were required to run MIXOR for the examples in this chapter, or for the SAS IML program to produce the pattern–mixture averaged results, can send an e-mail to the first author at hedeker@uic.edu.

ACKNOWLEDGMENTS

The authors thank Robin Mermelstein for many valuable discussions, comments, and suggestions prior to and during the preparation of this chapter. The authors also thank Clark Presson and Michael Seltzer for helpful and constructive comments on an earlier version of this article.

For use of the longitudinal data, the authors thank Steven J. Sherman, Laurie Chassin, and Clark Presson. Support for the longitudinal project was provided by National Institute of Child Health and Human Development Grant HD13449. Preparation of this article was supported by National Institutes of Mental Health Grant MH56146.

REFERENCES

Bock, R. D. (1975). *Multivariate statistical methods in behavioral research.* New York: McGraw-Hil.

Bock, R. D. (1983a). Within-subject experimentation in psychiatric research. In R. D. Gibbons & M. W. Dysken (Eds.), *Statistical and methodological advances in psychiatric research* (pp. 59–90). New York: Spectrum.

Bock, R. D. (1983b). The discrete Bayesian. In: H. Wainer & S. Messick (Eds.), *Modern advances in psychometric research* (pp.103–115). Hillsdale, NJ: Lawrence Erlbaum Associates.

Bock, R. D. (1989). Measurement of human variation: A two stage model. In R. D. Bock (Ed.), *Multilevel analysis of educational data* (pp. 319–342). New York: Academic Press.

Bryk, A. S., & Raudenbush, S. W. (1987). Application of hierarchical linear models to assessing change. *Psychological Bulletin, 101,* 147–158.

Bryk, A. S., & Raudenbush, S. W. (1992). *Hierarchical linear models: Applications and data analysis methods.* Newbury Park, CA: Sage.

Bryk, A. S., Raudenbush, S. W., & Congdon, R. T. (1996). *Hierarchical linear and nonlinear modeling with the HLM/2L and HLM/3L programs.* Chicago: Scientific Software, Inc.

Chassin, L., Presson, C. C., Rose, J. S., & Sherman, S. J. (1996). The natural history of cigarette smoking from adolescence to adulthood: Demographic predictors of continuity and change. *Health Psychology, 15,* 478–484.

Chassin, L., Presson, C. C., Sherman, S. J., Corty, E., & Olshavsky, R. (1984). Predicting the onset of cigarette smoking in adolescents: A longitudinal study. *Journal of Applied Social Psychology, 14,* 224–243

Chassin, L., Presson, C. C., Sherman, S. J., & Edwards, D. A. (1990). The natural history of cigarette smoking: Predicting young adult smoking outcomes from adolescent smoking patterns. *Health Psychology, 9,* 701–716.

Cohen, J. & Cohen, P. (1983). *Applied multiple regresion/correlation analysis for the behavioral sciences.* (2nd ed.). Hillsdale, NJ: Lawrence Erlbaum Associates.

DeLeeuw, J., & Kreft, I. (1986). Random coefficient models for multilevel analysis. *Journal of Educational Statistics, 11,* 57–85.

Dempster, A. P., Rubin, D. B., & Tsutakawa, R. K. (1981). Estimation in covariance component models. *Journal of the American Statistical Society, 76,* 341–353.

Diggle, P., & Kenward, M. G. (1994). Informative dropout in longitudinal data analysis (with discussion). *Applied Statistics, 43,* 49–94.

Diggle, P., Liang, K.-Y., & Zeger, S. L. (1994). *Analysis of Longitudinal Data.* New York: Oxford University Press.

EGRET. (1991). *Reference manual.* Seattle, WA: Statistics and Epidemiology Research Corporation.

Ezzet, F., & Whitehead, J. (1991). A random effects model for ordinal responses from a crossover trial. *Statistics in Medicine, 10,* 901–907.

Gibbons, R. D., & Bock, R. D. (1987). Trend in correlated proportions. *Psychometrika, 52,* 113–124.

Gibbons, R. D., & Hedeker, D. (1994). Application of random-effects probit regression models. *Journal of Consulting and Clinical Psychology, 62,* 285–296.

Gibbons, R. D., Hedeker, D., Elkin, I., Waternaux, C., Kraemer, H. C., Greenhouse, J. B., Shea, M. T., Imber, S. D., Sotsky, S. M., & Watkins, J. T. (1993). Some conceptual and statistical issues in analysis of longitudinal psychiatric data. *Archives of General Psychiatry, 50,* 739–750.

Glynn, R. J., Laird, N. M., & Rubin, D. B. (1986). Selection modeling versus mixture modeling with nonignorable nonresponse. In H. Wainer (Ed.), *Drawing inferences from self-selected samples* (pp. 115–142). New York: Springer-Verlag.

Goldstein, H. (1991). Nonlinear multilevel models, with an application to discrete response data. *Biometrika, 78*, 45–51.

Goldstein, H. (1995). *Multilevel statistical models, 2nd edition.* New York: Halsted Press.

Heckman, J. (1976). The common structure of statistical models of truncation, sample selection, and limited dependent variables and a simple estimator for such models. *Annals of Economic and Social Measurement, 5*, 475–492.

Hedeker, D., & Gibbons, R. D. (1994). A random-effects ordinal regression model for multilevel analysis. *Biometrics, 50*, 933–944.

Hedeker, D., & Gibbons, R. D. (1996a). MIXOR: A computer program for mixed-effects ordinal regression analysis. *Computer Methods and Programs in Biomedicine, 49*, 157–176.

Hedeker, D., & Gibbons, R. D. (1996b). MIXREG: A computer program for mixed-effects regression analysis with autocorrelated errors. *Computer Methods and Programs in Biomedicine, 49*, 229–252.

Hedeker, D., & Gibbons, R. D. (1997). Application of random-effects pattern–mixture models for missing data in longitudinal studies. *Psychological Methods, 2*, 64–78.

Hedeker, D., & Mermelstein, R.J. (1996). Application of random-effects regression models in relapse research. *Addiction, 91 (Supplement)*, S211–S229.

Heyting, A., Tolboom, J. T. B. M., & Essers, J. G. A. (1992). Statistical handling of drop-outs in longitudinal clinical trials. *Statistics in Medicine, 11*, 2043–2061.

Hogan, J. W., & Laird, N. M. (1997). Mixture models for the joint distribution of repeated measures and event times. *Statistics in Medicine, 16*, 239–258.

Jansen, J. (1990). On the statistical analysis of ordinal data when extravariation is present. *Applied Statistics, 39*, 75–84.

Jennrich, R. I., & Schluchter, M. D. (1986). Unbalanced repeated-measures models with structured covariance matrices. *Biometrics, 42*, 805–820.

Jones, R. H. (1993). *Longitudinal data analysis with serial correlation: A state-space approach.* New York: Chapman & Hall.

Kleinbaum, D. G. (1994). *Logistic regression: A self-learning text.* New York: Springer-Verlag.

Kreft, I. G., de Leeuw, J., & van der Leeden, R. (1994). Comparing five different statistical packages for hierarchical linear regression: BMDP-5V, GENMOD, HLM, ML3, and VARCL. *American Statistician, 48*, 324–335.

Laird, N. M., & Ware, J. H. (1982). Random-effects models for longitudinal data. *Biometrics, 38*, 963–974.

Laird, N. M. (1988). Missing data in longitudinal studies. *Statistics in Medicine, 7*, 305–315.

Leigh, J. P., Ward, M. M., & Fries, J. F. (1993). Reducing attention bias with an instrumental variable in a regression model: Results from a panel of rheumatoid arthritis patients. *Statistics in Medicine, 12*, 1005–1018.

Little, R. J. A. (1993). Pattern–mixture models for multivariate incomplete data. *Journal of the American Statistical Association, 88*, 125–133.

Little, R. J. A. (1994). A class of pattern–mixture models for normal incomplete data. *Biometrika, 81*, 471–483

Little, R. J. A. (1995). Modeling the drop-out mechanism in repeated-measures studies. *Journal of the American Statistical Association, 90*, 1112–1121.

Little, R. J. A., & Schenker, N. (1995). Missing Data. In G. Arminger, C. C. Clogg, & M. E. Sobel (Eds.), *Handbook of statistical modeling for the social and behavioral sciences* (pp. 39–75). New York: Plenum.

Long, J. S. (1997). *Regression models for categorical and limited dependent variables.* Thousand Oaks, CA: Sage.

Longford, N. T. (1986). VARCL–Interactive software for variance component analysis. *The Professsional Statistician, 74*, 817–827.

Longford, N. T. (1987). A fast scoring algorithm for maximum likelihood estimation in unbalanced mixed models with nested random effects. *Biometrika, 74*, 817–827.

Longford, N. T. (1993). *Random coefficient models.* New York: Oxford University Press.

McArdle, J. J., & Hamagami, F. (1991). Modeling incomplete longitudinal data using latent growth structural equation models. In L. Collins & J. L. Horn (Eds.), *Best methods for analysis of change* (pp. 276–304). Washington, DC: APA Press.
Muthén, B., Kaplan, D., & Hollis, M. (1987). On structural equation modeling with data that are not missing completely at random. *Psychometrika, 52*, 431–462.
Neuhaus, J. M., Kalbfleisch, J. D., & Hauck, W. W. (1991). A comparison of cluster-specific and population-averaged approaches for analyzing correlated binary data. *International Statistical Review 59*, 25–35.
Pendergast, J. F., Gange, S. J., Newton, M. A., Lindstrom, M. J., Palta, M., & Fisher, M. R. (1996). A survey of methods for analyzing clustered binary response data. *International Statistical Review 64*, 89–118.
Rasbash, J., Yang, M., Woodhouse, G., & Goldstein, H. (1995). *MLn: Command reference guide*. London: Institute of Education, University of London.
Schluchter, M. D. (1988). 5V: Unbalanced repeated measures models with structured covariance matrices. In W. J. Dixon (Chief Ed.), *BMDP statistical software manual* (vol. 2; pp. 1081–1114). Berkeley, CA: University of California Press.
Stiratelli, R., Laird, N. M., & Ware, J. H. (1984). Random-effects models for serial observations with binary response. *Biometrics, 40*, 961–971.
Strenio, J. F., Weisberg, H. I., & Bryk, A. S. (1983). Empirical Bayes estimation of individual growth curve parameters and their relationship to covariates. *Biometrics, 39*, 71–86.
Stroud, A. H., & Sechrest, D. (1966). *Gaussian quadrature formulas*. Englewood Cliffs, NJ: Prentice-Hall.
Tebes, J. K., Snow, D. L., & Arthur, M. W. (1992). Panel attrition and external validity in the short-term follow-up study of adolescent substance use. *Evaluation Review, 16*, 151–170.
Wald, A. (1943). Tests of statistical hypotheses concerning several parameters when the number of observations is large. *Transactions of the American Mathematical Society, 54*, 426–482.
Zeger, S. L., Liang, K.-Y., & Albert, P. S. (1988). Models for longitudinal data: A generalized estimating equation approach. *Biometrics, 44*, 1049–1060.

Appendix
Marginal and Subject-Specific Probabilities

As discussed in the chapter, subject-specific probabilities can be obtained using Equation 3.9, namely,

$$P(Y_{ij} = 1 \mid \boldsymbol{x}_{ij}, \upsilon_i) = \frac{1}{1 + \exp\left[-(\boldsymbol{x}'_{ij}\,\beta + \upsilon_i)\right]} \qquad (3.?)$$

This equation yields the probability of a response for particular groupings of subjects determined by covariate values (i.e., $\boldsymbol{x}_{ij}$) and specific values of the random subject effect υ_i. This is why the probabilities are termed subject-specific, they depend on values of the random subject effect. Alternatively, we might want to obtain probabilities not for a particular subject (i.e., for a specific value of the random effect υ_i), but averaged across all subjects (i.e., averaging across all values of the random effect υ_i). To obtain these marginal or population-averaged probabilities, the following integration over the random-effects distribution υ must be performed:

$$P(Y_{ij} = 1 \mid \boldsymbol{x}_{ij}, \upsilon_i) = \int_{\upsilon} \frac{1}{1+\exp\left[-(\boldsymbol{x}'_{ij}\beta + \upsilon_i)\right]} g(\upsilon)d\upsilon . \qquad (3.?)$$

Performing this integration may seem difficult, however, it can be approximated numerically to any practical degree of accuracy. Specifically, because the assumed distribution of υ is normal, Gauss–Hermite quadrature (Stroud & Sechrest, 1966) can be used to numerically perform the integration. Essentially, the numerical integration replaces the integral with a weighted summation over a finite number of points. Alternatively, another approximation described in Diggle, Liang, and Zeger (1994, p. 142), does not require use of numerical integration. Here, the regression coefficients β from the random-effects model are divided by $\sqrt{c^2\sigma_\upsilon^2+1}$ where $c = 16\sqrt{3}/(15\pi)$. The inverse of the constant c reflects the fact that the standard logistic distribution has variance $\pi\sqrt{3}$ and that multiplying this variance by 15/16 provides improved empirical fit of the data (see Long, 1997, pp. 40–48, or Zeger, Liang, & Albert, 1988). These marginalized coefficients, denoted β^*, can then be directly used to produce marginal proportions by applying

$$P(Y_{ij} = 1 \mid \boldsymbol{x}_{ij}) = \frac{1}{1+\exp\left[-(\boldsymbol{x}'_{ij}\beta^*)\right]} \qquad (3.?)$$

As an example, for the pattern–mixture analysis in Table 3.4, the marginalization factor is calculated as $\sqrt{(.346)(3.697)^2+1} = 2.394$ (note that $c^2 \approx .346$). Thus, the estimated proportion at adolescence for young completers is equal to

$$\frac{1}{1+\exp\{-[(-3.343+.513(-1)-.243(0))/2.394]\}} = \frac{1}{1+\exp[-(-1.611)]} = .166 . \qquad (3.?)$$

Alternatively, using quadrature yields .175 for this proportion. The actual observed proportion for this subgroup at adolescence is .173, so the quadrature estimate is a little closer to the observed data, though this is not always the case. The marginalization approach was used to produce the trend lines depicted in Figs. 3.1–3.3. A similar set of figures (not shown) was produced using quadrature. These two sets of figures differed very little. In general, because the marginalization approach is simpler and provides very similar results to the more involved quadrature approach, its use is reasonable for examining approximate model fit.

4

Methodological Issues in Random Coefficient Growth Modeling Using A Latent Variable Framework: Applications to the Development of Heavy Drinking Ages 18–37

Bengt Muthén
University of California, Los Angeles

This chapter discusses four methodological issues arising in growth modeling of alcohol use development with a national longitudinal data set. First, with development over a long period of time it is reasonable that the different background variables have different importance for the growth process during different segments of time. It is shown how this change in importance can be captured in the modeling using the concept of varying centering points. The variation of centering points is illustrated by both linear and quadratic growth models. Second, in a heterogeneous group of individuals such as obtained by a national sample, it is likely that individual trajectories belong to several different, unknown subpopulations corresponding to normative development and several forms of non-normative development. It will be shown how new modeling techniques can capture this notion. Third, the data are obtained in the form of multiple cohorts where nobody in the sample has observations on the outcome variables for all ages, leading to missing data modeling. Fourth, many of the households in the sample have multiple siblings and given that siblings within a household share the same home environment, this gives rise to intraclass correlation that violates customary analysis assumptions of independent observations. The chapter gives a nontechnical overview of solutions to these four complications, where the solutions reveal interesting information in the data. Emphasis is placed on the first two areas, whereas the last two areas are described briefly.

This chapter discusses methodological issues of interest to researchers who study developmental theories with longitudinal data. The discussion focuses on advanced growth modeling analyses. Although methodologically oriented, the discussion will be as nontechnical as possible, instead focusing on the interpretation of the analyses. The chapter by Curran also discusses growth modeling and is a suitable introduction to this chapter.

To motivate the methodological discussion and have a specific data source and growth analysis as a reference point, data from the National Longitudinal Survey of Youth (NLSY) will be used. This data set was collected by the Bureau of Labor Statistics with alcohol supplements from NIAAA. The NLSY is a nationally representative household survey of young adults living in the United States in 1979. It was obtained as a multistage probability sample with oversampling of Blacks, Hispanics, and economically disadvantaged non-Blacks and non-Hispanics. The NLSY includes eight birth-year cohorts from 1957 to 1964 with approximately 1,000 individuals per cohort. The sample includes every age-eligible household member. Alcohol measures are available from the years 1982, 1983, 1984, 1985, 1988, 1989, and 1994 covering ages 18 to 37.

The focus of this chapter is on the development of heavy drinking and alcohol-related problems. The heavy drinking variable is obtained from the question "How often have you had 6 or more drinks on one occasion during the last 30 days?" The response categories (and their scores) are never (0), once (1), 2 or 3 times (2), 4 or 5 times (3), 6 or 7 times (4), 8 or 9 times (5), 10 or more times (6). Also of interest is an outcome variable representing a severity score reflecting alcohol-related problems. The NLSY contains 22 symptom items designed to capture DSM-IV diagnoses of alcohol dependence and abuse. From a factor analysis of these items, a sum of the 17 items measuring the most severe dimension is created to form the severity score. Background variables, also referred to as time-invariant covariates, are gender, ethnicity (Black, Hispanic, other), family history of alcohol problems, early onset, dropping out of high school, and college education. Family history information is obtained by the question "Have any of your relatives listed on this card been alcoholics or problem drinkers at any time in their lives?", recorded as the three dummy variables FH1 (family history among first-degree relatives only), FH23 (family history among second- or third-degree relatives only), and FH123 (family history among first-and second- or third-degree relatives). Early onset information is obtained from the question "How old were you when you first started drinking?" with the probe "For example, having two or more drinks a week." This variable is scored 0/1 with 1 for onset at or before age 14. High school dropout is scored 0/1 with 1 for not having completed high school by age 22. College education is scored 0/1 with 1 for some college by age 22.

The sample means for frequency of heavy drinking are shown in Fig. 4.1 for ages 18 to 37. The mean curve shows the typical increase from age 18 to the peak around age 21 with a subsequent decline. The variation in individual trajectories is, however, large and it is of interest to assess the amount of variation, how much of this variation can be explained by the set of time-invariant covariates, and which covariates are particularly important.

The NLSY growth modeling gives rise to several interesting and common methodological issues, four of which are discussed here. The first two will be discussed at some length, and the last two are described more briefly in order to show the range of issues common in longitudinal analysis.

The first issue concerns the influence of background variables on the growth process. In the NLSY, development of problematic alcohol use is studied over an age span of 20 years. With development over such a long period of time, it is reasonable that different background variables have different importance for the growth process during different segments of time. For example, it is of interest to study if dropping out of high school has more or less of an impact on problematic alcohol use when an individual is in his/her 20s or 30s. It is shown how such changes in importance can be studied using modeling that applies the notion of varying centering points.

The second issue concerns the heterogeneity of development across individuals in the sample. In a heterogeneous group of individuals such as obtained by a national sample, it is likely that individual trajectories belong to several different, unknown subpopulations corresponding to normative development and several forms of non-normative development. It is shown how emerging modeling techniques can capture this type of heterogeneity. The new type of approach is very promising for developmental studies in that it provides estimates for all the different growth models for the subpopulations as well as estimates of a person's likelihood of belonging to a certain subpopulation.

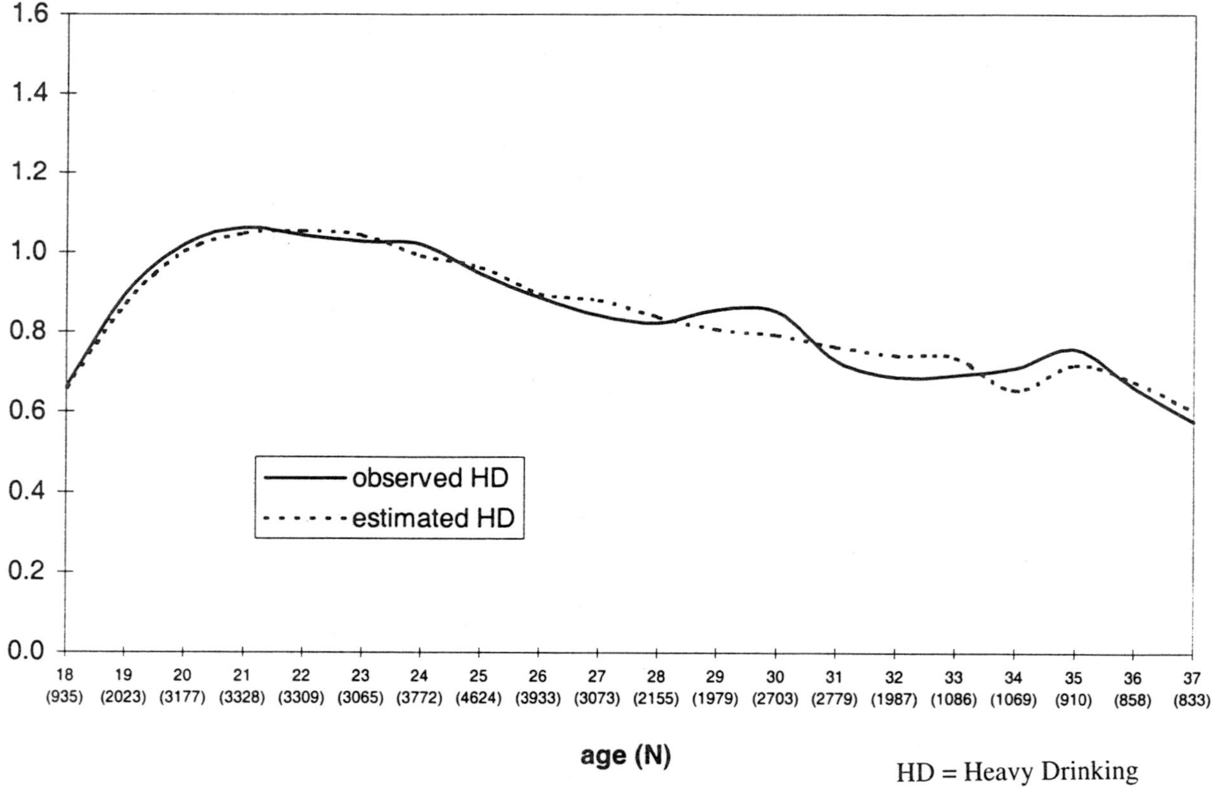

FIG. 4.1. Sample means for frequency of heavy drinking in the NLSY.

The third issue concerns the form of the data for studies covering a long time period. The data are often obtained in the form of multiple cohorts where nobody in the sample has observations on the outcome variables for all ages. This leads to modeling using missing data techniques. An overview is given of some key issues when modeling with missing data.

The fourth issue concerns nonindependence of sample units due to cluster sampling. In the NLSY, many of the households in the sample have multiple siblings and given that siblings within a household share the same home environment, this gives rise to intraclass correlation that violates customary analysis assumptions of independent observations. An overview is given of modeling with cluster samples.

The discussion of these four methodological issues draws on recent work in Muthén and Muthén (in press), Muthén and Shedden (in press), and Muthén and Khoo (1998). All analyses were performed using Mplus, a new comprehensive modeling program for applied researchers (Muthén & Muthén, 1998).

VARYING CENTERING POINTS

This section discusses the use of varying centering points to study changes over time in how a set of background variables influences the growth process. The concept of centering arises in longitudinal analyses carried out using random coefficient growth modeling, also known as multilevel or mixed linear modeling. Muthén and Curran (1997) presented a discussion of such modeling in the latent variable framework used here. An introduction is given in Curran (1998), in this book, and also in Muthén and Khoo (1998).

Briefly stated, growth modeling in the latent variable framework uses latent growth factors to describe the variation across individuals in development over time. For each individual, the growth factor values determine the systematic part of the development over time. This is the person's trajectory over time. Time-specific residuals add to this systematic development to give the observed outcome at each time point.

Conceptually, centering at a certain age can be seen as a particular way of defining the meaning of the growth factors. As an example, in a model of linear growth over time there is an intercept factor and a slope factor. Our discussion focuses on the intercept factor. The intercept factor determines the level of the systematic part of the development at a particular age. This age is chosen by the analyst as an age that is of particular substantive interest. If the lowest age is chosen, the intercept factor is referred to as initial status. Alternatively, the highest age may be of primary interest in which case the intercept factor concerns ending status. Relating the intercept factor to background variables assesses the influence of the background variables on the intercept defined at the age chosen. As is intuitively clear, however, the influence of the background variables need not be the same at different ages. The estimates will be different as will the standard errors and the t values. In this way, a background variable that is insignificant with a certain centering point may be significant with another. Here, it is proposed that it is valuable to look at the results for all possible centering points in order to study the change over age of the covariate influence.

Methodologically, centering at a certain age can be seen as a particular parameterization of the growth model. Different centering points are statistically equivalent in that they give the same model fit. Different choices correspond to different ways of looking at the same data. This is much like different rotations in exploratory factor analysis being different ways of representing the same sample correlation matrix. The growth analysis need in principle only be done for one centering point, whereupon a transformation of the solution to any other centering point can be carried out. Until this facility is commonly available in software, however, the user can simply reanalyze the model with the different centering points of interest.

Next, the principles of centering are shown for linear and quadratic growth. In the subsequent illustration, these centering types are applied, and are discussed in the context of joint growth modeling of two different outcomes.

Centering With Linear Growth

A simple linear growth model illustrates the formulas clearly. Consider an outcome y_{it} for individual i observed at time point t,

$$y_{it} = \eta_{0i} + \eta_{1i} x_t + \epsilon_{it} \tag{4.1}$$

where η_{0i} is the intercept factor, η_{1i} is the growth rate factor, x_t represents a time-related score such as age, and ϵ_{it} is the time-specific residual. A given individual, i, has the values η_{0i} and η_{1i} on the two growth factors. The growth process for this individual develops over time as x_t changes as $\eta_{0i} + \eta_{1i}\, x_t$. This is individual i's trajectory, describing the systematic part of the variation of the outcome at different time points. The individual's outcome at a certain time point t, y_{it}, is equal to the sum of the systematic part of the variation plus the time-specific residual ϵ_{it}.

Equation 4.1 is often referred to as the Level 1 equation, describing the repeated measures over time. The Level 2 equation describes the variation in the η_{0i} and η_{1i} factors as a function of covariates x,

$$\eta_{0i} = \alpha_0 + \sum_r \beta_{0r}\, x_{ri} + \zeta_{0i}, \tag{4.2}$$

$$\eta_{1i} = \alpha_1 + \sum_r \beta_{1r}\, x_{ri} + \zeta_{1i}, \tag{4.3}$$

where α coefficients are intercept parameters, β coefficients are regression weights for the covariates, and the ζ's represent residuals.

Centering relates to the choice of x_t scores in Equation 4.1, so it is important to consider how the x scores are chosen for different t values. For simplicity, say that t represents wave of measurement. With equidistant measurements, t would assume values 1, 2, ... , T, but nonequidistant time points are also possible. Or, t could represent year of measurement, such as 1998, 1999, 2000, 2001. Examples of x_t are age at time point t and school grade at time point t. To a certain degree, however, x_t values can be chosen by the researcher. For example, if age assumes the values 25, 27, 30, the same analysis would be obtained with age rescaled as 0, 1, 1.5. The key issue is that the distance between any two x_t values is comparable in the two scoring schemes. Centering is determined by choosing the time point t for which $x_t = 0$. Next, two alternative centering points are shown, defining the interpretation of the intercept factor as initial status and final status, respectively.

Initial Status Centering. Centering at the first time point ($t = 1$) implies that we set $x_1 = 0$ so that

$$y_{i1} = \eta_{0i} + \eta_{1i}\, 0 + \epsilon_{i1}, \tag{4.4}$$

which means that growth rate factor η_{1i} is not involved in the outcome y_{i1}, but the outcome is a function of the intercept factor η_{0i} and the time-specific residual ϵ_{i1}. The intercept factor η_{0i} is therefore the systematic part of the outcome y_{i1}, the part determined by the growth process. Because of this, η_{0i} is interpreted as the initial status growth factor, that is, the level of the growth process at the first time point, $t = 1$.

Final Status Centering. Centering at the last time point ($t = T$) implies that we set $x_T = 0$ so that $y_{iT} = \eta_{0i} + \epsilon_{iT}$. Here, the growth factor η_{0i} represents the level of the growth process at the last time point.

Any other centering point in between $t = 0$ and $t = T$ may be chosen. In this way, varying centering expresses the level of the growth process in terms of the η_{0i} factor at any time point t.

It may be noted that analysis of the regression coefficients for the intercept factor defined at different time points is not the same as doing regular regression analysis for the outcome y at these different time points. A set of regression analyses for each time point would give information about how the influence of covariates change over time. Growth modeling, however, considers the intercept factor as the more fundamental dependent variable given that it represents the systematic part of the developmental trajectory. There are two statistical reasons why it is advantageous to use the intercept factor as the dependent variable. First, the intercept factor avoids the time-specific influence of the residual contained in y, which results in a more powerful analysis. Second, in contrast to regular regression, the intercept factor draws on information for y from all time points, which also results in a more powerful analysis. Taken together, this means that a set of regression analyses might overlook important changes in covariate influence over time.

Centering With Quadratic Growth

In growth modeling with a quadratic growth function, three growth factors are involved: an intercept factor, a linear growth rate factor, and a quadratic growth rate factor. The quadratic growth model is written as

$$y_{it} = \eta_{0i} + x_t\eta_{1i} + x_t^2\eta_{2i} + \epsilon_{it}, \tag{4.5}$$

where η_{0i} is the intercept factor, η_{1i} is the linear rate factor, and η_{2i} is the quadratic rate factor. Consider centering at $t = 1$, using $x_1 = 0$. This means that the intercept factor η_{0i} is interpreted as an initial status factor. As in the linear case, choosing $x_t = 0$ defines the centering point at time point t. Even though the growth function is more complex in the quadratic case, the same simple interpretation as in the linear case is obtained for the intercept factor. With varying centering point corresponding to different time points, the factor is interpreted as the level of the growth process at that time point.

Application of Varying Centering to Heavy Drinking

Given the shape of the mean curve for heavy drinking in the NLSY shown in Fig. 4.1, a quadratic growth model with three random coefficients seems reasonable. Letting the subscript i denote individuals (i = 1, 2, ..., 7,933) and the subscript t time points (t = 1, 2, ..., 20), the model therefore expresses the development of the heavy drinking outcome variable over time as a quadratic function of three random coefficients η,

$$y_{it} = \eta_{0i} + \lambda_t \eta_{1i} + \lambda^2_t \eta_{2i} + \epsilon_{it}, \tag{4.6}$$

where η_{0i} is an intercept term, η_{1i} a coefficient for the linear term, and η_{2i} a coefficient for the quadratic term. In Equation 4.6, λ_t replaces x_t to show that these time steps are parameters that can be estimated. The centering point is defined as the time point t where the parameter λ_t is fixed to zero.

Letting the λ_t time steps be estimated gives a more flexible functional form for the growth model. This proved to be very important for the NLSY data where the heavy drinking sample means follow a curve that increases more rapidly than it decreases. Such a nonsymmetric form is not always well-fitted by a conventional quadratic function with fixed λ_t time steps. Corresponding to a centering at age 18, the first three λ_t values are fixed at 0, 1, and 2 for identification purposes, and the next 17 steps are estimated in the analysis.

The variation across individuals in the three growth factors is described by a set of covariates x, adding the model parts where α

$$\eta_{0i} = \alpha_0 + \sum_r \beta_{0r}\, x_{ri} + \zeta_{0i}, \tag{4.7}$$

$$\eta_{1i} = \alpha_1 + \sum_r \beta_{1r}\, x_{ri} + \zeta_{1i}, \tag{4.8}$$

$$\eta_{2i} = \alpha_2 + \sum_r \beta_{2r}\, x_{ri} + \zeta_{2i}, \tag{4.9}$$

and β are regression parameters and ζ are residuals. The model is estimated by maximum likelihood using the Mplus software (Muthén & Muthén, 1998).

Varying the Centering Over Ages 18 to 37: Estimated Effects of Covariates on Heavy Drinking. Figure 4.2 shows the results of varying centering points for the estimated growth model for the frequency of heavy drinking. The figure varies the centering point so that the random intercept is defined as the level of frequency of heavy drinking at ages 18, 21, 25, 27, 32, and 37. At each centering age, the figure shows the partial regression coefficients for the intercept factor regressed on the covariates, denoted β in Equations 4.7 through 4.9. Note that all covariates are 0/1 dummy variables. As usual in regression analysis, the partial regression coefficient for a 0/1 dummy variable represents the effect of going from status 0 to status 1, holding other background constant. The broken horizontal line represents a zero coefficient. Positive coefficients correspond to a higher frequency of heavy drinking and negative coefficients correspond to a lower frequency of heavy drinking. Covariates corresponding to risk factors are expected to have positive coefficients, and covariates corresponding to protective factors are expected to have negative coefficients.

Consider as an example centering at age 18. Here, the largest coefficient is for Onset with a value close to 0.9, saying that individuals who started drinking regularly before age 15 have a higher frequency of heavy drinking at age 18 than those who did not, holding other covariates constant. The legend shows that this coefficient is significantly different from zero (marked s as in significant). The second largest coefficient is for Black with a value close to -0.6, saying that Blacks have a lower frequency of heavy drinking than non-Blacks at age 18, holding other covariates constant. These findings are in line with the alcohol literature. The new findings of Muthén and Muthén (in press) are based on varying

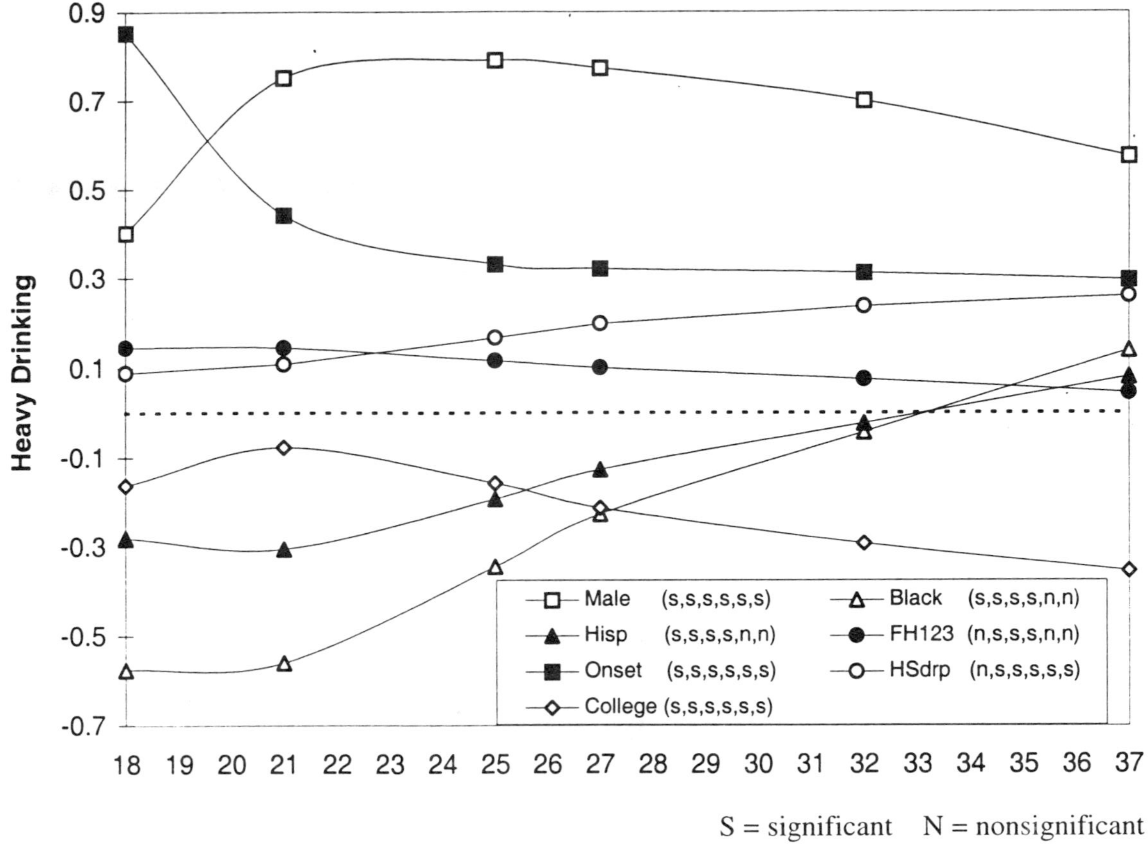

FIG. 4.2. Effects of background variables at different ages: Heavy drinking.

the centering point from 18 to 37. Following are some of the key findings for heavy drinking.

The heavy drinking coefficients in Figure 4.2 show that the largest partial regression coefficient at any age occurs for the early onset variable at age 18. This effect, however, diminishes over age. Figure 4.2 also shows that although Blacks and Hispanics have lower values than Whites at ages 18 and 21, their heavy drinking is not different from Whites at the later ages, 32 and 37. Furthermore, it is interesting to note that there is an increasing protective effect of college education.

Centering and Multiple Outcomes: Joint Analysis of Heavy Drinking and Severity Score

The severity score development in the NLSY is used to illustrate varying centering with a linear growth model. Before these results are presented, however, it is useful to discuss more generally some of the analysis considerations involved when studying two outcome variables. Furthermore, special considerations are needed for modeling the severity score given that, for this variable, only two repeated measures are available per individual.

Figure 4.3 shows the mean values for both of the outcome variables, heavy drinking and the severity score for alcohol-related problems. It is seen that the shape of the mean curve is rather similar for the ages where both are observed, 25 to 37. It is of interest to also study the individual variation in the severity trajectories and compare the relative importance of the time-invariant covariates for severity with that of heavy drinking. Using varying centering points, the comparison of relative importance of covariates can be made over age.

The severity score is based on items asked only in 1989 and 1994. An advantage of the multiple-cohort design is that this still results in coverage of a wide age range, 25 to 37 (this will be discussed later). The problem for growth modeling is, however, that for all individuals in the sample, the score is only available at two time points. With two time points, a general random coefficient growth model cannot be identified even when restricted to a linear model. This is clearly seen when noting that the growth model describes the correlations among the outcomes in terms of the variance and covariance parameters for the growth factors. With a linear model, there are three such parameters but with two measurement occasions per cohort there is only one observed

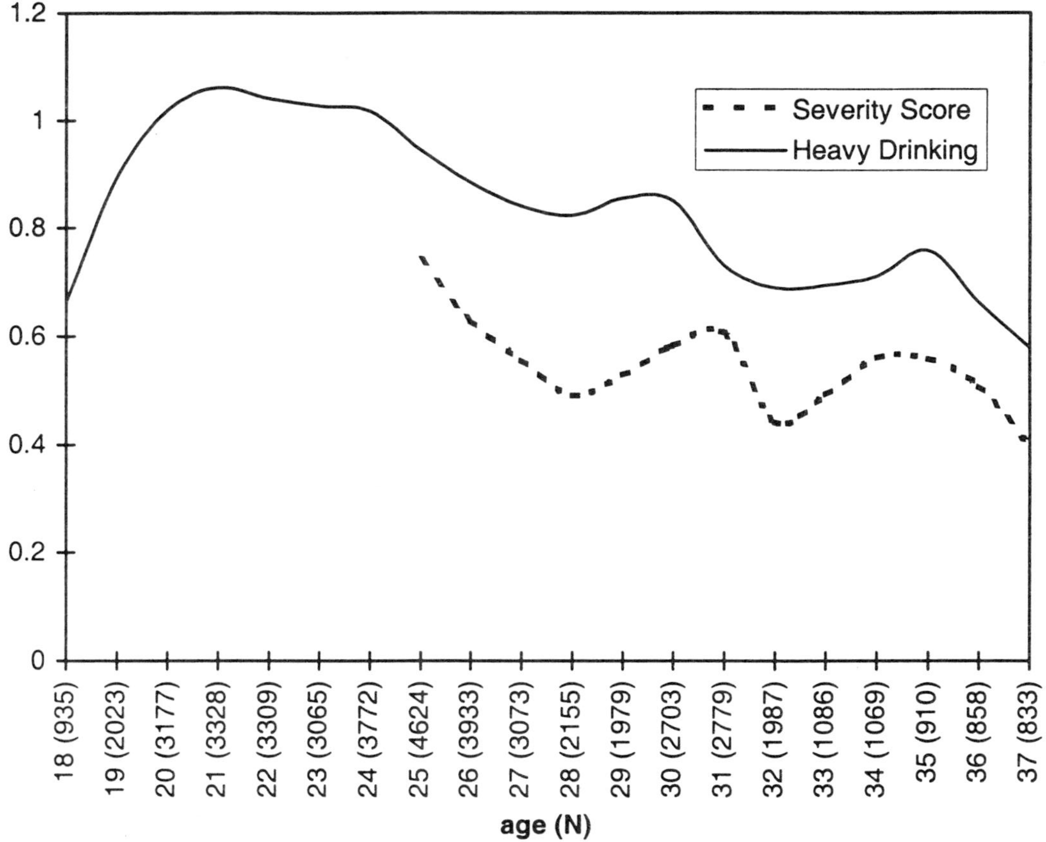

FIG 4.3. Sample means for heavy drinking and severity score in the NLSY.

covariance. This means that none of the cohorts supports the identification of the growth model. There are two possible solutions to this problem. One is to treat the two outcome variables as indicators of a single alcohol problem construct and the other is to specify a simplified linear model for the severity score. These two solutions will be discussed in turn.

Growth Modeling With Multiple Indicators. The severity score and the heavy drinking variable can be viewed as two indicators of the same unobserved construct, say alcohol problems. They reflect different aspects of this construct and it may be assumed that a factor analytic measurement model holds where the two indicators have different loadings and specific (error) variances. This formulation makes it possible to use the model in (Equation 4.1) with y replaced by this construct. The severity score indicator is only available in 1989 and 1994. The growth model can however be formulated for the full age range of 18 to 37, which means that an extrapolation of the severity score growth curve to younger ages than 25 is achieved. This multiple indicator approach does, however, carry with it certain assumptions. The most important assumption given our interest in the influence of covariates is that the model postulates that the regression coefficients for the two indicators on the covariates have the same proportionality factor across all covariates for all time points. This assumption would be violated if different covariates have different influence for the two indicators. Given that one key goal of the study is determining to which extent this holds, presupposing this in the modeling is not desirable.

Linear Growth Modeling With Two Time Points. A more straightforward approach is to specify a simplified linear model that is restricted to have a random coefficient only for the intercept and letting the slope be fixed, that is, not varying across individuals. This growth model is identified using only two time points. This fact was used in the modeling of Muthén and Muthén (in press). There are three reasons that the assumption of a linear growth model with fixed slope is plausible for the severity score in the 25 to 37 age range. First, this age range is past the normative peak age and shows roughly a linear decline. Second, estimates for heavy drinking show that most of the individual variation can be captured by variation in the intercept defined at 25. Third, the slope can still be allowed to vary as a function of the covariates so that its mean is different for different subgroups. The availability of only two time points does, however, limit the study of other growth forms than the linear. For example, it might have been interesting to study the significance of the bumps seen in Fig. 4.3.

The simplified linear model for the severity score can either be analyzed separately or together with the quadratic growth model for heavy drinking. When estimated together with heavy drinking, it is possible to study contemporaneous correlations between the severity score and heavy drinking. This is of interest to determine to which extent a high severity score occurs with high frequency of heavy drinking. Other types of joint growth modeling of several processes simultaneously were discussed in Muthén (1997) including comparisons with auto-regressive modeling with cross-lagged effects.

Varying the Centering Over Ages 18 to 37: Estimated Effects of Covariates on Severity Score. The results for varying centering points for the severity score are now presented. They are based on the growth model obtained by analyzing the two processes jointly using the simplified linear model. The results are given in Fig. 4.4, where the centering point is varied so that the random intercept is defined at ages 25, 27, 32, and 37. The figure shows the partial regression coefficients for the intercept on each covariate, denoted as β in Equations 4.7 to 4.9.

The severity score coefficients in Fig. 4.4 show an interesting trend of increasing importance of dropping out of high school. At age 37, the strongest partial regression coefficient is the lasting effect of having dropped out of high school. Other interesting trends are the diminishing effects of early onset and family history of alcoholism. The plot of the coefficients for varying centering points is clearly useful in that it indicates trends over age, thereby raising questions concerning future development.

MULTIPLE TRAJECTORY CLASSES

We now focus on modeling of heterogeneity across individuals in their development over time. Individuals may belong to different, unknown subpopulations corresponding to normative development and various forms of non-normative development. The applicability of new methods for studying such heterogeneity is summarized. The discussion indicates

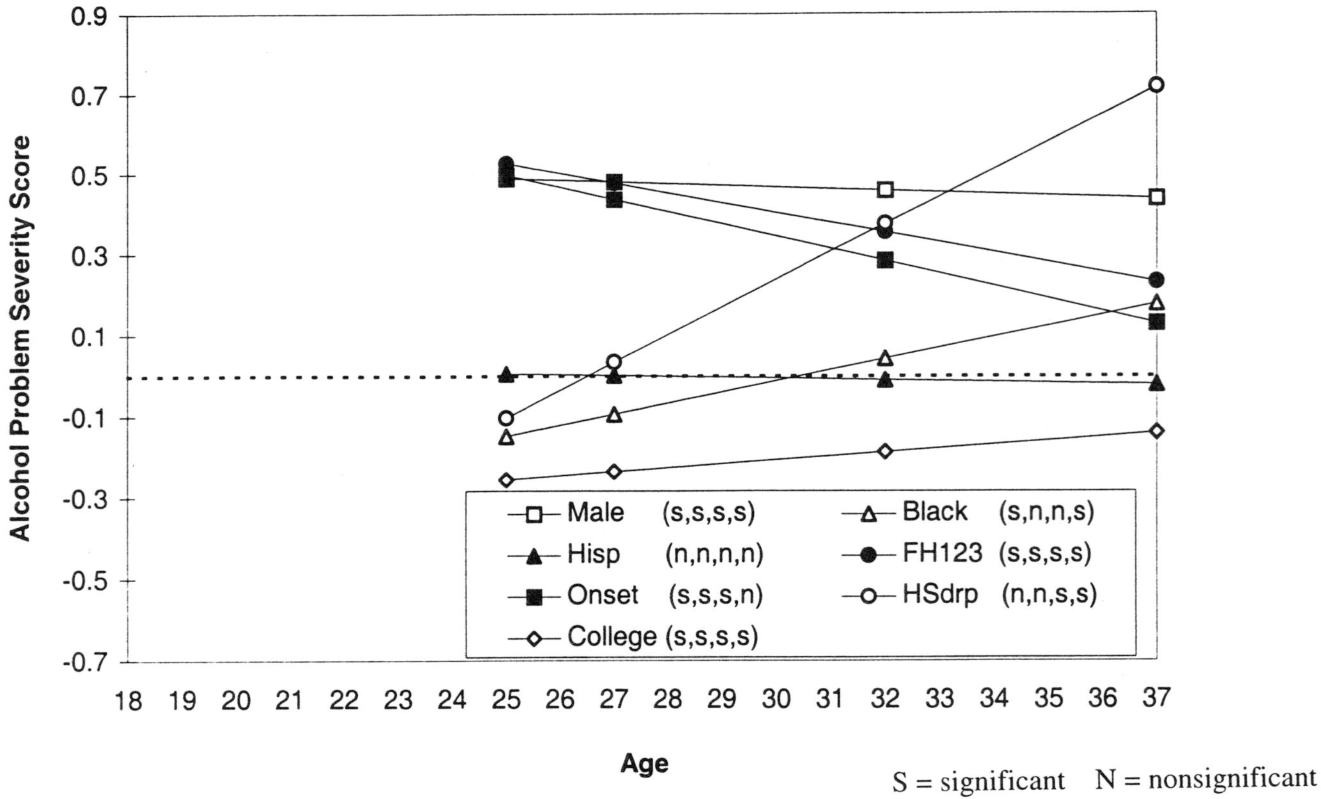

FIG 4.4. Effects of background variables at different ages: Alcohol problem severity score.

that growth mixture modeling is an exciting addition to the developmental analysis tools.

Figure 4.5 shows four different growth trajectories that can be derived from the estimated quadratic growth model for heavy drinking in the NLSY. They are obtained from different individual values on the growth factors for the intercept, linear, and quadratic terms. Given the large differences in the trajectories, one may ask if the individuals following these different trajectories really are from the same population or are better characterized as coming from different subpopulations. The different subpopulations may have different antecedents and different consequences. If this heterogeneity is ignored, incorrect conclusions may be drawn and useful information in the data overlooked.

Alcohol use development in terms of classes of trajectories has been discussed recently in the alcohol literature by Schulenberg, O'Malley, Bachman, Wadsworth, and Johnston (1996) and Bates and Labouvie (1997). Drawing on the Zucker (1994) distinction between different pathways to alcohol problems and alcoholism, one may hypothesize that trajectory Class 4 corresponds to "antisocial alcoholism", and that

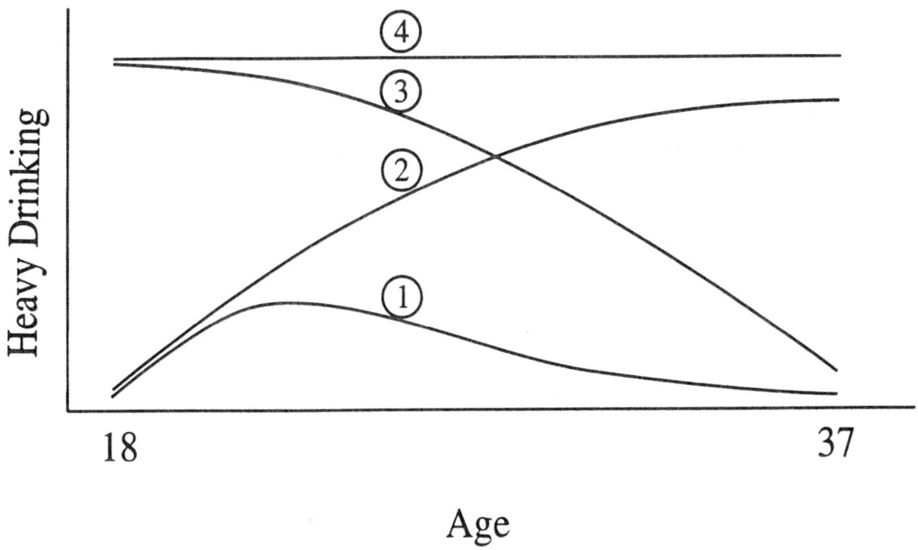

FIG 4.5. Four hypothetical trajectory classes of alcohol development.

trajectory Class 3 corresponds to "developmentally limited alcoholism." Trajectory Class 1 is the normative class of typical development. Class 4 may be more common in males and associated with family history of alcoholism, early onset, and antisocial behavior. Class 4 represents an early start of heavy drinking, which in contrast to Class 3 does not subside in early adulthood. Whereas Class 3 shows a transition from excessive use to more moderate use, Class 2 shows a transition from normative use to problematic use at the typical peak age. An important task of alcohol research is to find predictors of these transitions. The classes also differ in that Class 4 and Class 2 may be associated with alcohol-related problem consequences in young adulthood, whereas Class 1 and Class 3 are not.

Growth Mixture Modeling

The idea of multiple unobserved subpopulations or latent classes can be incorporated into growth modeling using the statistical feature of finite mixture modeling. This is referred to as growth mixture modeling. Muthén and Shedden (in press) used growth mixture modeling to formulate a latent trajectory class model for heavy drinking in the youngest NLSY cohort for ages 18 to 25. Three latent trajectory classes were obtained. The three classes are those of Fig. 4.5 except that Class 3 and Class 4 of Fig. 4.5 are combined.

There are two interesting aspects of growth mixture modeling with latent trajectory classes, studying growth process antecedents and studying growth process consequences.

Growth Mixture Modeling of Antecedents. Growth model covariates play the role of antecedents to the growth process in that they can be used to predict the growth factor values for an individual. With growth mixture modeling, the analysis of antecedents is strengthened by allowing different means and different covariate influence across the different latent trajectory classes.

The quadratic growth model used for NLSY heavy drinking analysis is

$$y_{it} = \eta_{0i} + \lambda_t \eta_{1i} + \lambda^2_t \eta_{2i} + \epsilon_{it}, \tag{4.10}$$

$$\eta_{0i} = \alpha_0 + \sum_r \beta_{0r}\, x_{ri} + \varsigma_{0i}, \tag{4.11}$$

$$\eta_{1i} = \alpha_1 + \sum_r \beta_{1r}\, x_{ri} + \varsigma_{1i}, \tag{4.12}$$

$$\eta_{2i} = \alpha_2 + \sum_r \beta_{2r}\, x_{ri} + \varsigma_{2i}, \tag{4.13}$$

Growth mixture modeling allows the α and β coefficients in these equations to vary across the latent trajectory classes. Consider first the means of the three growth factors, where for simplicity covariates are ignored here so that the means are represented by the α coefficients. The trajectory Classes 2 and 3 in Fig. 4.5 may be taken as an example. The mean of the intercept factor at age 18 (α_0) is lower for Class 2 than for Class 3. The mean of the linear growth rate factor (α_1) is positive for Class 2 and negative for Class 3. The quadratic growth rate factor (α_2) is slightly negative for Class 2 reflecting a downturn at higher ages, whereas it is slightly positive for Class 3. Class-varying β coefficients are also likely across the four classes in Fig. 4.5. Normative development may for example be influenced by different covariate coefficients than non-normative development.

Once trajectory class membership has been found to influence the growth process, the interest turns to finding predictors of trajectory class membership. These might be sought among the growth model covariates but there may also be additional variables. Prediction of class membership using the growth model covariates was studied for the three-class model of Muthén and Shedden (in press). This showed that compared to the normative class, the likelihood of membership in the non-normative class of escalating use was increased by being male, being a high school dropout and having positive family history. The likelihood of membership in the non-normative class of high use already at age 18 was increased by early onset, being male, not being Black, and having positive family history. As discussed in Muthén and Shedden (in press), this analysis can be taken one step further by the inclusion of early latent class indicators, for example, adolescent measures such as early onset and antisocial behavior, in order to provide an early warning system for problematic alcohol development.

Growth Mixture Modeling of Consequences. Conventional growth modeling considers the growth curve as an outcome predicted by the covariates. Growth curves need not, however, be used only as outcomes but can also be used as predictors. The growth mixture modeling used in Muthén and Shedden (in press) also served a predictive purpose. Muthén and Shedden (in press) were interested in using the heavy drinking growth model to predict alcohol dependence 5 years later. Alcohol dependence was formulated as a binary variable based on the symptom items discussed earlier forming DSM-IV diagnostic criteria to give dependence diagnoses. They argued that it was difficult to use the conventional growth model for this prediction because the growth factors have a nonlinear interplay in their determination of curve shape. For example, curve Classes 4 and 2 in Fig. 4.5 may be hypothesized to be associated with a higher probability of later alcohol dependence than curve Class 1. Using the intercept factor and the linear factor value for this prediction, however, does not fit this hypothesis. An increasing intercept factor value is not always predictive of dependence under this assumption given that a low intercept value is associated with both curve Class 1 and curve Class 2. Likewise, although curve Class 2 has a high linear slope factor value, curve Class 4 has a zero value. This shows that it is hard to use curve shape for prediction in the conventional growth model. The growth mixture model solves this dilemma in a simple way.

The growth mixture model allows for different curve classes such as those in Fig. 4.5 by letting different latent classes of individuals have different growth factor means. In this way, a categorical latent variable, the latent class variable, is introduced in addition to the continuous latent variables corresponding to the growth factors. The categories (classes) of the categorical latent variable correspond directly to the curve shapes and the model lets the dependence probability vary over these categories. The estimated three-class model in Muthén and Shedden (in press) showed that the normative Class 1 had a dependence probability of 0.08. The two non-normative classes, however, had probabilities of 0.23 and 0.39. The corresponding odds ratios for dependence in each of the two non-normative classes relative to the normative class are 3.43 and 7.35. This clearly shows the elevated risk of developing alcohol dependence for individuals who are in the non-normative classes. It is therefore of interest to study the possibility of predicting class membership as early as possible.

MULTIPLE-COHORT ANALYSIS

This section discusses methodological issues related to the use of multiple cohorts in longitudinal studies. The data structure of multiple-cohort analysis is similar to that of missing data and some general missing data issues are also discussed.

Figure 4.6 shows the multiple-cohort design of the NLSY for the 18 to 37 age range when alcohol measurements were included in the survey. Nonshaded areas correspond to years when alcohol measures were obtained. In our analyses, 1985 is excluded given that a question format change was made that may have made across-time comparisons invalid. Figure 4.6 shows that different ages are covered by different cohorts and are represented by different number of observations. For example, age 18 is only represented by the approximately 1,000 individuals in cohort 64, yet age 25 is represented by several cohorts. For each cohort only six ages/occasions are represented (excluding the 1985 measure). The design is sometimes referred to as an *accelerated cohort design* given that there are only 12 years between the first measurement year of 1982 and the last measurement year of 1994, whereas 20 years of development from 18 to 37 is covered.

Multiple Cohorts Viewed as Missing Data

Figure 4.6 may be viewed in terms of missing data. The figure may be seen as a data matrix with rows corresponding to (groups of) people and columns corresponding to variables. For example, the 20 columns may be seen as corresponding to the heavy drinking outcome variable measured at ages 18 to 37. In this way, the shaded areas in the matrix represent missing data. Data are missing by design, which means that if the cohorts can be assumed to be drawn from the same population, we can assume that data are missing completely at random (MCAR) to use a term introduced in Little and Rubin (1987). Simply put, MCAR means that the probabilities of the values being missing are not related to the values that would have been observed. Here, the missing data cannot be handled by the common approach of listwise deletion because nobody in the sample has complete data.

The missing data structure in Fig. 4. 6 can be handled analytically in several ways using a latent variable framework. Using maximum-likelihood estimation with continuous-normal outcome variables, a multiple-group mean and covariance structure analysis of the eight cohorts may be carried out as in Muthén, Kaplan, and Hollis (1987). In this analysis, the groups are assumed to come from one and the same population so that across-group equality constraints are imposed on all parameters

	Age																			
Cohort	18	19	20	21	22	23	24	25	26	27	28	29	30	31	32	33	34	35	36	37
57								82	83	84	85	86	87	88	89	90	91	92	93	94
58							82	83	84	85	86	87	88	89	90	91	92	93	94	
59						82	83	84	85	86	87	88	89	90	91	92	93	94		
60					82	83	84	85	86	87	88	89	90	91	92	93	94			
61				82	83	84	85	86	87	88	89	90	91	92	93	94				
62			82	83	84	85	86	87	88	89	90	91	92	93	94					
63		82	83	84	85	86	87	88	89	90	91	92	93	94						
64	82	83	84	85	86	87	88	89	90	91	92	93	94							

FIG. 4.6. NLSY alcohol multiple-cohort data structure.

that are in common. The analysis needs to fill in dummy elements in the sample means, variances and covariances corresponding to variables that are missing for the group (cohort). Newer latent variable software can handle the missing data feature directly without such an artificial setup by allowing for a single-group, single-population analysis where eight missing data patterns are present, one corresponding to each cohort.

Hidden Cohort Design

It should be noted that the missing data structure represented by Fig. 4.6 often arises in circumstances where it is not planned, which implies hidden modeling opportunities. A typical example is where, at a given wave of measurement, all individuals cannot be followed up at the same point in time given practical constraints on interview load and so on. Often these differences in timing can be ignored but in many cases they are important and should be taken into account in the growth modeling. For example, in terms of reading development, early and late grade-one measures do not refer to the same developmental age and college drinking experiences may not be same at the beginning and end of the freshman year. As yet another example, consider a study where drinking is to be examined across a 10-year period assuming a starting age that ranges from 20 to 40 in the sample. For individuals starting at age 20, the 10-year development is in a different developmental phase than for individuals starting at age 40. Ignoring these differences in developmental phases and analyzing the data with time point (1, 2, ..., 10), not age (20, 21, ..., 50), as the time axis would amount to analysis of heterogeneous subpopulations and would give misleading results.

This data situation may be referred to as a *hidden cohort* design. There is a direct analogy between these examples and the missing data structure of Fig. 4.6. For example, to make a distinction between early and late in the year, one may consider the age points as fractions of a year, or use the month of the year, which results in no individual being observed more than once in the year. Letting age define the time axis has the useful side effect of "spreading out the time axis" so that more time points are available for growth modeling. This is highly desirable in order to capture more features in the data and avoid restrictive modeling assumptions. Frequently, it is quite reasonable to assume that MCAR is at hand, given that the time point at which an individual is interviewed is likely to be uncorrelated with his/her observed values.

Attrition

In passing, it should be pointed out that in addition to missingness by design, longitudinal data typically show attrition over time with loss of individuals to follow up. It is often a reasonable approximation that such missingness satisfies the assumption of "missing at random" (MAR) in Little and Rubin (1987) terms. In contrast to MCAR, MAR makes the more realistic assumption that the probabilities of values being missing can be related to variables that are not missing, for instance covariates and variables observed at the first time point of the study. When MAR holds, correct maximum-likelihood estimates are obtained by using all available data, including observations on individuals who have missing data on some of the analysis variables.

Cohort Differences

With multiple cohorts it is important to test for across-cohort parameter invariance. Societal changes may have taken place that influence drinking practices in the 7 years between the births of the youngest and oldest cohort in the NLSY. At the one extreme, one may assume that all cohorts come from the same population as just discussed. At the other extreme, one may assume that each cohort corresponds to its own population. Model tests of fit can be compared for models corresponding to these two extremes. If there is a significant difference in model fit, it is of interest to consider the eight-population analysis with full across-population invariance and study model modification indices that point to invariance assumptions that do not match the data well. In such an eight-population analysis of the Fig. 4.6 data on heavy drinking, the growth modeling of Muthén and Muthén (in press) found that full invariance could not be rejected in important ways except for the mean of one of the time-invariant covariates, early onset. The mean of this variable increased steadily from cohort 57 to cohort 64.

INTRACLASS CORRELATION DUE TO MULTIPLE SIBLINGS PER HOUSEHOLD

This section discusses methodological aspects of analyzing data obtained by cluster sampling. Cluster sampling is a common feature in national samples such as the NLSY. In the NLSY, clustering occurs because for each household all siblings in a given age range are interviewed. Clustered data causes nonindependence among the observations that needs to be taken into account. The methods in this area range from those focusing on nonindependence corrections of standard errors and chi-square tests of model fit to methods that express the nonindependence explicitly in the models.

The left part of Table 4.1 shows the number of interviewed siblings per household in the NLSY sample. It is seen that for almost a quarter of the households more than one sibling was interviewed. The average number of siblings per household is 1.4. The right part of Table 4.1 shows the intraclass correlation, that is, the correlation between siblings within households, for the heavy drinking variable. The intraclass correlation is measured as the ratio of the between-household variance divided by the sum of the within- and between-household variance. If siblings within a household have similar heavy drinking, the within variance is small and the intraclass correlation is large. Table 4.1 shows that the intraclass correlation ranges from 0.19 in 1982 to 0.06 in 1989. It is reasonable that the value decreases by year as the siblings grow older, move out of their original households, and lead more independent lives.

Design Effects

A positive value of the intraclass correlation indicates that the conventional analysis assumption of independence is violated. The seriousness of this violation, however, may not be large. Intraclass correlations are

TABLE 4.1

NLSY Household Clusters

Household Type (Number of respondents)	*Number of Households**	*Intraclass Year*	*Correlations for Siblings Heavy Drinking*
Single	5,944	1982	0.19
Two	1,985	1983	0.18
Three	634	1984	0.12
Four	170	1985	0.09
Five	32	1988	0.04
Six	5	1989	0.06

Total number of households: 8,770

Total number of respondents: 12,686

Average number of respondents per household: 1.4

*Source: NLS User's Guide, 1994, p. 247

discussed in the context of cluster sampling where the ratio of the variance of an estimator computed under cluster sampling (correlated observations) to that computed under simple random sampling (independent observations) is described as the design effect. Cluster sampling gives a larger variance than when using simple random sampling formulas. The degree to which the design effect exceeds 1 reflects the error in the variance assessment when ignoring the non-independence of observations obtained under cluster sampling. In other words, ignoring the intraclass correlation results in standard errors of estimates that are too small by a factor corresponding to the square root of the design effect. For an estimator of the mean and clusters of equal size, the design effect is $1 + \rho(c - 1)$, where ρ is the intraclass correlation and c is the common cluster size. This shows that it is the product of the intraclass correlation and the cluster size that is the important factor.

The NLSY data does not have equal household sizes and the estimates in growth modeling concern more complex parameters than means, but as Muthén and Satorra (1995) pointed out in their simulations' the product formula just shown may give a rough indication of the potential problem at hand. From Table 4.1, one finds that the product is maximally $0.19\,(1.4 - 1) = 0.076$, which assuming equal household sizes would result in a design effect of 1.076, a very small value that indicates that the lack of independence can probably be ignored.

Latent Variable Analysis With Cluster Samples

Muthén and Satorra (1995) discussed ways to compute correct standard errors and chi-square tests of model fit when the design effects are not ignorable. Even when they are ignorable, however, interesting modeling opportunities exist for describing the correlations among siblings within households as long as these correlations are not zero. As the discussion in Muthén and Satorra (1995) pointed out, sampling statisticians sometimes make a distinction between aggregated and disaggregated analysis.

Aggregated analysis amounts to a conventional analysis that ignores the lack of independence when computing parameter estimates, but uses special formulas to compute standard errors and chi-square tests of fit.

Disaggregated analysis instead deals with the lack of independence directly by modeling it. Multilevel modeling is an example of disaggregated analysis, where in this case one would consider a three-level model for repeated measures over time for each sibling for each household. Here, independence of observations is only assumed for clusters (households), not individuals. Three-level modeling in a latent variable framework was discussed in Muthén (1997). Muthén and Khoo (1998), see also Khoo in this volume, discuss disaggregated analysis alternatives for household (family) data.

CONCLUSIONS

This chapter discussed four analysis complications arising in growth modeling of a national longitudinal sample and analysis approaches were presented that handled these complications. More importantly, each of the four complications can be seen as representing an opportunity for understanding deeper features in the data that are not usually uncovered by conventional techniques. The discussion pointed to solutions that use modeling with a richer set of parameters in order to answer new research questions.

All techniques proposed here are available in the new latent variable software package Mplus (Muthén & Muthén, 1998b). Hopefully, this chapter will stimulate the use of these techniques for new, interesting longitudinal studies in the alcohol area and developmental research in general.

ACKNOWLEDGMENTS

This research was supported by Grant 1 K02 AA 00230-01 from NIAAA, and by Grant 1 R21 AA10948-01A1 from NIAAA. I thank Tom Harford for suggestions regarding the NLSY application.

REFERENCES

Bates, M. E., & Labouvie, E. W. (1997). Adolescent risk factors and the prediction of persistent alcohol and drug use into adulthood. *Alcoholism Clinical and Experimental Research, 21,* 944-950.

Little, R. J. A., & Rubin, D. B. (1987). *Statistical analysis with missing data.* New York: Wiley.

Muthén, B.(1997). Latent variable modeling with longitudinal and multilevel data. In A. Raftery (Ed.), *Sociological methodology 1997* (pp. 453-480). Boston: Blackwell.

Muthén, B., & Curran, P. (1997). General longitudinal modeling of individual differences in experimental designs: A latent variable framework for analysis and power estimation. *Psychological Methods, 2*, 371–402.

Muthén, B., Kaplan, D., & Hollis, M. (1987). On structural equation modeling with data that are not missing completely at random. *Psychometrika, 42*, 431–462.

Muthén, B., & Khoo, S. T. (1998). Longitudinal studies of achievement growth using latent variable modeling. *Learning and individual differences, Special issue: Latent growth curve analysis, 10*, 73–101.

Muthén, B., & Khoo, S. T. (1998). *Growth modeling of family data*. In preparation.

Muthén, B., & Muthén, L. (In press). The development of heavy drinking from age 18 to 37 in a U.S. national sample. Forthcoming in *Journal of Studies on Alcohol.*

Muthén, L., & Muthén, B. (1998b). *Mplus user's guide.* Los Angeles: Muthén & Muthén.

Muthén, B., & Satorra, A. (1995). Complex sample data in structural equation modeling. In P. V. Marsden (Ed.), *Sociological methodology* (pp. 267–316). Washington, DC: American Sociological Association.

Muthén, B.,& Shedden, K. (in press). Finite mixture modeling with mixture outcomes using the EM algorithm. Forthcoming in *Biometrics.*

Schulenberg, J., O'Malley, P. M., Bachman, J. G., Wadsworth, K. N., & Johnston, L. D. (1996). Getting drunk and growing up: Trajectories of frequent binge drinking during the transition to young adulthood. *Journal of Studies on Alcohol, 57*, 289–304.

Zucker, R. A. (1994). Pathways to alcohol problems and alcoholism: A developmental account of the evidence for multiple alcoholisms and for contextual contributions to risk. In R. A. Zucker, G. M. Boyd, & J. Howards (Eds.), *The development of alcohol problems: Exploring the biopsychosocial matrix of risk* (pp. 225–289). NIAAA Research Monograph 26.

5

Contrasts in Multiple Mediator Models

David P. MacKinnon
Arizona State University

This chapter describes methods to test contrasts among mediated effects in multiple mediator models. First, multiple mediator models are described as a way to summarize drug prevention programs. Second, technical material on testing of specific mediated effects and the standard errors of specific mediated effects is presented. A methodology to test contrasts among mediated and direct effects in multiple mediator models is then presented and illustrated in the evaluation of a multicomponent program to prevent anabolic steroid use among high school football players. Limitations of multiple mediator models and methods to test contrasts in these models are discussed.

A *mediator* is a variable that accounts for all or part of the relation between a predictor and an outcome because the mediator is intermediate in the causal pathway from the independent variable to the dependent variable (Baron & Kenny, 1986; Sobel, 1990). The mediation effect is also called the *indirect effect* because it represents the independent variable effect on the dependent variable through a mediating variable. Mediating variables are sometimes called *intervening variables* because they are intermediate between the independent and dependent variable.

Theory regarding mediated or indirect effects is an important part of social science research (Alwin & Hauser, 1975) and drug abuse research as well (Judd & Kenny, 1981). For example, the effects of socioeconomic status on adolescent drug use may be mediated through failure to bond with school and parents (Hawkins, Lishner, Jensen, & Catalano, 1987). Chassin and colleagues (Chassin et al., 1993) postulated several mediating mechanisms for the effects of parental alcoholism on children's early initiation of drug use. Model estimates were consistent with parental alcoholism leading to increased exposure to life stress, which increases negative affect, which then leads to association with drug-using peers. Another mediational pathway was consistent with parental alcoholism leading to impaired parenting ability, which leads to association with peers who use drugs.

One of the best examples of mediation analysis is the estimation of mediated effects in the experimental evaluation of health-related prevention programs. Prevention and intervention programs are designed to change putative mediating constructs thought to be causally related to health outcomes (MacKinnon & Dwyer, 1993). For example, the Multiple Risk Factor Intervention Trial (MRFIT) was designed to change smoking, cholesterol, and blood pressure in order to prevent heart disease (Multiple Risk Factor Intervention Trial Research Group, 1990). In the prevention case, mediation analysis assesses the process by which the program changed the mediator, which in turn, changed the outcome variable. Mediational analysis in prevention studies is important because the processes that lead to behavior change are studied (MacKinnon, 1994). If the critical mediating processes are identified, then prevention programs can be improved by focusing on the most effective components and removing ineffective or even counterproductive program components.

Hansen (1992) reviewed the 12 mediators most commonly targeted by drug prevention curriculum: information, decision making, pledges, values clarification, goal setting, stress management, self-esteem, resistance skills training, life skills training, norm setting, assistance, and alternatives. Different programs target different groups of mediators. Because drug prevention programs typically target several mediators, the multiple mediator model is the correct model for the evaluation of such programs. Hansen concluded that social influence and comprehensive programs had the most evidence for success, although there were successful programs that did not target social influence mediators, and some programs targeting social influence mediators were not successful. The detailed examination of the contributions of multiple mediators to changes in drug use may clarify the critical mediators for successful drug prevention programming as well as help resolve the discrepancies among studies. Mediation analyses in drug prevention research are most compelling when alternative theories predict different mediational pathways for program effects on drug use. Hansen and Graham (1991), for example, found evidence for the norm setting mediator, but not the resistance skills mediator, following an experimental manipulation of these mediators. Similarly, social influence approaches have been more successful than affective based programs (Flay, 1985).

Mediation Analysis

In the most common method to identify mediated effects, three causal hypotheses are tested based on estimates from regression equations (Baron & Kenny, 1986; Judd & Kenny, 1981). The single mediator model is shown in Fig. 5.1. The first test requires that there is a significant effect of the independent variable (X) on the dependent variable (Y) as shown in Eqn. 5.1. The logic here is that in order for a relationship to be mediated, it is first necessary to show that a relationship exists. However, as described later in this chapter, mediation can occur even when there is not a relationship between the independent and dependent variable.

$$Y = \tau X + \epsilon_1 \tag{5.1}$$

where Y is the outcome variable, X is the program or independent variable, and τ codes the regression coefficient between the program and the dependent variable, which is the total effect, ϵ_1 codes unexplained variability, and the intercept is not shown to simplify the presentation.

The second causal hypothesis requires that there is a significant program effect on the mediating variable, X_M, as summarized in the following equation where the regression coefficient relating the program to the mediating variable is α, and X is the same as described previously, ϵ_2 codes unexplained variability, and the intercept is not shown.

$$X_M = \alpha X + \epsilon_2 \tag{5.2}$$

The third causal hypothesis is a test of whether the mediator is a significant predictor of the dependent variable when both the independent variable and the mediator simultaneously predict the dependent variable.

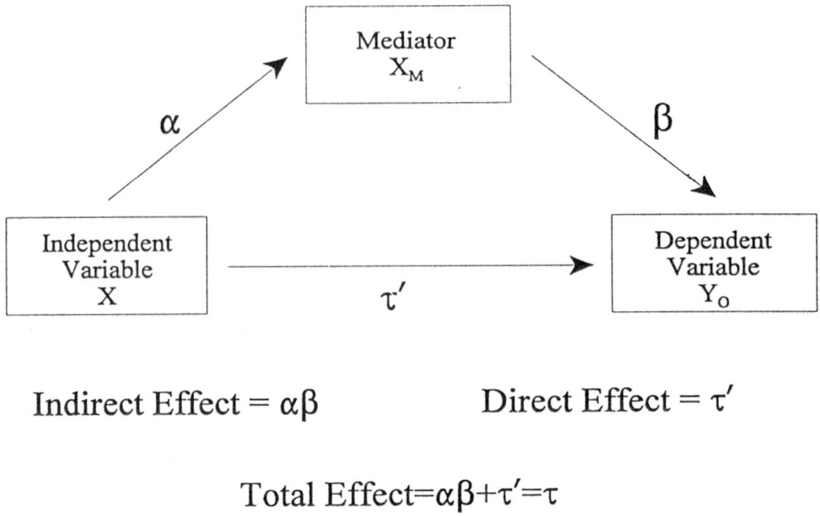

Fig. 5.1. Single mediator model.

$$Y = \tau' X + \beta X_M + \epsilon_3 \tag{5.3}$$

where τ' is the regression coefficient relating the independent variable to the dependent variable adjusted for the effects of the mediator, β is the regression coefficient relating the mediator to the dependent variable adjusted for the independent variable, ϵ_3 codes unexplained variability, and the intercept is not shown.

There are several limitations of this causal hypotheses test for mediation effects. First, no estimate of the mediated effect is available, although the causal hypotheses method can be combined with other approaches that do provide estimates of the mediated effect. An estimate of the mediated effect can be obtained by either $\tau - \tau'$ or $\alpha\beta$, which are equivalent (MacKinnon, Warsi, & Dwyer, 1995). The $\tau - \tau'$ estimate is based on the difference between the independent variable effect on the dependent variable unadjusted (τ) and adjusted (τ') for the mediator. The $\alpha\beta$ estimate is based on the notion that the mediated effect (also called the indirect effect) is equal to the relationship between the independent variable to mediator, α, times the relationship between mediator and the dependent variable, β.

A second limitation of the causal hypotheses method is the lack of an estimate of the standard error of the mediated effect to construct confidence limits and test the mediated effect for statistical significance. Sobel (1982; see also Bobko & Rieck, 1980; Folmer, 1981) noted that the variance of the mediated effect could be derived based on the multivariate delta method and is equal to:

$$\sigma^2_{\alpha\beta} = \sigma_\alpha{}^2\beta^2 + \sigma_\beta{}^2\alpha^2 \tag{5.4}$$

where $\sigma_\alpha{}^2$ is the variance of the regression coefficient α, β^2 is the squared β regression coefficient, $\sigma_\beta{}^2$ is the variance of the regression coefficient β, and α^2 is the squared α regression coefficient. MacKinnon (1994) proposed that the estimate of the mediated effect and its standard error can be used to provide a fourth test, a test of whether the mediated effect is larger than expected by chance based on the confidence limits of the mediated effect. Simulation studies of the standard error suggest that it is accurate for sample sizes of 50 or larger for multivariate normally distributed data (MacKinnon et al., 1995).

A third limitation of the causal hypotheses method, most important for this chapter, is that multiple mediators are difficult to incorporate in the analysis. Only the total mediated effect can be obtained from $\tau - \tau'$. There is not a procedure in the causal hypotheses method to obtain adjusted estimates of the individual mediated effects. It is possible to obtain estimates of the individual mediated effects and standard errors for multiple mediators as will be described next (Bollen, 1987; Sobel, 1982).

A fourth limitation of the causal hypotheses method is the requirement that there must be a significant total effect, τ, for mediation analysis to proceed. Actually, significant mediation effects may be present when the total effect, τ, is equal to zero, but this requires that the mediated effect, $\alpha\beta$, and the direct effect, τ', have opposite signs. Such inconsistent effects are possible in drug prevention programs where changing some mediators leads to counterproductive effects. Models with positive and negative mediated effects are called inconsistent models (Blalock, 1969; Davis, 1985). Inconsistent effects are likely with multiple mediators. As the number of mediators increases, the number of possible combinations of consistent and inconsistent mediation effects increases. As a result, the requirement of a significant total effect, τ, may be even less realistic in the case of multiple mediator models.

Mediation Analysis in Multiple Mediator Models

The statistical methods for multiple mediators is a straightforward extension of the single mediator case. Figure 5.2 depicts the four mediator model. The regression equations for a four mediator model are described next. The first equation is the same as in the single mediator case, where the total effect of the independent variable on the dependent variable is estimated.

$$Y = \tau X + \epsilon_1 \quad (5.5)$$

where Y is the dependent variable, X is the independent variable, τ is total effect, ϵ_1 is unexplained variability, and the intercept is not shown to simplify presentation.

An estimate of the effect of the independent variable on each mediator can be obtained by estimating the following regression equations:

$$X_{M1} = \alpha_1 X + \epsilon_2 \quad (5.6)$$

$$X_{M2} = \alpha_2 X + \epsilon_3 \quad (5.7)$$

$$X_{M3} = \alpha_3 X + \epsilon_4 \quad (5.8)$$

$$X_{M4} = \alpha_4 X + \epsilon_5 \quad (5.9)$$

where regression coefficients α_1, α_2, α_3, and α_4 are the program effects on each mediator, X_{M1} is the first mediator, X_{M2} is the second mediator, X_{M3} is the third mediator, X_{M4} is the fourth mediator, ϵ_2, ϵ_3, ϵ_4, and ϵ_5 code unexplained variability and the intercepts are not presented to simplify presentation. The equation to test whether the mediators are significant predictors when both the independent variable and the mediators are included is shown in Eqn. 5.10.

$$Y = \tau' X + \beta_1 X_{M1} + \beta_2 X_{M2} + \beta_3 X_{M3} + \beta_4 X_{M4} + \epsilon_6 \quad (5.10)$$

where regression coefficients β_1, β_2, β_3, and β_4 are the adjusted effects of each of the four mediators, τ' is the regression coefficient relating the

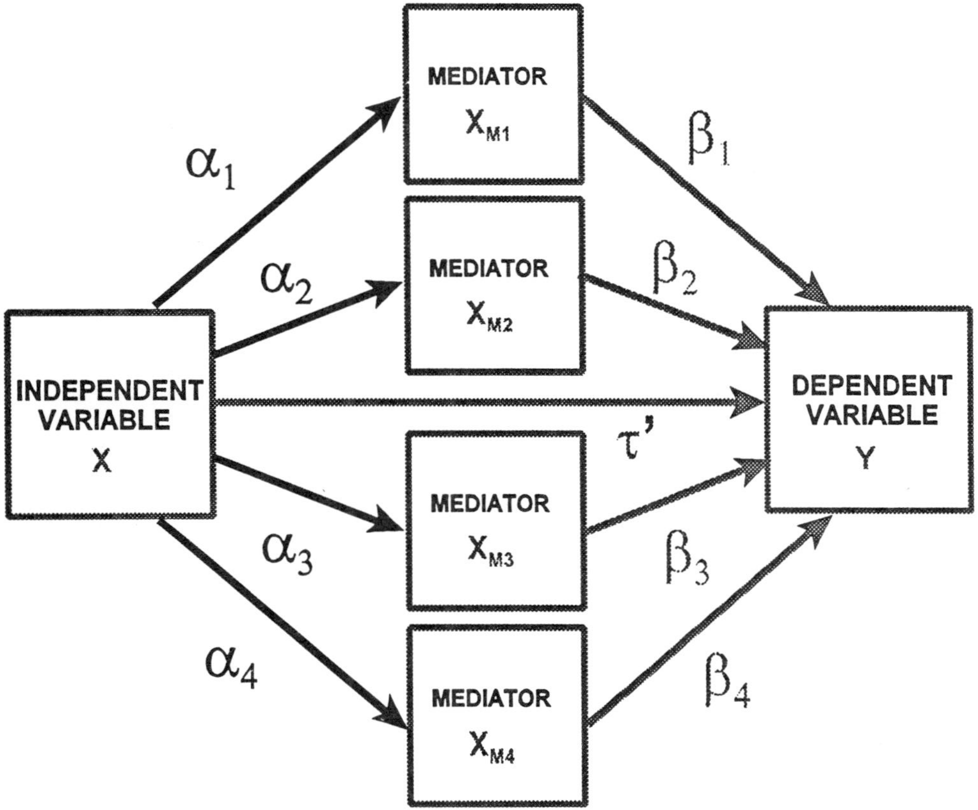

Fig. 5.2. Four mediator model.

independent variable to the dependent variable adjusted for the effects of the four mediators, and $\in_6$ codes unexplained variability.

The value of the total mediated or total indirect effect equals the difference in the independent variable coefficients $\tau - \tau'$, which is equal to the sum of four mediated effects $\alpha_1\beta_1$, $\alpha_2\beta_2$, $\alpha_3\beta_3$, and $\alpha_4\beta_4$. If a significant total effect (τ) is reduced to zero (i.e., $\tau' = 0$) when the four mediators are included in the model, then the program effect is entirely mediated by the four mediators.

Bollen (1987) called each mediated effect, $\alpha_1\beta_1$, $\alpha_2\beta_2$, $\alpha_3\beta_3$, $\alpha_4\beta_4$, a specific indirect effect to clarify the distinction between the total mediated effect and the individual mediated effects. Bollen described the application of the multivariate delta method to find the variance of several types of specific mediated effects and the total mediated effects for a variety of models, including several models more complicated than the model described in this chapter. The standard error of each mediated effect in the multiple mediator model is the same as the standard error of the single mediated effect in Eqn 5.4. The variance of the total mediated effect based on the multivariate delta method is equal to:

$$\begin{aligned} \text{var}\,(\alpha_1\beta_1 + \alpha_2\beta_2 + \alpha_3\beta_3 + \alpha_4\beta_4) &= \sigma_{\alpha 1}^2\beta_1^2 + \sigma_{\beta 1}^2\alpha_1^2 + \sigma_{\alpha 2}^2\beta_2^2 \\ &+ \sigma_{\beta 2}^2\alpha_2^2 + \sigma_{\alpha 3}^2\beta_3^2 + \sigma_{\beta 3}^2\alpha_3^2 + \sigma_{\alpha 4}^2\beta_4^2 + \sigma_{\beta 4}^2\alpha_4^2 + 2\,\alpha_1\alpha_2\sigma_{\beta 1\,\beta 2} \\ &+ 2\,\alpha_1\alpha_3\sigma_{\beta 1\,\beta 3} + 2\,\alpha_1\alpha_4\sigma_{\beta 1\,\beta 4} + 2\,\alpha_2\alpha_3\sigma_{\beta 2\,\beta 3} \\ &+ 2\,\alpha_2\alpha_4\sigma_{\beta 2\,\beta 4} + 2\,\alpha_3\alpha_4\sigma_{\beta 3\,\beta 4} \end{aligned} \quad (5.11)$$

The EQS (Bentler, 1989) and LISREL (Jöreskog & Sörbom, 1988) covariance structure modeling programs include routines to compute total mediated effects and total direct effects and their standard errors. Procedures to test specific mediated effects are not available in these computer programs although Bollen (1987) gives matrix routines to get these quantities. In this chapter, I show how to compute specific mediated effects and standard errors for a four mediator model.

Derivation of Variances Using the Multivariate Delta Method

In this section, an overview of the application of the delta method to find the variance of functions of random variables is given, from which the variance is used to obtain the large sample standard errors of contrasts among mediated effects. Detailed information on the delta method can be found in Bishop, Feinberg, and Holland (1975; chapter 14), Dunn (1989; Appendix 3) and Rao (1973).

The delta method is most easily described in the one variable case known as the single delta method. The single delta method states that the variance of a function of a random variable is equal to the square of its derivative times the variance of the original variable. For example, the variance of 5X is equal to the squared derivative of 5X with respect to X times the variance of X, yielding the well-known result that the variance of 5X is equal to 25 $\sigma^2 x$.

The multivariate delta method is a general method to determine the variance of functions of two or more random variables that follow a multivariate normal distribution (Bishop, Feinberg, & Holland, 1975). The method consists of pre- and postmultiplying the covariance matrix among parameters in the function by a vector of partial derivatives of the function. Note that the multivariate delta method is undefined when the parameters in the function approach zero (Sobel, 1982). The multivariate delta method has been used by several researchers to find the variance of the mediated effect and related functions (Bobko & Rieck, 1980; Bollen, 1987; Sobel, 1982; 1986). For the case of the mediated effect, the function is $f = \alpha\beta$, with partial derivatives,

$$\frac{\partial f}{\partial \alpha} = \beta \quad \text{and} \quad \frac{\partial f}{\partial \beta} = \alpha \tag{5.12}$$

Pre- and postmultiplying the vector of partial derivatives by the covariance matrix

$$V_{2 \times 2} = \begin{bmatrix} \sigma^2_\alpha \; 0 \\ \hline 0 \; \sigma^2_\beta \end{bmatrix} \tag{5.13}$$

yields the estimate of the variance of the mediated effect in Equation 5.4,

$$\sigma^2_{\alpha\beta} = \sigma^2_\alpha \, \beta^2 + \sigma^2_\beta \, \alpha^2 \tag{5.14}$$

Note the variance–covariance matrix among the parameter estimates is available in the output from covariance structure and regression programs. The vector of partial derivatives, however, must be derived for each function of regression estimates. Any calculus textbook (e.g.,

Loomis, 1974) contains the background necessary to find the partial derivatives for most functions. Some elements of the covariance matrix among parameter estimates may be equal to zero due to the independence of parameters. Bollen (1987) and Sobel (1982) give several examples of partial derivatives useful for testing specific indirect effects, but not contrasts, presented here. The multivariate delta method can be used to find the variance of a large variety of functions of mediated effects, but the adequacy of the formulas depends on accuracy of the large sample approximation of the multivariate delta method based on first derivatives and the distribution of the function of the regression coefficients (Springer, 1979).

Contrasts in Multiple Mediator Models

Because each mediated effect and the direct effect are in the metric of the dependent variable, the scale of the mediator does not affect the size of the mediated effect or direct effect. Any mediated effect is the product of a α and β, but the metric of the mediator is in the numerator of one term, β, and the denominator of the other term, α, which cancels out in the mediated effect estimate. For this reason, it is justifiable to test comparisons among mediated and direct effects. In the evaluation of a prevention program, researchers may want to know which mediated effect is largest or to determine if a specific mediated effect is larger than the direct effect. Contrasts can be used to test comparisons among pairs of specific mediated effects or between a specific mediated effect and the direct effect.

The multivariate delta method can be used to find the variance for any pairwise comparison among mediated effects. For a contrast comparing the first mediated effect to the second mediated effect in a multiple mediator model, the function is $f = \alpha_1 \beta_1 - \alpha_2 \beta_2$ with four partial derivatives,

$$\frac{\partial f}{\partial \alpha_1} = \beta_1, \frac{\partial f}{\partial \alpha_2} = -\beta_2, \frac{\partial f}{\partial \beta_1} = \alpha_1, \frac{\partial f}{\partial \beta_2} = -\alpha_2 \qquad (5.15a)$$

Pre and postmultiplying the vector of partial derivatives,

$[\beta_1 \ -\beta_2 \ \alpha_1 \ -\alpha_2]$ by the 4 by 4 covariance matrix,

$$\begin{bmatrix} \sigma^2_{\alpha_1} \; 0 \; 0 \; 0 \\ 0 \; \sigma^2_{\alpha_2} \; 0 \; 0 \\ 0 \; 0 \; \sigma^2_{\beta_1} \; \sigma_{\beta_1\beta_2} \\ 0 \; 0 \; \sigma_{\beta_1\beta_2} \; \sigma^2_{\beta_2} \end{bmatrix} \quad (5.15b)$$

yields the variance of the difference between two mediated effects:

$$\text{var}(\alpha_1 \beta_1 - \alpha_2 \beta_2) = \sigma^2_{\alpha 1} \beta_1^2 + \sigma_{\beta 1}^2 \alpha^2_1 + \sigma^2_{\alpha 2} \beta^2_2 + \sigma^2_{\beta 2} \alpha^2_2 - 2\, \alpha_1 \alpha_2 \sigma_{\beta 1\, \beta 2} \quad (5.16)$$

A specific mediated effect could be compared to the direct effect $f = \alpha_1 \beta_1 - \tau'$ or the average of two specific mediated effects could be compared to a direct effect, $f = \alpha_1 \beta_1 + \alpha_2 \beta_2 - 2\tau'$. An example of a complex comparison is the comparison between the average of the first two mediated effects and the average of two other mediated effects, $f = \alpha_1 \beta_1 + \alpha_2 \beta_2 - \alpha_3 \beta_3 - \alpha_4 \beta_4$.

The variances based on the multivariate delta method for these contrasts are shown as follows:

$$\text{var}(\alpha_1 \beta_1 - \tau') = \beta_1^2 \sigma_{\alpha 1}^2 + \alpha^2_1 \sigma_{\beta 1}^2 - 2\, \alpha_1 \sigma_{\beta 1 \tau'} + \sigma^2_{\tau'} \quad (5.17)$$

$$\text{var}(\alpha_1 \beta_1 + \alpha_2 \beta_2 - 2\tau') = \beta_1^2 \sigma_{\alpha 1}^2 + \beta_2^2 \sigma_{\alpha 2}^2 + \alpha_1^2 \sigma_{\beta 1}^2 + 2\, \alpha_1 \alpha_2 \sigma_{\beta 1\, \beta 2} - 4\, \sigma_{\beta 1 \tau'} \alpha_1 + \alpha_2^2 \sigma_{\beta 2}^2 - 4\, \sigma_{\beta 2 \tau'} \alpha_2 + 4\, \sigma_{\tau'}^2 \quad (5.18)$$

$$\begin{aligned} \text{var}(\alpha_1 \beta_1 + \alpha_2 \beta_2 - \alpha_3 \beta_3 - \alpha_4 \beta_4) = {} & \sigma_{\alpha 1}^2 \beta_1^2 + \sigma_{\beta 1}^2 \alpha_1^2 + \\ & \sigma_{\alpha 2}^2 \beta_2^2 + \sigma_{\beta 2}^2 \alpha_2^2 + \sigma_{\alpha 3}^2 \beta_3^2 + \sigma_{\beta 3}^2 \alpha_3^2 + \\ & \sigma_{\alpha 4}^2 \beta_4^2 + \sigma_{\beta 4}^2 \alpha_4^2 + 2\, \alpha_1 \alpha_2 \sigma_{\beta 1\, \beta 2} - 2\, \alpha_1 \alpha_3 \sigma_{\beta 1\, \beta 3} \\ & - 2\, \alpha_1 \alpha_4 \sigma_{\beta 1\, \beta 4} - 2\, \alpha_2 \alpha_3 \sigma_{\beta 2\, \beta 3} - 2\, \alpha_2 \alpha_4 \sigma_{\beta 2\, \beta 4} + 2\, \alpha_3 \alpha_4 \sigma_{\beta 3\, \beta 4} \end{aligned} \quad (5.19)$$

Confidence limits for the contrast effect are obtained from the formula $\Psi \pm 1.96\, (\sigma_\Psi)$, where Ψ is the value of the contrast, 1.96 is the value for the 95% confidence limits assuming a normal distribution, and σ_Ψ is the square root of the variance of the contrast. The confidence limits assume a normal distribution for the contrast effect, which may not necessarily be accurate given that functions of random variables may only be normal asymptotically.

Contrasts in the Evaluation of Mediation Effects

The Adolescents Training and Learning to Avoid Steroids (ATLAS) program is a school-based prevention program based on a theoretical model derived from androgenic and anabolic steroid (AAS) risk-factor research (Goldberg et al., 1996a, 1996b). The program aims to modify behavioral intent and risk factors of AAS use, improve healthy behaviors and attitudes, and ultimately reduce steroid use. Based on earlier studies (Bents, Young, Bosworth, Boyea, Elliot, & Goldberg, 1990; Goldberg & Elliot, 1996), sports nutrition and effective strength training were presented as alternatives to AAS use. Because of the success of peer educators in other drug prevention programs (Klepp, Halper, & Perry 1986) and the social influences in the team setting, a peer-taught, team-based approach was used.

An evaluation of the ATLAS program delivered to 15 of 31 high school football teams showed that experimental subjects were significantly less interested in trying AAS after intervention, and other risk factors promoting AAS use were also reduced. The 16 teams in the control condition were given an antisteroid informational pamphlet. A recent paper examined the mediating mechanisms responsible for the reductions in intentions to use steroids (MacKinnon et. al., 1999). There was considerable evidence that attitudinal measures such as perceived severity of AAS use and normative measures, such as team norms about nutrition, were important mediators of the program effect on intentions to use anabolic steroids.

Four potential mediators were selected to illustrate the application of the contrasts developed in this chapter: knowledge of the effects of anabolic steroid use, perceived risks of using steroids, perception of the team as a source of health and other information, and the number of reasons to use steroids. There were statistically significant program effects on each of these potential mediators. The multiple mediator model shown in Fig. 5.3 was estimated. The figure includes standardized estimates in bold, as well as unstandardized parameter estimates and standard errors. The figure has been simplified so that it is a multiple mediator model for one wave, even though the estimates are based on pretest and posttest measurements. The actual model included paths from Time One measures of each mediator and intentions to use anabolic steroids to posttest measures of all mediators and intentions to use steroids. All pretest measures, including group, were allowed to correlate. In addition, the errors among all the mediators were allowed

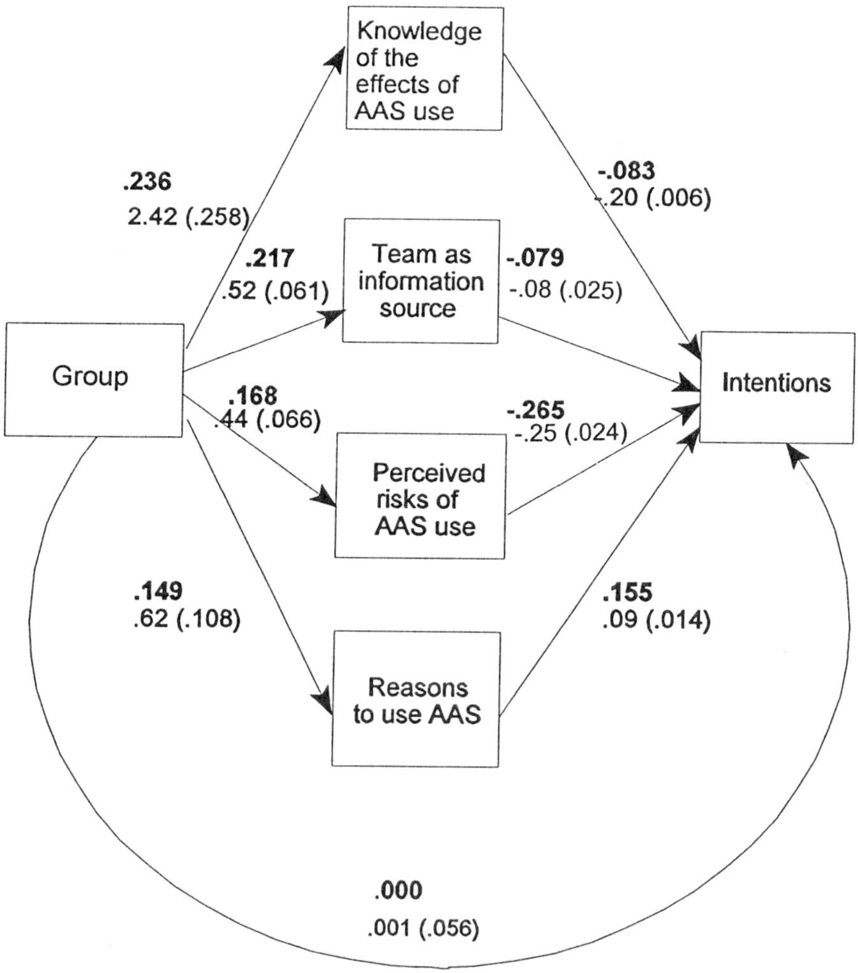

Fig. 5.3. Four mediator model for the effects of an anabolic steroid prevention program on intentions to use steroids

to correlate at posttest. The model was based on data from 1,144 football players measured before and after the 1994 football season, representing an 81% retention rate. The model was an adequate fit (χ^2 (4) = 14.20) with a normed fit index of .995 (where a fit of .9 or greater is usually considered to be a good fit). The parameters and standard errors in the model were estimated using the AMOS (Arbuckle, 1997) computer program.

All mediated effects were statistically significant, but there were considerable differences in the size of the mediated effect estimates. As shown in Table 5.1, the confidence limits for the total mediated effect did not include zero, suggesting that the effect was larger than expected by chance. One of the mediated effects, through reasons to use anabolic steroids was counterproductive in that subjects exposed to the ATLAS program were better able to identify the reasons to use steroids than control subjects and that, in turn, led to greater intentions to use steroids. This inconsistent mediation effect is opposite to the desired effects.

The results of testing several contrasts are summarized in Table 5.1. Pairwise comparisons suggested that the mediated effects for knowledge of AAS and team as information source were not significantly different from each other, but there were significant differences between each of these mediated effects and both the perceived severity of steroid use and reasons for using steroids mediated effects. The average of the mediated effects through knowledge and team as information source was not significantly different from the average of the mediated effects for perceived severity of and reasons to use steroids, although this test is presented to illustrate a complex comparison and does not have a compelling substantive motivation. The knowledge mediated effect was not significantly different from the direct effect. The average of the knowledge and team as information source mediated effects was not significantly different from the direct effect, primarily due to the large standard error of the direct effect.

In order to clarify the use of these formulas to compare mediated effects, the computations necessary to compare the mediated effect through knowledge to the mediated effect through perceived severity are given here. The regression coefficients, standard errors, and covariances are inserted into Eqn 5.16, giving a variance estimate of .00061 and a standard error estimate of .0247. The 95% confidence interval of .013 (lower confidence limit) and .109 (upper confidence limit) does not contain zero, suggesting that the mediated effect through perceived severity is larger than the mediated effect through knowledge.

$$\begin{aligned} \text{var}(\alpha_1\beta_1 - \alpha_2\beta_2) &= (.258^2)(-.019^2) + (.006^2)(2.423^2) + (.066^2)(.246^2) + (.024^2)(.437^2) \\ &\quad - (2)(2.423)(.437)(.0000) \\ &= .00061 \end{aligned} \tag{5.20}$$

These contrasts suggest that our revised ATLAS curriculum reduce focus on reasons to use anabolic steroids. Of course, description of the

TABLE 5.1

Mediated Effects and Contrasts Among Mediated Effects

Effect	*Estimate/ (Standard Error)*	*Estimate/ (Standard Error)*	*LCL*	*UCL*
Knowledge Mediated Effect	-.046 (0.15)	-3.00	-.075	-.017
Team as Information Mediated Effect	-.041 (.014)	-2.97	-.068	-.014
Perceived Severity Mediated Effect	-.108 (.019)	-5.56	-.145	-.071
Reasons to Use Anabolic Steroid Mediated Effect	.056 (.013)	4.29	.031	.081
Total Mediated Effect	-.139 (.031)	- 4.54	-.199	-.078
Direct Effect of Program on Intentions to Use AAS	.001 (.056)	0.18	-.109	.111
Pairwise Contrast of Knowledge vs. Team as Information Mediated Effects	-.005 (.021)	-0.22	-.046	.036
Pairwise Contrast of Knowledge vs. Perceived Severity Mediated Effects	.061 (.025)	2.49	.012	.110
Pairwise Contrast of Team as Information Source and Perceived Severity	.066 (.025)	2.67	.017	.115
Complex Comparison of Knowledge and Team as Information Mediated Effect vs. Perceived Severity and Reasons to Use AAS Mediated Effects	-.036 (.032)	-1.12	-.099	.027
Complex Comparison of Knowledge and Reasons to Use vs. Team as Information Source and Perceived Severity Mediated Effects	.159 (.031)	5.18	.098	.220
Comparison of Knowledge Mediated Effect and the Direct Effect	-.047 (.062)	-.746	-.168	.074
Comparison of Knowledge and Team as Information Source Mediated Effects and the Direct Effect	-.089 (.120)	-.746	-.324	.146

Note: LCL is lower confidence limit and UCL is the upper confidence limit. The standard error of the mediated effect is based on the multivariate delta method.

reasons for, as well as reasons against, steroids may be necessary to retain credibility and may also contribute to the positive effects of other mediators. That perceived severity appears to be a substantial mediator is surprising given that perception of severity is not typically an important predictor of health behavior in other studies (Janz & Becker, 1984). Perceived severity may be important in the case of anabolic steroids because the risks of anabolic steroid use are not as well known and the program is designed to increase perceptions of severity. Similarly, knowledge is not often found to be a significant mediator of program effects but the risks of anabolic steroids are not well known unlike other risky behaviors such as smoking whose risks are well known. Norms are often significant mediators of program effects in drug prevention research (Mackinnon, Johnson, Pentz, et al., 1991). Team as information source is a measure of team norms regarding nutrition, strength training, and steroid information. Nevertheless, the mediated effect of knowledge was not significantly different from the team as information source mediated effect.

Summary and Future Directions

Multiple mediator models were described in this chapter and procedures to test contrasts among mediated effects were developed and applied to the evaluation of an anabolic steroid drug prevention program. The assessment of mediated effects in multiple mediator models should improve understanding of the effects of drug prevention programs. Tests of the mechanisms of prevention program effects are only recently appearing in the research literature (MacKinnon et al. 1991; Vinokur & Schul, 1997). The methods to test contrasts described in this chapter will be most useful as drug prevention programs advance to the state where hypotheses regarding mediation effects are the focus of research studies. It is best to think of mediation analysis as extracting additional information in the evaluation of prevention programs rather than as final tests. The next experimental study is then based on the results of the mediation analysis.

Simulation studies of the contrasts proposed in this chapter suggest that the standard errors are accurate for sample sizes of at least 100 (Taborga, MacKinnon, & Krull, 1999). There is evidence that the statistical tests of the effects are conservative, however. In this regard, it is most sensible to use the standard errors to construct confidence limits, rather than to compute tests of statistical significance. Information on effect size in multiple mediator models can be obtained by examining the standardized regression coefficients and partial correlation values for the parameters in the multiple mediator model.

The methods described here were for two-path mediated effects, that is, two regression coefficients. The multivariate delta method is easily extended to more complicated models and contrasts such as a contrast testing a three-path mediation effect (three regression coefficients) to a two-path mediation effect. The mediated effects are in the same metric so the comparison between a two-path mediated effect to a three-path mediated effect is justified as long as the same dependent variable is examined. If there is more than one dependent variable, these methods are not accurate for testing mediated effects between dependent variables because of the different metric for each dependent variable.

There are several assumptions of the mediated effects tests proposed here. Multivariate normality is assumed for the data and regression coefficients. It is also assumed that the contrast estimate and the contrast divided by its standard error are normally distributed. In the ATLAS example, the nesting of football players in teams was ignored, which is likely to make standard error estimates smaller than they should be. No adjustments for multiple comparisons were made in this research although Type 1 error inflation and distinctions between planned and post-hoc comparisons are important for these tests (Keppel, 1991). It was also assumed that each mediator is measured well and that the estimate of the mediated effect is not spurious because of unreliability in the measurement of the mediator. One way to adjust for unreliability is to develop measurement models for the constructs in the mediation model using covariance structure modeling (Bentler, 1989; Bollen, 1989). Contrasts among mediated effects in the structural model relating these constructs can be tested using the same methods described previously.

It is also assumed that the direction of causality is correct, that is, the program changed the mediator, which in turn changed the outcome variable (Holland, 1988). It is possible that the direction of effects was such that the change in the dependent variable led to the change in the mediator. Interpreting effects is more complicated in multiple mediator models because the relationships among the mediators and the dependent variable are correlational. Similarly, the methods described here also apply to nonexperimental studies but such designs have more alternative explanations of effects because the relationship between the independent variable and the mediator is correlational. This ambiguity suggests that multiple mediator analysis is not a substitute for experimental evaluation of treatments designed to test hypotheses about mediators (West & Aiken, 1997).

Despite the difficulty of determining causality in mediation models, multiple mediator models may actually increase ability to identify the most important mediators of program effects. Assuming that the measurement of each mediator is comparable, a significant mediation effect through one mediator but not through another mediator provides evidence for the specificity of effects. Such differences between mediated effects provide additional evidence that the significant mediational pathway is the important one. Future experimental studies are then necessary to provide more compelling evidence for one mediating mechanism rather than another.

ACKNOWLEDGEMENT

This chapter was supported in part by National Institute on Drug Abuse grants DA09757 and DA07356. Some of this research was presented at the 1996 meeting of the Prevention Research Methodology Group in Tempe, Arizona. I thank Jeanne Hoffman, Jennifer Krull, and Chondra Lockwood for their help in this research.

REFERENCES

Alwin, D. F., & Hauser, R. M. (1975). The decomposition of effects in path analysis. *American Sociological Review, 40,* 37–47.

Arbuckle, J. L. (1997). *AMOS Publication.* Chicago, IL: SmallWaters.

Baron, R. M., & Kenny, D. A. (1986). The moderator–mediator distinction in social psychological research: Conceptual, strategic, and statistical considerations. *Journal of Personality and Social Psychology, 51*(6), 1173–1182.

Bentler, P. M. (1989). *Theory and implementation of EQS: A structural relations program.* Los Angeles: BMDP Statistical Software.

Bents, R., Young, J., Bosworth, E., Boyea, S., Elliot, D., & Goldberg, L. (1990). An effective educational program alters attitudes toward anabolic steroid use among adolescent athletes. *Medicine and Science in Sports Exercise, 22,* S64.

Bishop, Y. M., Feinberg, S. E., & Holland, P. W. (1975). *Discrete multivariate analysis: Theory and practice.* Cambridge, MA: MIT Press.

Blalock, H. M. (1969). *Theory construction: From verbal to mathematical formulations.* Englewood Cliffs, NJ: Prentice-Hall.

Bobko, P., & Rieck, A. (1980). Large sample estimators for standard errors of functions of correlation coefficients. *Applied Psychological Measurements, 4*(3), 385–398.

Bollen, K. A. (1987). Total direct and indirect effects in structural equation models. In C. C. Clogg (Ed.), *Sociological Methodology* (pp. 37–69). Washington, DC: American Sociological Association.

Bollen, K. A. (1989). *Structural equations with latent variables.* New York: Wiley.

Chassin, L., Pillow, D. R., Curran, P. J., Molina, B. S. G., & Barrera, M., Jr. (1993). Relation of parental alcoholism to early adolescent substance use: A test of three mediating mechanisms. *Journal of Abnormal Psychology, 102*(1), 3–19.

Davis, J. A. (1985). The logic of causal order. *Sage University Paper Series on Quantitative Applications in the Social Sciences* (pp. 07–055). Beverly Hills, CA: Sage.

Dunn, G. (1989). *Design and analysis of reliability studies: The statistical evaluation of measurement errors.* New York: Oxford University Press.

Flay, B. R. (1985). Psychosocial approaches to smoking prevention: A review of findings. *Health Psychology, 4*(5), 449–488.

Folmer, H. (1981). Measurement of the effects of regional policy instruments by means of linear structural equation models and panel data. *Environmental and Planning Annual, 13,* 1435–1448.

Goldberg, L., & Elliot, D. L. (1996). Offering alternatives to anabolic steroids: Effects on adolescent intentions. *Children and adolescents in athletic competition: Rewards and adversities* (pp. 103–108). Eugene, OR: International Institute for Sport and Human Performance.

Goldberg, L., Elliot, D. L., Clarke, G. MacKinnon, D. P., Moe, E., Zoref, L., Green, C., & Wolf, S. L. (1996a). The Adolescents Training and Learning to Avoid Steroids (ATLAS) prevention program: Background and results of a model intervention. *Archives of Pediatric and Adolescent Medicine, 150,* 713–721.

Goldberg, L. Elliot, D., Clarke, G. N., MacKinnon, D. P., Moe, E., Zoref, L., Green, C., Wolf, S. L., Greffrath, E., Miller, D. J., & Lapin, A. (1996b). Effects of a multidimensional anabolic steroid prevention intervention. *Journal of the American Medical Association, 276*(19), 1555– 1562.

Hansen, W. B., (1992). School-based substance abuse prevention: A review of the state-of-the-art in curriculum, 1980–1990. *Health Education Research, 7,* 403,430.

Hansen, W. B., & Graham, J. W. (1991). Preventing alcohol, marijuana, and cigarette use among adolescents: Peer pressure resistance training versus establishing conservative norms. *Preventive Medicine, 20,* 414–430.

Hawkins, J. D., Lishner, D. M., Jensen, J. M., & Catalano, R. F. (1987). Delinquents and drugs: What the evidence suggests about prevention and treatment programming. *Youth at High Risk for Substance Abuse,* DHHS No. (ADM) 87–1537, 81–133.

Holland, P. W. (1988). Causal inference, path analysis, and recursive structural equations models. *Sociological Methodology, 18,* 449–484.

Janz, N. K., & Becker, M. H. (1984). The health belief model: A decade later. *Health Education Quarterly, 11,* 1–47.

Jöreskog, K. G., & Sörbom, D. (1988). *LISREL 7: A guide to the program and applications.* Chicago: SPSS, Inc.

Judd, C. M., & Kenny, D. A. (1981). Process Analysis: Estimating mediation in treatment evaluations. *Evaluation Review, 5*(5), 602–619.

Keppel, G. (1991). *Design and analysis: A researcher's handbook.* Englewood Cliffs, NJ: Prentice-Hall.

Klepp, K. L., Halper, A., & Perry, C. L. (1986). The efficacy of peer leaders in drug abuse prevention. *Journal of School Health, 56,* 407–411.

Loomis, L. (1974). *Calculus,* Reading, MA: Addison-Wesley.

MacKinnon, D. P. (1994). Analysis of mediating variables in prevention and intervention studies. In A. Cazares and L. A. Beatty (Eds.), *Scientific Methods in Prevention Research, NIDA Research Monograph 139,* pp. 127–153, DHHS Pub. 94–3631, Washington, DC.

MacKinnon, D.P., & Dwyer, J. H. (1993). Estimating medicated effects in prevention studies. *Evaluation Review, 17,* 144–158.

MacKinnon, D. P., Goldberg, L., Clarke, G. N., Elliot, D. L., Cheong, J., Lapin, A., & Moe, E. L., Krull, J. (1999). Mediating mechanisms in a program to reduce intentions to use anabolic steroids and improve strength training self-efficacy, and dietary behavior. Manuscript submitted for publication.

MacKinnon, D. P., Johnson, C. A., Pentz, M. A., Dwyer, J. H., Hansen, W. B., Flay, B. R., & Wang, E. (1991). Mediating mechanisms in a school-based drug prevention program: First-year effects of the Midwestern Prevention Project. *Health Psychology, 10*(3), 164- 172.

MacKinnon, D. P., Warsi, G., & Dwyer, J. H. (1995). A simulation study of indirect effect measures. *Multivariate Behavioral Research, 30*(1), 41-62.

The Multiple Risk Factor Intervention Trial Research Group. (1990). Mortality rates after 10.5 years for participants in the Multiple Risk Factor Intervention Trial: Findings relate to a priori hypotheses of the trial. *Journal of the American Medical Association, 263*(13), 1795-1801.

Rao, C. R. (1973). *Linear statistical inference and its applications.* New York: Wiley.

Sobel, M. E. (1982). Asymptotic confidence intervals for indirect effects in structural equation models. In S. Leinhardt (Ed.), *Sociological methodology 1982* (pp. 290-312). Washington, DC: American Sociological Association.

Sobel, M. E. (1986). Some new results on indirect effects and their standard errors in covariance structure models. In N. Tuma (Ed.), *Sociological methodology 1986* (pp. 159-186). Washington, DC: American Sociological Association.

Sobel, M. E. (1990). Effect analysis and causation in linear structural equation models. *Psychometrika, 55*(3), 495-515.

Springer, M. D. (1979). *The algebra of random variables.* New York: Wiley.

Taborga, M. P., MacKinnon, D. P., & Krull, J. L. (1999). A simulation study of effect size measures in mediation models. Poster presented at the 7th Annual Meeting of the Society for Prevention Research, New Orleans, LA.

Vinokur, A. D., & Schul, Y. (1997). Mastery and inoculation against setbacks as active ingredients in JOBS Intervention for the Unemployed. *Journal of Consulting and Clinical Psychology, 65*(5), 867-877.

West, S. G. & Aiken, L. S., (1997). Toward understanding individual effects in multicomponent prevention programs: Design and analysis strategies. In B. Kendall, M. Windle, & S. West (Eds.), *The science of prevention: Methodological advances from alcohol and substance abuse research* (pp. 167-209). Washington, DC: American Psychological Association.

6

Estimation and Equivalent Models for Quadratic and Interactive Latent Variable Models

Phillip K. Wood
University of Missouri–Columbia

Estimation of interactive and quadratic effects using the original factor model of Kenny and Judd (1984) is difficult because specification of models with interactive or quadratic effects is cumbersome and prone to human error. In addition, even models with correct specifications often fail to converge or do not converge to proper solutions. Although standard remedies in the literature for remediating improper solutions can be used for the Kenny and Judd model, such changes imply an entirely different set of constraints on the solution. Finally, on a conceptual level, published discussion of the Kenny and Judd model has used latent variables with simple structure. As a remedy to these issues, two solutions are proposed. First, it is recommended that models with interactive and/or quadratic effects be fit by first fitting the usual measurement model and using the resulting solution as the basis for start values for all terms in the model. Second, to reduce the cumbersome path diagrams from such models, it is recommended that effects associated with cross-product terms that are not of theoretical interest be fit by means of nonlinear constraints. Although such procedures can be used with all structural modeling programs that allow estimation of nonlinear effects, a short SAS™ program, GENINT, is described that takes, as input, the PROC CALIS solution for the traditional measurement model and automatically produces a second PROC CALIS program for use in modeling interactive or quadratic effects. Extensions of the polynomial model to other models are presented, including mean level information and/or complex factor structure as well as longitudinal models, which allow for correlated errors. Applications of the technique to substance abuse research are illustrated using prospective data of alcohol expectancies and alcohol use disorder in a sample of young adults. Although an interaction effect had been claimed for this model in earlier work, follow-up graphical techniques for these data reveal that the presence of a significant interaction term may be liable to alternative interpretations.

ALTERNATIVE MODELS FOR POLYNOMIAL AND INTERACTIVE EFFECTS IN LATENT VARIABLES

Substance abuse researchers often adopt sophisticated prospective and multivariate research designs in order to better understand the complex way that two or more constructs relate to a criterion of interest. When the effects of interest are only linear and the researcher is interested in determining which multivariate effects contribute uniquely to the criterion, structural models using latent variables have been attractive because they allow the researcher to estimate the relation of a construct to the criterion within the absence of the confounding bias introduced by differential measurement error.

Conceptually, however, there are many cases when a researcher has reason to believe that a good model of substance abuse behavior involves interactive or even nonlinear relationships. Such relationships are especially relevant to the identification of systematic individual differences in substance abuse. For example, Cooper, Russell, Skinner, Frone, & Mudar (1992) documented a moderator effect of prior alcohol use and alcohol outcome expectancies on subsequent alcohol use. Other researchers (e.g., Gotham, 1995) have explored the degree to which an interaction between personality variables and expectancy measures are related to problematic alcohol use. Although examples from other aspects of alcohol use are easily cited, examples of interaction in substance abuse research may be grossly classified as attempts to identify how individual difference variables affect the relationship of prior substance use patterns on subsequent substance use or, more generally, how one construct moderates the effect of another construct on the behavior of interest. The first rationale for such interaction models of individual differences seems particularly useful as researchers integrate their models of substance abuse within a life-span developmental approach. For example, in outlining the goals of lifespan developmental psychology, Baltes, Reese, and Nesselroade (1977) described its focus as the "description, explanation, and modification _ of intraindividual change in behavior across the life span, and on interindividual differences (and similarities) in intraindividual change." Within the context of substance abuse research, interindividual differences (such as personality variables or alcohol outcome expectations) are assumed to result in different prospective relationships in substance abuse.

Even though structural models are used widely in substance abuse research, and even though interactions constitute logical and exciting tests of psychological theory, researchers often avoid structural models entirely when modeling interactions. Instead, composite measures across manifest variables are calculated, which enable the use of more familiar regression, analysis of variance, or other general linear models. Such an approach has advantages in that the relevant models are easier to specify, more familiar to reviewers, and do not require the researcher to make several choices during the course of model specification. The problems with such an approach are, however, well known. Specifically, parameter estimates from such models may be significantly biased and/or inconsistent given the presence of measurement error in the predictor variables (Busemeyer & Jones, 1983).

Kenny and Judd (1984) have proposed one model for estimation of interaction and quadratic effects in latent variables using traditional structural equation modeling software. This approach, in its original form, involves calculation of appropriate product variables, which, under a given measurement model and the assumption that error and exogenous factor variances are distributed normally, allows specification of the desired quadratic or product latent variable(s). Implementation of the model, however, also requires the specification of several additional latent variables in order to account for cross-product effects associated with factors and uniquenesses of the extended model. Such cross-product effects, however, are often merely nuisance parameters in that they are unrelated to the conceptual hypotheses of interest. Larger discussions of this basic model have made their way into several texts on structural equation models (Bollen, 1989; Hayduk, 1987; Loehlin, 1992).

Since that time, several structural equation modeling programs now allow specification of nonlinear constraints, allowing researchers more opportunity to apply and evaluate such nonlinear effects. Although the elegance of the Kenny and Judd model is attractive, writing the structural equation program setups for such models remains tedious and prone to human programming error (Jöreskog & Yang, 1996; Ping, 1995, 1996). In addition, Ping (1995) noted that latent variable models seemed especially prone to convergence problems and unfeasible solutions. Remedies to such unfeasible solutions in traditional structural models involve respecification of the original model. For example, McDonald (1980) discussed how observed negative error variances may be dealt with by respecifying the error variance as unity, with a free path from the error variance to the manifest variable. Such an approach is not often employed in nonlinear factor models, however, because such respecification involves an entirely different set of constraints in the model, thereby doubling the program tedium necessary to specify the model. Even if a researcher desired to expend the effort to program this new set of constraints, there is no guarantee that any particular respecification will converge on a correct solution, leaving the researcher with a large menu of tedious programs to write, most of which may not achieve the desired result of representing a feasible solution.

At this point, it seems unclear as to whether the convergence and unfeasible solution difficulties are due to the specification of additional variables, as suggested by Ping (1995, p. 338), or whether the optimization terrain for polynomial latent variable models contains several local minima, which often preclude arrival at the true local minimum value of the fit function. As a word of explanation, local minima result when the computer program produces a set of parameter estimates for the data that are "best" in terms of the convergence criteria for the computer program, but that are not actually those parameter estimates that actually best reproduce the observed variance–covariance matrix. In either situation, it seems reasonable to believe, however, that the algorithms used by a structural equation programs for generating start values are less than optimal for polynomial applications. Such algorithms often do not incorporate the additional information about relationships of factor loadings and uniquenesses of the linear measurement model with factor variances, loadings, and error variances associated with product or interaction terms. It seems reasonable, therefore, to believe that a latent variable model with polynomial effects has a greater chance of converging if start values taken from the linear measurement model were used to specify start values for all elements of the corresponding polynomial model. Unfortunately, using information from the final estimates of the linear model to generate feasible start values for polynomial effects involves a large amount of hand calculation and increases the programming tedium yet again by another order of magnitude.

Outline

Original and Reduced Models for Polynomial Effects. This chapter introduces a small SAS program that takes as input a PROC CALIS program setup for the usual measurement model of linear effects (excluding the criterion). This program then automatically generates the CALIS program setup for a Kenny and Judd-style model for desired latent variable model of interactive and/or quadratic effects. The program allows researchers to generate the interactive or quadratic effects under common model respecifications such as those described by McDonald (1980) in order to remediate improper estimated solutions. In addition, the program will also automatically generate a more parsimonious, but mathematically equivalent, structural model for interaction effects that does not contain the cross-product nuisance effects of the original Kenny and Judd formulation. Extensions of such a model to interactive models that include the mean intercept follow the basic logic of the Kenny and Judd model, but are based on an adjoined Sum of Squares and Cross-Products (SSCP) matrix and include mean level effects (discussed in more detail later). The rationale for the model fitting approach outlined here is illustrated using Kenny and Judd's original simulated data examples for quadratic and interactive effects.

Analysis of Real-World Data. In practice, fitting interactive and quadratic factor models using real-world data has often proved more difficult than presentations using generated data would suggest. To show how the SAS program can be used in such situations, data from a prospective study is used to investigate a prior research claim that heavy alcohol use is a function of prior alcohol expectancies, prior heavy alcohol use, and the interaction of prior alcohol expectancies with prior heavy alcohol use. Two approaches to testing for interaction effects are compared: a traditional multiple regression model that includes an interaction variable and an interactive model that assumes a complex factor structure and serial correlation over time. Models involving interactive latent variables for the real-world data set are estimated and compared under the generalized least squares (GLS) criterion originally recommended by Kenny and Judd, as well as the maximum likelihood (ML) and weighted least squares (WLS) criteria as suggested in more recent papers. Parameter estimates and fit from the interactive model are compared under the original scale, as well as a rank normalized version of the data.

Prior Methods for Quadratic and Interactive Effects[1]

Before proceeding, it is appropriate to discuss briefly some of the recent work on interactive and quadratic effects in latent variables. Recently, Bollen (1995) proposed a two-stage least squares approach to the estimation of nonlinear effects that makes no assumptions regarding the underlying distribution of unobserved variables. Although the use of limited information estimators constitutes an interesting and less restrictive alternative to the factor model of Kenny and Judd, a comparison of such Bollen's approach with the factor models is beyond the scope of this chapter and the subsequent discussion is limited to full information factor models of interaction and quadratic effects. Several variants of the Kenny and Judd model have been proposed that are less laborious than the original. Specifically, Jaccard and Wan (1995) specified a slight modification of the original Kenny and Judd model in that the interaction model they describe imposes the correct nonlinear constraints on interaction factor loadings and factor variances, but allows covariances between manifest product variables to be freely estimated.

Although the example analysis they present in their appendix contains nonlinear constraints similar to those calculated by the proposed program, their model is not strictly equivalent to either the Kenny and Judd approach nor to that proposed here.

In an attempt to limit the tedium associated with modeling all variables, Jaccard and Wan calculated only some of the possible product variables in the model. In addition, (and because of the omission of some product variables) error variances and covariances in the model were not constrained to be nonlinear functions of previously estimated parameters. Although one may view this example as a reduced version of the original Kenny and Judd model, it may also be argued that it constitutes a relaxed version of the Kenny and Judd approach because it does not specifically constrain error variances and covariances. To the extent that measurement error and/or nuisance factors do not meet the normality assumption assumed by Kenny and Judd, such models may better fit the data. This topic is revisited in the discussion of Jöreskog and Yang's (1996) chapter later.

[1]Parts of the literature review and real-world data example are excerpted from Wood and Erickson (1998).

Ping (1995) proposed two simple alternatives to the original Kenny and Judd model that are less tedious and appear to perform well in simulated data. One approach, described as a two-step estimation technique, involved solving the linear measurement model as a way of recovering factor loadings, and then using these estimates as fixed values in the larger polynomial model. A second, more parsimonious, model was developed based on the fact that the product or quadratic expressions of the sum of variables in the linear measurement model should have a known relationship to the product or quadratic latent variable of interest. Such an approach has the advantage that only one additional manifest and latent variable is needed to estimate the desired interactive or quadratic effect. Unfortunately, the approach also assumes that the manifest variables load on only one latent variable (although Ping also proposes an indirect two-step estimation technique for complex factor models).

Ping's refinements of the original Kenny and Judd approach are simple, elegant, and may constitute a more preferable avenue for many researchers, especially given the fact that the ratio of observed variables and estimated parameters to sample size may be unfavorable for many models. The advantages of using the program proposed here are that convergence rates and the likelihood of a proper solution appear to be quite good using the proposed computer program. Also, use of the program permits for simultaneous estimation of cross-loadings and correlated errors without using Ping's proposed two-step approach.

Finally, Jöreskog and Yang (1996) proposed that researchers analyze an adjoined matrix that incorporates mean level information in the structural model of polynomial effects. Like Jaccard and Wan, Jöreskog and Yang also noted that the advent of the ability to program nonlinear constraints in structural equation modeling programs eliminates the need to specify the nuisance parameters. As such, the restated model is an example of the reduced model proposed here for estimating interactive effects. Jöreskog and Yang (1996) provided an example respecification of the original Kenny and Judd model and their own restatement based on the adjoined variance–covariance matrix. In addition, they consider the effects of nonnormality by allowing the factor variances associated with the interaction latent variable to be free, as opposed to their constrained estimates under normal theory. They also explore whether product terms of the error variances are the product of normal variates. One of the points not discussed explicitly in the presentation, however, is the way in which the expression of the model (particularly in reduced form) must be changed in response to such nonlinearity. As discussed next, the adjoined model proposed by Jöreskog and Yang can also be generated using the SAS program described here, with slight modifications.

COMPLETE AND REDUCED STRUCTURAL EQUATION MODEL REPRESENTATIONS FOR QUADRATIC AND INTERACTIVE EFFECTS

The key to a more compact notation for specification of quadratic and interactive effects results from the fact that it is possible to specify a structural equation model. This requires only those latent variables of interest (such as squared and interaction factor models) and correctly specifying variance components of a manifest variable due to nuisance factors such as cross-product terms by including them in that part of the model associated with error variance. Similarly, covariances between manifest product variables that arise due to these nuisance factors can be expressed as an additional covariance between error variances. A simulated data example of this helps to illustrate the approach.

A Simulated Data Example. Consider the interaction model proposed by Kenny and Judd (1984) in which two variables, X_1 and X_2, are indicators of a factor, F_1 and two other variables, Z_1 and Z_2, are indicators of factor F_2. Factors F_1 and F_2 are thought to influence a third variable, *Y*, through linear effects as well as an interaction. In the notation that follows, upper case letters denote latent and manifest variables, *e* is used to index *uniquenesses*, and lower case is used for estimated *factor loadings*. The linear measurement model in equation form for this model is expressed as:

$$\begin{bmatrix} X_1 \\ X_2 \\ X_3 \\ X_4 \end{bmatrix} = \begin{bmatrix} F_1 + e_1 \\ gF_1 + e_2 \\ F_2 + e_3 \\ hF_2 + e_4 \end{bmatrix}, \tag{6.1}$$

where *g* and *h* represent factor loadings associated with manifest variables X_2 and Z_2 respectively and the factor loadings associated with X_1

and Z_1 are set to 1 to produce an identified solution. The CALIS program used to estimate such a measurement model is given in Table 6.1. Note for ease of presentation that the values given in parentheses are given as start values for the program and represent final parameter estimates for the ML solution for simulated data to be described later.

To test for interaction effects, the Kenny and Judd model considers the implied measurement model associated with all product variables associated with the manifest variables;

$$\begin{bmatrix} X_1 \\ X_2 \\ Z_1 \\ Z_2 \end{bmatrix} \begin{bmatrix} X_1 \\ X_2 \\ Z_1 \\ Z_2 \end{bmatrix}^T = \begin{bmatrix} F_1 + e_1 \\ gF_1 + e_2 \\ F_2 + e_3 \\ hF_2 + e_4 \end{bmatrix} \begin{bmatrix} F_1 + e_1 \\ gF_1 + e_2 \\ F_2 + e_3 \\ hF_2 + e_4 \end{bmatrix}^T, \tag{6.2}$$

where T denotes the transpose operator. Carrying out this multiplication yields the following symmetric matrix:

$$\begin{bmatrix} F_1^2 + 2F_1e_1 + e_1^2 & - & - & - \\ gF_1^2 + F_1e_2 + gF_1e_1 + e_1e_2 & g^2F_1^2 + F_1e_2 + e_2^2 & - & - \\ F_2F_1 + F_2e_1 + F_1e_3 + e_3e_1 & gF_2F_1 + F_2e_2 + gF_1e_3 + e_3e_2 & F_2^2 + F_2e_3 + e_3^2 & - \\ hF_1F_2 + F_1e_4 + hF_2e_1 + e_1e_4 & ghF_1F_2 + gF_1e_4 + hF_2e_2 + e_2e_4 & hF_2^2 + F_2e_4 + hF_2e_3 + e_3e_4 & h^2F_2^2 + 2hF_2e_4 + e_4^2 \end{bmatrix} \tag{6.3}$$

for the respective elements of the matrix of product variables,

$$\begin{bmatrix} X_1^2 & \cdots & \cdots & \cdots \\ X_2X_1 & X_2^2 & \cdots & \cdots \\ X_1Z_1 & X_2Z_1 & Z_1^2 & \cdots \\ X_1Z_2 & X_2Z_2 & Z_2Z_1 & Z_2^2 \end{bmatrix} \tag{6.4}$$

To test for the presence of interactions, the measurement model associated with these product variables is added to the base measurement model along with information concerning the measurement model associated with the criterion or criteria of interest. The basic Kenny and Judd model assumes that the factors and error terms associated with manifest variables of the base measurement model are distrib-

TABLE 6.1

Basic Measurement Model for Interaction Effects

```
proc calis method=ml cov; lineqs
x1 = 1 f1 + 1 e1,
x2 = g (0.6000) f1 + 1 e2,
z1 = 1 f2 + 1 e3,
z2 = h (0.6000) f2 + 1 e4;
std
f1 = fv1 (2.154309),
f2 = fv2 (1.603206),
e1 = ev1 (0.360721),
e2 = ev2 (0.811623),
e3 = ev3 (0.490982),
e4 = ev4 (0.641283);
cov
f2 f1 = c12 (0.371688);
```

uted normally with a mean of zero. Under this assumption, nonlinear constraints may be imposed on the factor loadings as well as the error and factor variances associated with the product variables. These are described at length in Kenny and Judd (1984) as well as Jöreskog and Yang (1996). Figure 6.1 shows the path model associated with the original Kenny and Judd (1984) model. In this figure, the original linear model is indicated within the box in the upper left hand corner. In order to specify how the latent variable associated with the interaction is related to the criterion, it is necessary to generate four new variables that are the products of the manifest variables (X_1Z_1 through X_2Z_2 in the figure) and to specify the corresponding measurement model involving (among other terms) the latent variable interaction term F_1F_2.

Estimation Using SAS Program. It is quite easy to use the GENINT program to generate the CALIS setup for the original Kenny and Judd model. To generate the PROC CALIS program for the interactive model, the GENINT program requires two inputs: the base measurement model (Table 6.1, for example) and a list of relevant

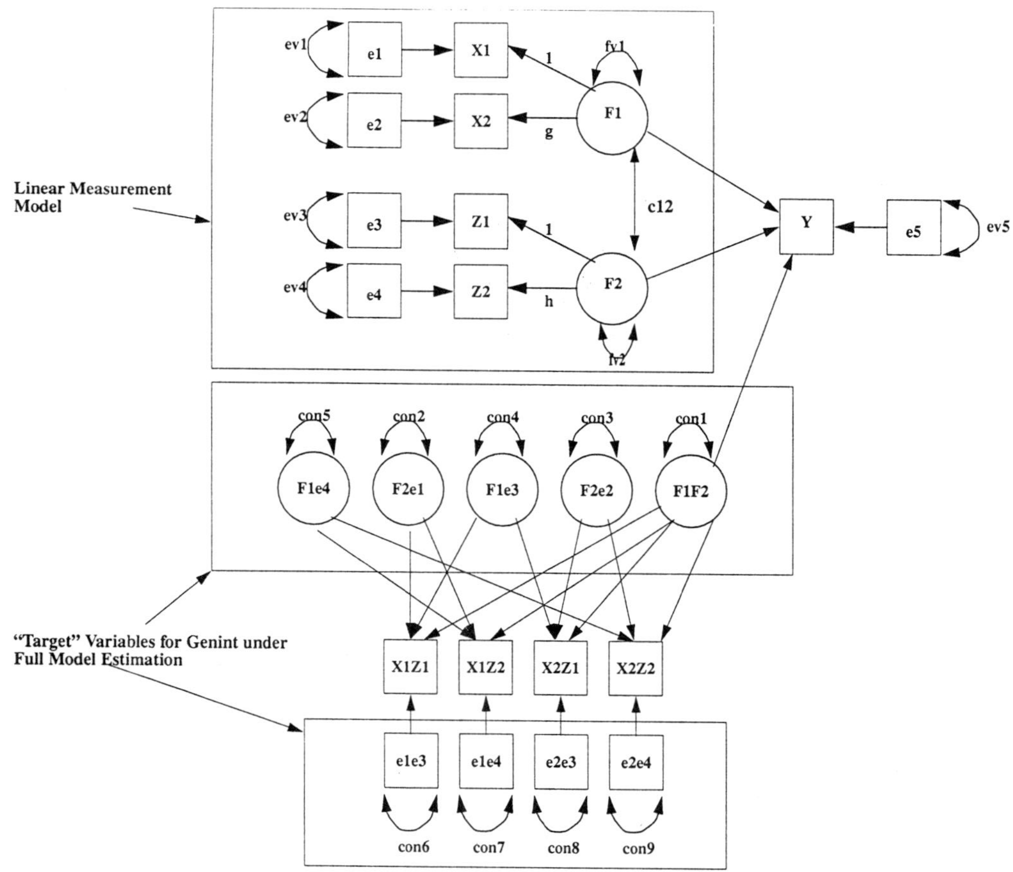

FIG. 6.1. Path diagram for Kenny and Judd (1984) latent variable interaction example.

product variables terms that the researcher would like to see in the resulting CALIS program. For ease of presentation, all unmeasured variables associated with the interaction model are termed *targets*. For these data, the relevant targets are shown in the boxes indicated in the lower left side of Fig. 6.1. These implied cross-product terms can also be found by examining Equation 6.3. Specifically, when running the GENINT program, it is necessary to supply a CALIS setup describing the linear measurement model and to specify which cross-product variables are of interest, as well as the number of cross-product variables. For the Kenny and Judd example, the following SAS code in Table 6.2 needs to be inserted:

To illustrate this, a sample size of 500 observations was randomly generated using SAS with parameter values equal to those described in the original Kenny and Judd article. For ease of presentation, factor loadings associated with the base measurement model were generated to be exactly their population values. The GENINT program automatically generates a CALIS program (shown in Table 6.3) containing the original (linear) measurement model, as well as the measurement model associated with the relevant product variables under the Kenny and Judd model. To fit the model, the researcher must add the measurement model associated with the criterion variable(s) of interest (indicated as underlined text in Table 6.3):

TABLE 6.2

Target Information to Produce Original Kenny and Judd Interaction Model

```
target1 = 'f1f2';
target2 ='f1e3';
target3 = 'f2e1';
target4 = 'e1e3';
target5 = 'f1e4';
target6 = 'e1e4';
target7 = 'f2e2';
target8= 'e2e3';
target9 = 'e2e4';
tindex = 9;
```

TABLE 6.3

Calis Program for Original Kenny and Judd Interaction Model

```
proc calis method=ml cov omethod=quanew;
lineqs y = c (-0.1324) f1 + d (0.3085) f2 + e (0.6589) f1f2 + e5,
x1z1 = 1 f1e3 + 1 f2e1 +1 f1f2 + e1e3,
x1z2 = 1 f1e4 + h (0.6000) f2e1 + h (0.6000) f1f2 + e1e4,
x2z1 = g (0.6000) f1e3 + 1 f2e2 + g (0.6000) f1f2 + e2e3,
x2z2 = g (0.6000) f1e4 + h (0.6000) f2e2 + gh (0.36) f1f2 + e2e4,
x1 = 1 f1 + 1 e1,
x2 = g (0.6000 ) f1 + 1 e2,
z1 = 1 f2 + 1 e3,
z2 = h (0.6000 ) f2 + 1 e4;
STD
f1 = fv1 (2.154309),
f2 = fv2 (1.603206),
f1f2 = con1 (3.5919471371),
f2e1 = con2,
f2e2 = con3,
f1e3 = con4,
f1e4 = con5,
e1e3 = con6,
e1e4 = con7,
e2e3 = con8,
e2e4 = con9,
e1 = ev1 (0.360721),
e2 = ev2 (0.811623),
e3 = ev3 (0.490982),
e4 = ev4 (0.641283),
e5 = ev5 (0.189198);
COV
f1 f2 = c12 (0.37168) ;
con1 = fv2*fv1+(c12*c12);
con2 = fv2*ev1;
con3 = fv2*ev2;
con4 = fv1*ev3;
con5 = fv1*ev4;
con6 = ev1*ev3;
con7 = ev1*ev4;
con8 = ev2*ev3;
con9 = ev2*ev4;
gh = g*h;
```

Note in Table 6.3 that the GENINT program generates the nonlinear constraints associated with the specified targets and that start values for all factor product terms and covariances are automatically calculated and inserted by the GENINT program. In addition, that factor loadings that are products of other factor loadings (such as the loading *gh* associated with product variable X_2Z_2) are calculated and inserted as well. Calculation of start values based on the final solution of the linear measurement model has been found to greatly reduce the incidence of nonconvergence–local minimum problems in parameter estimation for such models.

Alternate Specifications of Measurement Models. Although not discussed in the original Kenny and Judd article, it is quite possible that researchers may wish to specify the linear measurement model in some alternative form. For example, a researcher may wish to specify that the error variance of a manifest variable is fixed to unity, while the path from the error term to the manifest variable is free to vary. This modification is frequently encountered when a variance term is negative (McDonald, 1980). Such a modification, although somewhat more tedious for the researcher to program, can be easily generated with the GENINT Program. As an example, suppose that the researcher wished to specify the base measurement model for our example as is shown in Table 6.4, in which the error variances associated with variables X1 and Z1 are set to unity and the paths from these terms to the manifest variables are free to vary. The CALIS program statements for the original Kenny and Judd model under such a measurement model are calculated by specifying the original nine target variables in the GENINT program and are shown in Table 6.5 as follows:

It seems, however, somewhat unnecessary to consider a CALIS program (or path diagram) that contains nuisance latent variables (in this example, F_1e_4–F_2e_2) when only one latent variable, F_1F_2 is conceptually of interest. Because of this, a researcher may wish to consider only the latent variable F_1F_2 as a target, in which case the GENINT program will generate nonlinear constraints on error variances and covariances of the product variables so that the effects of these nuisance parameters are taken into account. As can be seen in Table 6.6, the resulting CALIS program generates the same predicted variance–covariance matrix as

TABLE 6.4

Alternative Base Measurement Model for Interactive Effects

```
proc calis omethod=quanew cov method=ml;
lineqs
x1 = 1 f1 + 1 e1,
x2 = g (.6 ) f1 + s2 (0.9009 ) e2,
z1 = 1 f2 + 1 e3,
z2 = h (.6 ) f2 + s4 (0.8008 ) e4;
STD
f1 = fv1 (2.154309),
f2 = fv2 (1.603206),
e1 = ev1 (0.360721),
e2 = 1,
e3 = ev3 (0.490982),
e4 = 1;
COV
f1 f2 = c12 (.371688);
```

Kenny and Judd's original model shown in Fig. 6.2, but with only the latent variables of interest (F_1, F_2, and F_1F_2).

When these two programs are used on the example data set described previously, the fit statistics and parameter estimates are identical.

Quadratic Models in Reduced Form. The procedures for generating quadratic models according to the Kenny and Judd specifications is the same as for the interactive models. Specifically, the measurement model for the Kenny and Judd example for two manifest variables, A and B can be written as:

$$\begin{bmatrix} A \\ B \end{bmatrix} [A \; B] = \begin{bmatrix} A^2 & AB \\ AB & B^2 \end{bmatrix} = \begin{bmatrix} F_1 + e_1 \\ bF_1 + e_2 \end{bmatrix} [F_1 + e_1 \;\; bF_1 + e_2] = \begin{bmatrix} F_1^2 + 2F_1e_1 + e_1^2 & bF_1^2 + F_1e_2 + bF_1e_1 + e_1e_2 \\ bF_1^2 + F_1e_2 + bF_1e_1 + e_1e_2 & b^2F_1^2 + 2bF_1e_2 + e_2^2 \end{bmatrix}, \tag{6.5}$$

TABLE 6.5
Original Kenny and Judd Model involving Error Standard Deviations

```
proc calis cov omethod=quanew method=ml;
lineqs
y = c f1 + d f2 + e f1f2 + e5,
x1z1= 1 f1e3 + 1 f2e1 +1 f1f2 + e1e3,
x1z2 = 1 f1e4 + h (.6) f2e1 + h (.6) f1f2 + e1e4,
x2z1 = g (.6) f1e3 + 1 f2e2 + g (.6) f1f2 + e2e3,
x2z2 = g (.6) f1e4 + h (.6) f2e2 + gh (0.36) f1f2 + e2e4,
x1 = 1 f1 + 1 e1,
x2 = g (.6 ) f1 + s2 (0.9009) e2,
z1 = 1 f2 + 1 e3,
z2 = h (.6 ) f2 + s4 (0.8008) e4;
STD
f1 = fv1 (2.154309),
f2 = fv2 (1.603206),
f1f2 = con1 (3.591953084),
f2e1 = con2,
f2e2 = con3,
f1e3 = con4,
f1e4 = con5,
e1e3 = con6,
e1e4 = con7,
F1F2 is
e2e3 = con8,
e2e4 = con9,
e1 = ev1 (0.360721),
e2 = 1,
e3 = ev3 (0.490982),
e5 = e5,
e4 = 1;
COV
f1 f2 = c12 (.371688) ;
con1 = fv2*fv1+(c12*c12);
con2 = fv2*ev1;
con3 = fv2*s2*s2;
con4 = fv1*ev3;
con5 = fv1*s4*s4;
con6 = ev1*ev3;
con7 = ev1*s4*s4;
con8 = s2*s2*ev3;
con9 = s2*s2*s4*s4;
gh = g*h;
```

TABLE 6.6
Reduced Form for Interaction Effects Assuming Error Standard Deviations

```
proc calis omethod=quanew cov method=ml;
lineqs
y = c (-0.1324) f1 + d (0.3085) f2 + e (0.6589) f1f2 + e5,
x1z1 =1 f1f2 + e1e3,
x1z2 = h (.6) f1f2 + e1e4,
x2z1 = g (.6) f1f2 + e2e3,
x2z2 = gh (0.36) f1f2 + e2e4,
x1 = 1 f1 + 1 e1,
x2 = g (.6 ) f1 + s2 (0.9009 ) e2,
z1 = 1 f2 + 1 e3,
z2 = h (.6 ) f2 + s4 (0.8008 ) e4;
STD
f1 = fv1 (2.154309),
f2 = fv2 (1.603206),
f1f2 = con1 (3.591953084),
e1e3 = con2 (3.093340),
e1e4 = con3 (1.467620),
e2e3 = con4 (2.888367),
e2e4 = con5 (1.486253),
e1 = ev1 (0.360721),
e2 = 1,
e3 = ev3 (0.490982),
e5 = e5 (0.189198),
e4 = 1;
COV
f1 f2 = c12 (.371688) ,
e1e3 e1e4 = con6 (0.360721),
e1e3 e2e3 = con7 (0.466263),
e1e4 e2e4 = con8 (0.828910),
e2e3 e2e4 = con9 (1.049089);
con1 = fv2*fv1+(c12*c12);
con2 = ev1*ev3 + fv1*ev3 + fv2*ev1;
con3 = ev1*s4*s4 + fv1*s4*s4 + h*h*fv2*ev1;
con4 = s2*s2*ev3 + g*g*fv1*ev3 + fv2*s2*s2;
con5 = s2*s2*s4*s4 + g*g*fv1*s4*s4 + h*h*fv2*s2*s2;
con6 = h*fv2*ev1;
con7 = g*fv1*ev3;
con8 = g*fv1*s4*s4;
con9 = h*fv2*s2*s2;
gh = g*h;
```

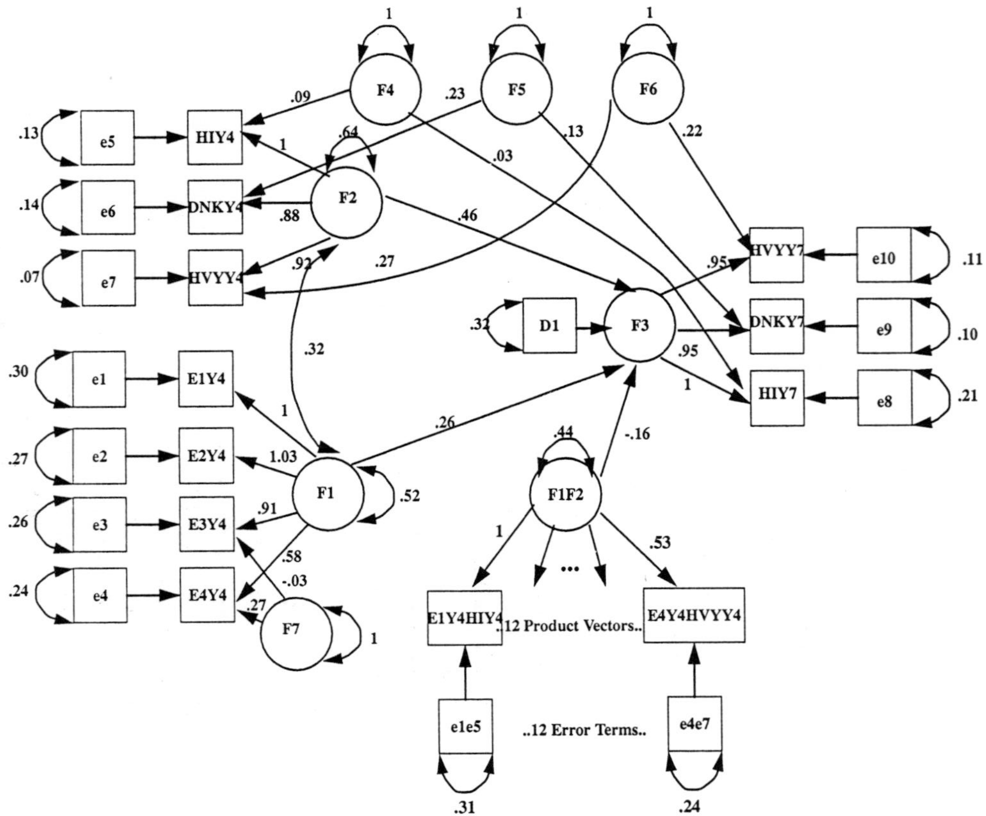

FIG 6.2. Path diagram for predicting problematic alcohol use at Year 7 from alcohol expectancies and problematic alcohol use at Year 1.

which then yields the following measurement model:

$$\begin{bmatrix} A \\ B \\ A^2 \\ AB \\ B^2 \\ Y \end{bmatrix} = \begin{bmatrix} F_1 + e_1 \\ bF_1 + e_2 \\ F_1^{\,2} + 2F_1e_1 + e_1^2 \\ bF_1^{\,2} + F_1e_2 + bF_1 + e_1e_2 \\ b^2F_1^{\,2} + 2bF_1e_2 + e_2^2 \\ cF_1 + dF_1^2 + e_3 \end{bmatrix}. \tag{6.6}$$

For the GENINT program, it is necessary to specify the CALIS program for the base linear model, a specification of the target product terms in the measurement model of interest, and the number of target terms. When these are supplied to the GENINT program, requesting the full and reduced programs, the following CALIS setups are produced in Tables 6.7 and 6.8.

Additional Analyses and Models. The fact that the GENINT program can produce the appropriate CALIS program so quickly and the fact that use of final parameter estimates taken from the base linear model have resulted in almost no convergence failures or improper solutions means that researchers are much freer to explore whether their conclusions regarding a proposed interactive or quadratic effect are perhaps due to a confound with some other interaction or quadratic effect. For example, it is possible to specify both quadratic and interactive effects merely by specifying that the targets *f1f2*, *f1f1*, and *f2f2* are of interest to the researcher and including these targets in the model for the criterion of interest.

However, note that, in such situations, it will be necessary to specify additionally the covariances between quadratic and interaction effects. For example, one additional constraint is: $Cov(F_1F_2,F_1F_1) = 2(Var(F_1)Cov(F_1,F_2)$, see Kendall and Stuart (1958). Additionally, many of the published examples of interactive and quadratic effects have assumed that the measurement models for latent variables involve simple structure. Presumably, these examples have made this assumption because of the programming tedium involved in creating all of the necessary cross-product variables and constraints implied by a complex

TABLE 6.7

Quadratic Effects for Original Kenny and Judd Model

```
proc calis cov method=ml;
lineqs y= a f1 + b f1f1 + e3,
x1x1 =1 f1f1 +2 f1e1 +e1e1,
x2x2 = gg (0.66683556) f1f1 +_2g (1.6332) f1e2 +e2e2,
x1x2 = 1 f1e2 + g (0.8166) f1e1 + g (0.8166) f1f1 + e1e2,
x1 = 1 f1 + 1 e1,
x2 = g (0.8166 ) f1 + 1 e2;
STD
f1 = fv1 (0.793227),
f1f1 = con1 (1.2584181471),
f1e1 = con2 (0.4079621987),
f1e2 = con3 (0.3036758518),
e1e1 = con4,
e2e2 = con5,
e1e2 = con6,
e1 = ev1 (0.514307),
e2 = ev2 (0.382836),
e3 = ev3 (0.154195);
_2g = 2*g;
con1 = fv1*fv1*2;
con2 = fv1*ev1;
con3 = fv1*ev2;
con4 = 2*ev1*ev1;
con5 = 2*ev2*ev2;
con6 = ev1*ev2;
gg = g*g;
```

factor model. In the GENINT program, such models can be generated automatically. Additionally, although many authors have cautioned that generalized least squares, maximum likelihood, and ADF estimates may be quite different, few researchers have been inclined to conduct alternate models in an attempt to examine how choice of minimization function affects the final estimates of interaction or quadratic effects.

TABLE 6.8

Quadratic Effects for Reduced Model

```
proc calis cov method=ml;
lineqs y= a f1 + b f1f1 + e3,
x1x1 =1 f1f1 +e1e1,
x2x2 = gg (0.66683556) f1f1 +e2e2,
x1x2 = g (0.8166) f1f1 + e1e2,
x1 = 1 f1 + 1 e1,
x2 = g (0.8166 ) f1 + 1 e2;
STD
f1 = fv1 (0.793227),
f1f1 = con1 (1.2584181471),
e1e1 = con2,
e2e2 = con3,
e1e2 = con4,
e1 = ev1 (0.514307),
e2 = ev2 (0.382836),
e3 = ev3 (0.154195);
COV
e1e1 e1e2 = con5 ,
e1e2 e2e2 = con6 ;
con1 = fv1*fv1*2;
con2 = 2*ev1*ev1 + 4*1*1*fv1*ev1;
con3 = 2*ev2*ev2 + 4*g*g*fv1*ev2;
con4 = ev1*ev2 + 1*1*fv1*ev2 + g*g*fv1*ev1;
con5 = 1*g*fv1*ev1 + 1*g*fv1*ev1;
con6 = 1*g*fv1*ev2 + 1*g*fv1*ev2;
gg = g*g;
```

As discussed in greater detail in Wood and Erickson (1998), detection of significant interactions using the Kenny and Judd technique appears sensitive to the presence of skew, in that variance estimates of interaction effects appear inflated, thereby inflating the estimated variance of product terms, and reducing patterns of statistical significance for interaction effects.

Although, in one respect, it may be tempting to transform manifest variables in order to reduce such skew, such a decision is particularly ill-advised for polynomial models because interaction effects may be made to appear or disappear under suitable choice of transformation. One transform, however, which does remediate some effects of skew and retain the conceptual meaning of the variables (at least as relative standing within the population under study) is the rank transform. As discussed next, the rank transform appears to produce a more coherent set of analyses across estimation techniques and to produce estimates of interaction effects that are more similar to those obtained by multiple regression.

Finally, it is also possible to explore the degree to which the multivariate normality of error and latent variables affects key parameters of the model (as implied by the constraints on error and latent variable product terms). In such situations when the researcher has specified the full Kenny and Judd model, such explorations may be made by merely commenting out constraints involving a given error or factor variance.

A Real-World Data Example. As mentioned before, comparison of GLS and ML estimates and standard errors have appeared roughly comparable in examples that use generated data. Given the ease with that interactive models can be generated and tested using the SAS program described here, it seems appropriate to compare estimated parameters under GLS, ML, and WLS estimation. It is also instructive to compare the estimates of the unique proportion of variability accounted for by the interaction under the traditional multiple regression approach with the Kenny and Judd factor model. Following this, a simple extension of the Kenny and Judd model to a factorially complex model is presented. Before discussing the statistical models further, it is appropriate to discuss a rudimentary rationale for the substantive questions explored.

As mentioned initially, prior research has noted that the prediction of heavy alcohol use appears to be moderated by both prior heavy drinking and alcohol outcome expectancies (e.g., Cooper et al., 1992). In a test of this finding within a larger prospective study, Gotham (1995) examined the prospective moderator relationships of heavy drinking, personality traits, sex, and four positive alcohol outcome expectancy scales on subsequent heavy drinking in a college-aged sample of 446 individuals. (Analyses proposed here, however, are based on 443 individuals because three individuals possessed missing data values on some of the manifest variables of interest.) Gotham's analysis considered complete data records gathered from participants who were assessed 4 years following their freshman year in college and again 3 years later. See Kushner, Sher, Wood, & Wood (1994), Sher, Wallitzer, Wood, & Brent (1991), and Gotham (1995) for a further description of the sample and instruments. Gotham et al. used the traditional general linear models approach to test for the presence of interactions involving only composite manifest variables and a much larger set of predictor variables than only the alcohol expectation and heavy alcohol use constructs considered here. For the analyses presented here, we will consider only the relationship of Year 4 alcohol expectancies and heavy alcohol use to prediction of heavy alcohol use 3 years later.

Within a larger model containing personality and gender effects, Gotham et al. found that subsequent heavy alcohol use was predicted significantly by prior alcohol expectancies, prior heavy alcohol use, and the interaction of prior alcohol expectancies and heavy alcohol use. In the interest of brevity, the significant main effects for sex, interactions of sex with heavy alcohol use, and interactions of personality variables with previous heavy alcohol use are not addressed in these example analyses. The purpose of the present analysis is to examine the magnitude of structural relationships predicting heavy alcohol use from prior heavy alcohol use, alcohol expectancies and the interaction of prior heavy alcohol use and alcohol expectancies.

Multiple Regression Models. As a measure of the practical benefit, if any, of conducting factor models to test for the interaction effect of alcohol expectancies and prior problematic alcohol use on subsequent problematic alcohol use, a multiple regression was conducted on centered data predicting a composite measure of Year 7 alcohol use from Year 4 alcohol use, four measures of alcohol expectancies, and the interaction of Year 4 alcohol use and alcohol expectancies. (The variables comprising these composites are discussed under the measurement model.) The results of this regression model with and without the interaction term are shown in the two left-hand columns of Table 6.9.

As can be seen, the standardized estimates associated with the expectancy and initial alcohol use composites do not change appreciably between a model that contains only main effects and that which contains

TABLE 6.9

Comparison of Estimates Predicting Year 7 Heavy Alcohol Use from Heavy Alcohol Use and Alcohol Expectancies at Year 4

	Parameter Estimated	Model: Multiple Regression[a] — Main Effects Only	Main Effects & Interactions	Polynomial Factors Assuming Serial Correlation and Complex Factor Structure — GLS	ML	WLS
Paths Predicting Year 2 Heavy Alcohol Use	Heavy Alcohol Use	.5026[b]	.5189	.3628[c]	.5496	.6174
	Year 4	.4871***	.5132***	.3477 *	.4408*	.5477*
	Alcohol Expectancy	.1258	.1177	.2410	.1006	.1301
	Year 4	.1212	.1133	.0489*	.0189*	.0199*
	Alcohol Use x Alcohol		-.0895	-.1066	-.1398	.0009
	Expectancy		-.0794**	-.0248	-.0209*	.0002
Variances	Heavy Alcohol Use	1	1	1	1	1
	Year 4	.9396	.9396	.6454	1.1694	.2943
	Alcohol Expectancy	1	1	1	1	1
	Year 4	.9910	.9910	14.3837	21.3842	9.8867
	Alcohol Use x Alcohol		1	1	1	1
	Expectancy		1.1662	10.9533	33.6955	3.3836
Covariances	Alcohol Use,	.5588	.5588	.4241	.5894	.4036
	Alcohol Expectancy	.5392***	.5392***	1.2921	2.9475	.6884*
R^2 Year 2b Heavy Alcohol Use		.3119	.3267	.2752	.3969	.4628

[a]Multiple regression models and models correcting for attenuation employ composites of relevant constructs. Polynomial factor models employ the measurement model described in text.

* = $p < .05$; ** = $p < .01$; *** = $p < .001$

[b]First number in each cell denotes standardized value. Second number denotes unstandardized value.

[c]First number in each cell denotes standardized parameter estimate. Second number denotes unstandardized estimate.

the interaction effect, suggesting that the interaction effect is not highly collinear with main effects due to skew or other non-normality of the data. In addition, the regression weight associated with the interaction vector is statistically significant, although the unique variability in the criterion accounted for by the interaction is somewhat small.

Latent Variable Models of Interaction. Turning now to the factor model, it becomes necessary to introduce the manifest variables associated with problematic alcohol use and alcohol expectancies. Specifically, the construct of heavy alcohol use on both occasions consisted of three variables: Subjects were asked to consider their drinking over the past month and to estimate the mean number of occasions per week that the individual: Had five or more drinks in a single session (HVY); felt a little high or light-headed from alcohol (HI); or became drunk (DNK). Alcohol expectancies, by contrast, were measured by four manifest variables. Specifically, items from the expectancy questionnaire explored the degree to which individuals endorsed items related to Tension Reduction (F1Y4), Social Lubrication (F2Y4), Activity Enhancement (F3Y4), and Performance Enhancement (F4Y4).

Anderson and Gerbing (1988) and others have stressed the need to first fit the basic measurement model for a given data set before proceeding to a structural analysis. Examination of fit measures associated with the simple structure model revealed a generally acceptable degree of fit to the data (e.g., NFI = .97; NNFI = .96; RMSEA = .06).

Although it is possible to fit a latent variable model predicting problematic alcohol use from initial problematic alcohol use, alcohol expectancies, and the interaction of the two constructs, it was decided to first investigate whether a simple structure factor model was reasonable for these data. Three conceptual reasons also motivated this decision: (1) Such effects may constitute a class of alternative explanations for a proposed nonlinear effect; (2) assuming a simple factor structure where a pattern of complex loadings exists may result in an overestimate of covariances between predictor factors; and (3) such violations of the measurement model may be one reason that the variances of product variables involving either error or factor variances do not conform to the normal theory values predicted under the Kenny and Judd model. To see how the program may be used to estimate interactive and quadratic effects for complex models, interactive effects in the real-world data example are presented under a factorially complex model that assumes serial autocorrelation and a complex factor structure.

In order to improve the measurement model with interactive effects, the measurement model for linear effects was re-examined on conceptual and empirical grounds. Specifically, it seemed reasonable to assume that the manifest variables associated with heavy alcohol use would demonstrate serial correlation over time (i.e., it seemed reasonable to think that each alcohol use item at the first assessment would be systematically related to the same item at a subsequent occasion due to, for example, the fact that the items share the same format). The measurement model for the heavy alcohol use latent variable at Years 4 and 7 was therefore respecified as:

$$\begin{bmatrix} \text{HIY4} \\ \text{DNKY4} \\ \text{HVYY4} \end{bmatrix} = \begin{bmatrix} F_2 + hF_4 + e_5 \\ aF_2 + iF_5 + e_6 \\ bF_2 + jF_6 + e_7 \end{bmatrix} \tag{6.7}$$

and

$$\begin{bmatrix} \text{HIY5} \\ \text{DNKY5} \\ \text{HVYY5} \end{bmatrix} = \begin{bmatrix} F_3 + kF_4 + e_8 \\ fF_3 + lF_5 + e_9 \\ gF_3 + mF_6 + e_{10} \end{bmatrix}, \tag{6.8}$$

where F_2 and F_3 represent the latent variable of problematic alcohol use in Years 4 and 7, respectively, and F_4, F_5, and F_6 represent method factors that account for serial correlation in the manifest variables over time. It is worth noting that all of the serial autocorrelation loadings were statistically significant under GLS and ML analyses of the linear measurement model and these associations remained significant under the interactive models described later.

Changes were also made to the four alcohol outcome expectancies as well. A similar measurement model was used for heavy alcohol use at Year 7. Examination of the Lagrange multipliers for the linear measurement model revealed significant unmodeled covariation between the Activity Enhancement and Performance Enhancement scales not explained by the general expectancy variables. Accordingly, this association was modeled as an additional method factor with two indicators. The measurement model for the four alcohol expectation manifest variables can be expressed as:

$$\begin{bmatrix} \text{F1Y4} \\ \text{F2Y4} \\ \text{F3Y4} \\ \text{F4Y4} \end{bmatrix} = \begin{bmatrix} F_1 + e_1 \\ cF_1 + e_2 \\ dF_1 + nF_7 + e_3 \\ eF_1 + oF_7 + e_4 \end{bmatrix}, \tag{6.9}$$

where F_1 is, as before, the general latent variable of alcohol expectancies, and F_7 is the additional method factor associated with the third and fourth expectancy scales. Fit indices for this revised linear measurement model are given in Table 6.10. Although the increment in fit between the simple and complex measurement models is significant (chi-square difference = 19.28; 8 df; $p = .01$), examination of overall fit statistics for the model reveals only a modest improvement in an already high degree of fit to the data.

As can be seen from Table 6.10, the fit of the measurement models associated with the three latent variables is quite good, although the RMSEA estimate reveals (as before) a slightly less acceptable report of model fit. GLS and WLS solutions were also computed for the data and were, as expected, much lower, due to the theory underlying the fit indices.

The structural models for cross-products variables, which are used as indicators of the interactive latent variable are then calculated by multiplying the appropriate measurement equations, that is,

TABLE 6.10

Selected Fit Indices for ML Solution

Fit Index	*Value*
Model χ^2	68.7224
df = 24	
AIC	20.7224
RMSEA Estimate	.0649
(90% CI)	(.0472 - .0832)
SBC	-77.3252
CFI	.9789
NNFI	.9791
NFI	.9683

Note:
AIC = Akaike's Information Criterion
RMSEA = Root Mean Square Error of Approximation
SBC = Schwarz's Bayesian Information Criterion
CFI = Comparative Fit Index
NNFI = Non-Normed Fit Index
NFI = Normed Fit Index

$$\begin{bmatrix} \text{F1Y4} \\ \text{F2Y4} \\ \text{F3Y4} \\ \text{F4Y4} \end{bmatrix} [\text{HIY4 DNKY4}] = \begin{bmatrix} \text{F}_1 + e_1 \\ c\text{F}_1 + e_2 \\ d\text{F}_1 + n\text{F}_7 + e_3 \\ e\text{F}_1 + o\text{F}_7 + e_4 \end{bmatrix} [F_2 + hF_4 + eF_2 + iF_5 + e_6 \; bF_2 + jF_6 + e_7] \quad (6.10)$$

The resulting measurement models associated with the product variables are somewhat large and cumbersome. For example, cross-product terms for the four expectancy variables when multiplied by HIY4 are:

$$\begin{array}{c} F_2\text{F}_1 + h\text{F}_1F_4 + e_5\text{F}_1 + F_2e_1 + he_1F_4 + e_1e_5 \\ cF_2\text{F}_1 + hF_4c\text{F}_1 + ce_5\text{F}_1 + F_2e_2 + he_2F_4 + e_2e_5 \\ dF_2\text{F}_1 + dhF_4\text{F}_1 + de_5\text{F}_1 + nF_2\text{F}_7 + hn\text{F}_7F_4 + ne_5\text{F}_7 + F_2e_3 + he_3 F_4 + e_3e_5 \\ eF_2\text{F}_1 + ehF_4\text{F}_1 + ee_5\text{F}_1 + oF_2\text{F}_7 + ho\text{F}_7F_4 + oe_5\text{F}_7 + F_2e_4 + he_4F_4 + e_4e_5 \end{array} \quad (6.11)$$

Cross-product terms of the four expectancy measures with DNKY4 and HVYY4 were calculated in an analogous manner.

Estimation Using SAS Program. Generation of the SAS program proceeded as described previously for the simple structure interaction model, except that the corresponding measurement models from the factorially complex linear model were used. As an additional check on the accuracy of estimation, this model was run twice, once specifying only the targets of interest (f1f2), and again, specifying all cross-product variables as described by Kenny and Judd (1984). Both programs converged to the same parameter estimates and fit function values.

As can be seen from Table 6.11, the fit of the full interactive model is considerably poorer when based on the original raw data than when based on rank transformed versions of the manifest variables.

Selected parameters estimates from the raw and rank-transformed version of the data set are presented in Table 6.12.

Patterns of statistical significance, however, were different with only the ML solution reaching statistical significance. This effect is perhaps due to chance variation, given the similar magnitude of the parameter estimates. When the analyses were based on ranked data, both the GLS and ML solutions demonstrated a significant interaction effect, whereas this was found only for the ML solution before. To some extent, the convergence of parameter estimates across GLS and ML may be an artifact of the approximately equal variance assigned to all variables under the rank-normalized transform. In spite of this, however, for both

TABLE 6.11

Selected Fit Measures for Complex Measurement Model Including Interactive Effects: ML Solutions.

Structure	*Complex (df = 231)*	
Fit Index	*Raw Data*	*Ranked Data*
Model X^2	1981.2827	684.1800
AIC	1539.2827	242.1800
RMSEA Estimate (90% CI)	.1342 (.1288 - 13.97)	.0689 (.0630 -.0748)
SBC	634.6038	-662.4989
CFI	.8086	.9521
NNFI	.7999	.9499
NFI	.7898	.9309

of these analyses, the WLS estimates associated with the interaction effect were minimal, in contrast to the statistically significant ML and GLS estimates.

Supplemental Analyses

Although the GENINT program is useful in producing convergent solutions for factorially simple and complex models, the utility of the program is perhaps most clear when considering competing models for the data that a reasonable skeptic could raise regarding the reasonableness of a proposed interactive effect. Space prohibits a presentation and extended discussion of such alternatives, but a discussion of the general characteristics of such analyses seems appropriate. Exploration of the Functional Relationship Between Predictors and Criteria.

Although the sign associated with the interactive or quadratic effect is helpful in interpreting the nature of the interaction, it would seem reasonable to also explore the relationship between predictors and criteria by graphical means. In addition to providing valuable clues as to whether the interaction effect is an artifact of relatively few influential observations, graphic approaches can also be used to explore other model violations that may not readily come to mind a priori. In the context of the present model, for example, factor scores were generated from the basic linear measurement model for initial alcohol abuse and

TABLE 6.12

Polynomial Factor Models Assuming Serial Correlation in Manifest Alcohol Use Variables and Complex Factor Structure for Alcohol Expectancies.

	Parameter Estimated	*Polynomial Factors*			*Ranked Polynomial Factors*		
		GLS	*ML*	*WLS*	*GLS*	*ML*	*WLS*
Paths Predicting Year 2 Heavy Alcohol Use	Heavy Alcohol Use	.3628[a]	.5496	.6174	.4855	.5969	.6134
	Year 4	.3477*	.4408*	.5477*	.4651*	.5388*	.5941*
	Alcohol Expectancy	.2410	.1006	.1301	.2476	.0525	.0323
	Year 4	.0489*	.0189*	.0199*	.2633*	.0523	.0276
	Alcohol Use x	-.1066	-.1398	.0009	-.1399	-.0951	.0054
	Alcohol Expectancy	-.0248	-.0209*	.0002	-.1620*	-.0941*	.0050
Variances	Heavy Alcohol	1	1	1	1	1	1
	Use Year 4	.6454	1.1694	.2943	.6422	.7733	.6405
	Alcohol Expectancy	1	1	1	1	1	1
	Year 4	14.3837	21.3842	9.8867	.5214	.6327	.8223
	Heavy Alcohol Use	1	1	1	1	1	1
	x Alcohol Expectancy	10.9533	33.6955	3.3836	.4395	.6434	.6880
Covariances	Alcohol Use	.4241	.5894	.4036	.5587	.5612	.5535
	Alcohol Expectancy	1.2921	2.9475	.6884*	.3233*	.3925	.4017
R^2 Year 7 Heavy Alcohol Use		.2752	.3969	.4628	.4509	.4032.	.3992

[a]First number in each cell denotes standardized parameter estimate. Second number denotes unstandardized estimate.

* = $p < .05$

alcohol expectation factors as well as the subsequent alcohol abuse factor. The scoring coefficients from these factors were then used to generate factor scores, which can then be used to explore the nature of the interaction effect.

To ease presentation, the scatterplots of the initial alcohol abuse factor are plotted against the subsequent alcohol abuse factor score separately for quintiles based on their alcohol expectation scores, and are shown in Fig. 6.3.

Although conclusions regarding the scatterplots based on approximately 50 individuals are tentative at best, it seems reasonable to believe

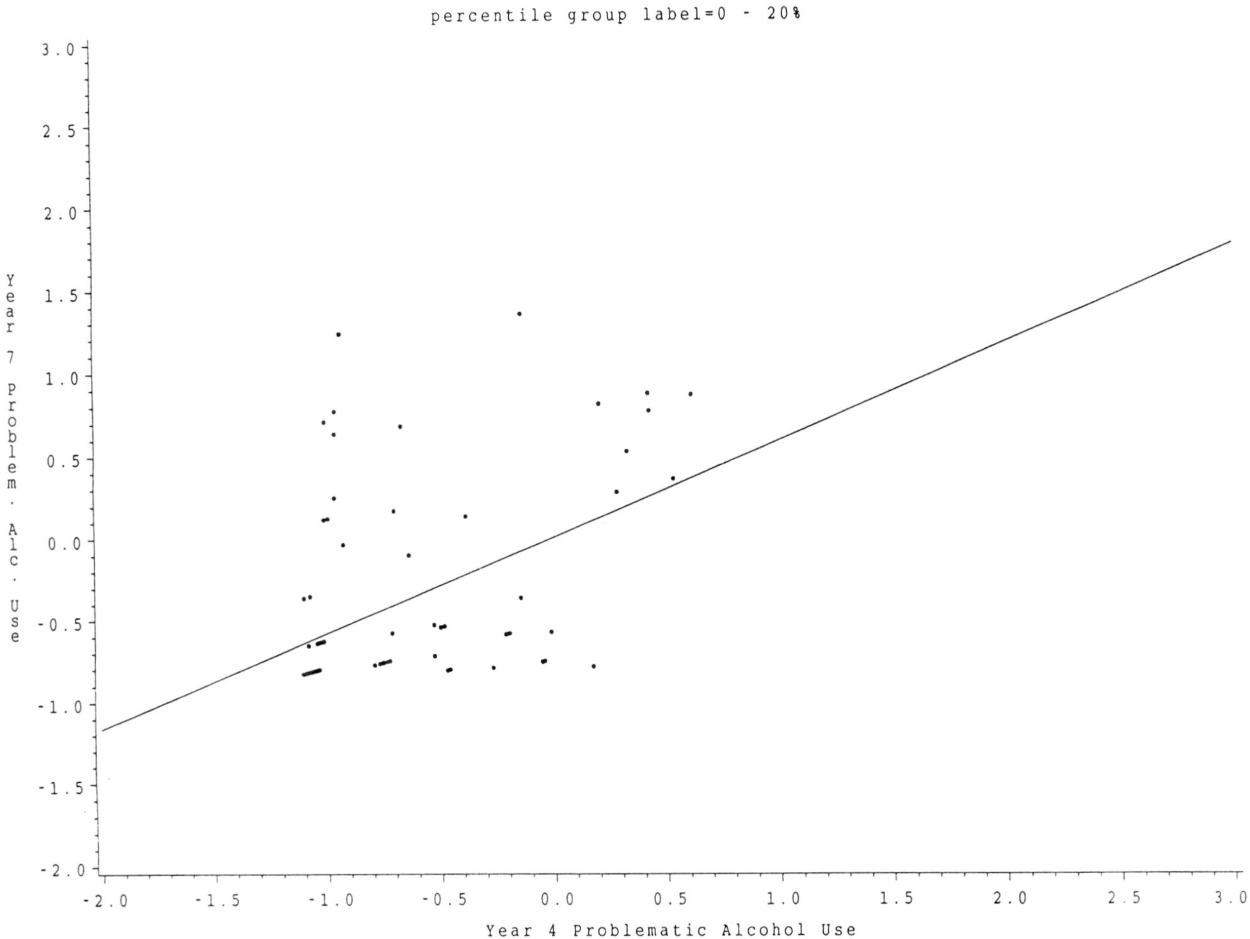

FIG. 6.3. Scatterplots of factor scores for problematic alcohol use at Years 4 and 7 grouped by quintiles of alcohol expectancies at Year 4.

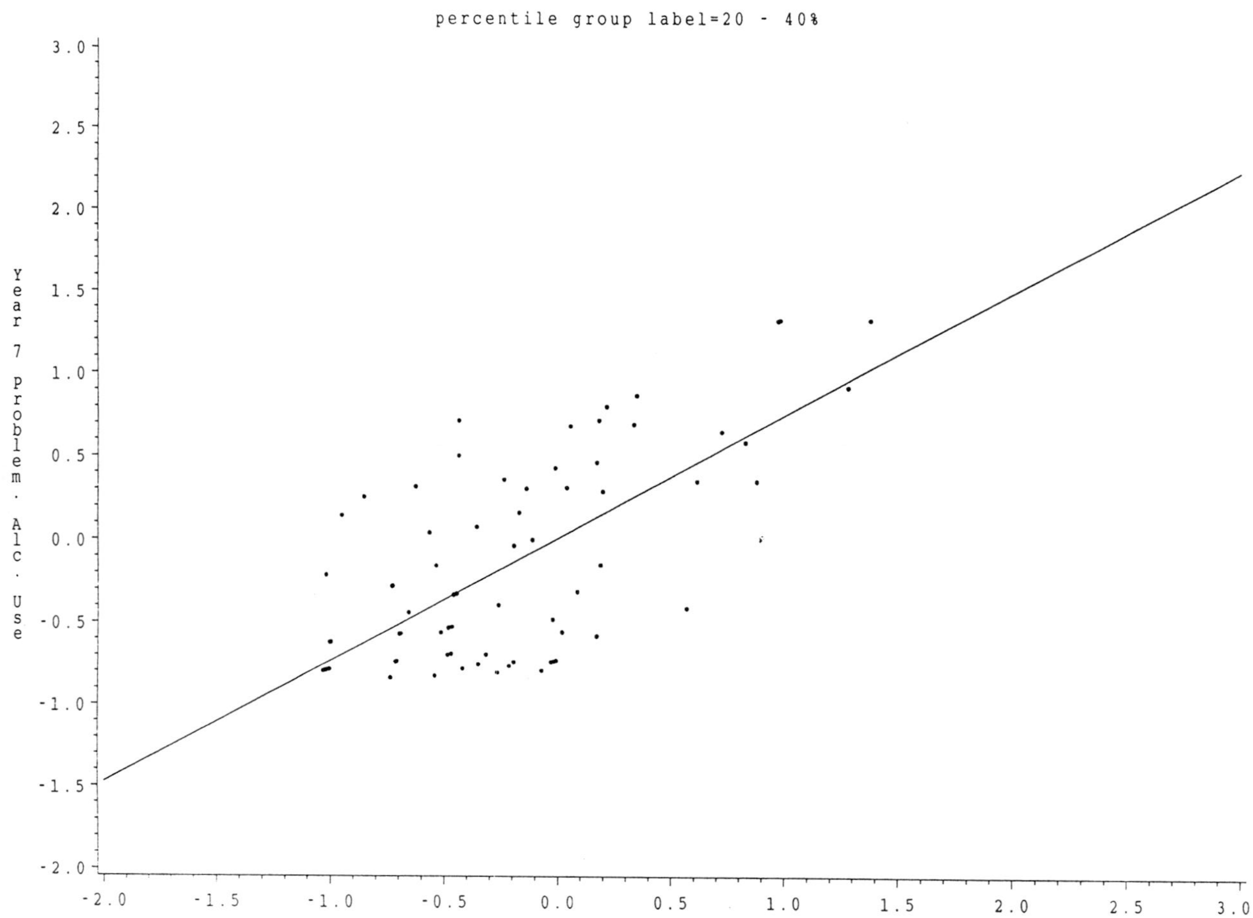
percentile group label=20 - 40%
Year 7 Problem. Alc. Use
3.0
2.5
2.0
1.5
1.0
0.5
0.0
-0.5
-1.0
-1.5
-2.0
-2.0 -1.5 -1.0 -0.5 0.0 0.5 1.0 1.5 2.0 2.5 3.0

FIG. 6.3B.

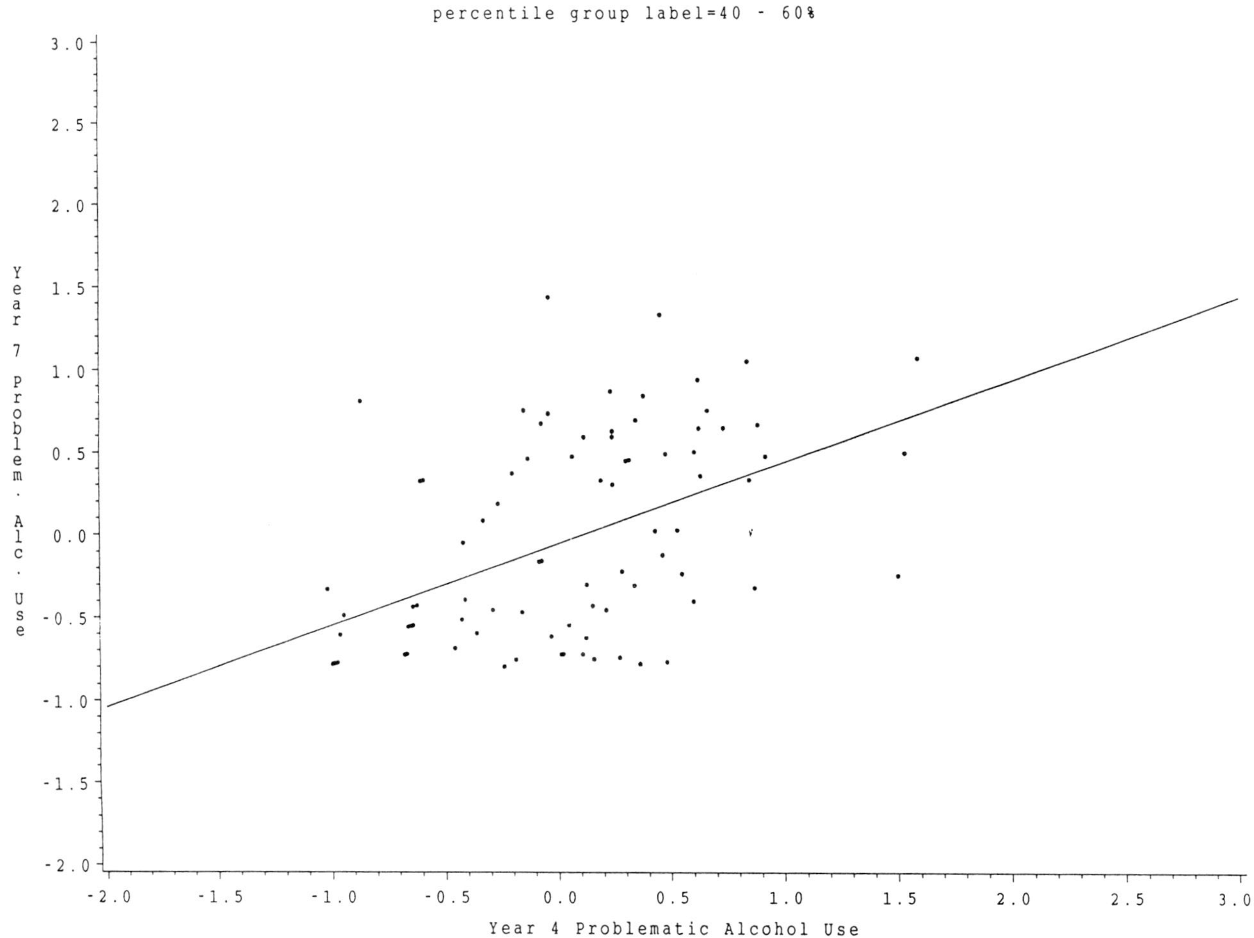

percentile group label=40 - 60%
Year 7 Problem. Alc. Use
3.0
2.5
2.0
1.5
1.0
0.5
0.0
-0.5
-1.0
-1.5
-2.0
-2.0
-1.5
-1.0
-0.5
0.0
0.5
1.0
1.5
2.0
2.5
3.0
Year 4 Problematic Alcohol Use

FIG. 6.3C.

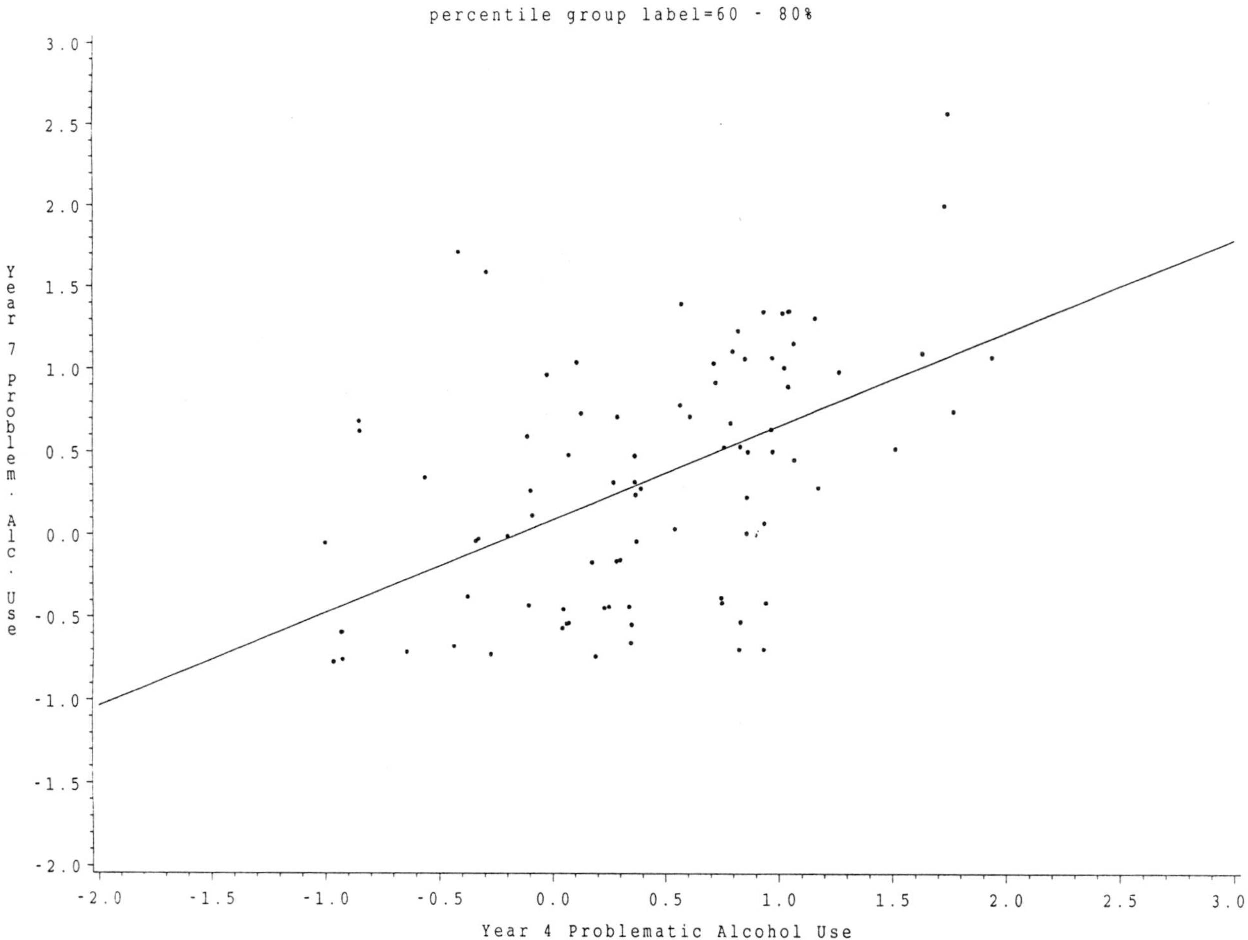
percentile group label=60 - 80%
3.0
2.5
2.0
1.5
1.0
0.5
0.0
-0.5
-1.0
-1.5
-2.0
Year 7 Problem. Alc. Use
-2.0
-1.5
-1.0
-0.5
0.0
0.5
1.0
1.5
2.0
2.5
3.0
Year 4 Problematic Alcohol Use

FIG. 6.3D.

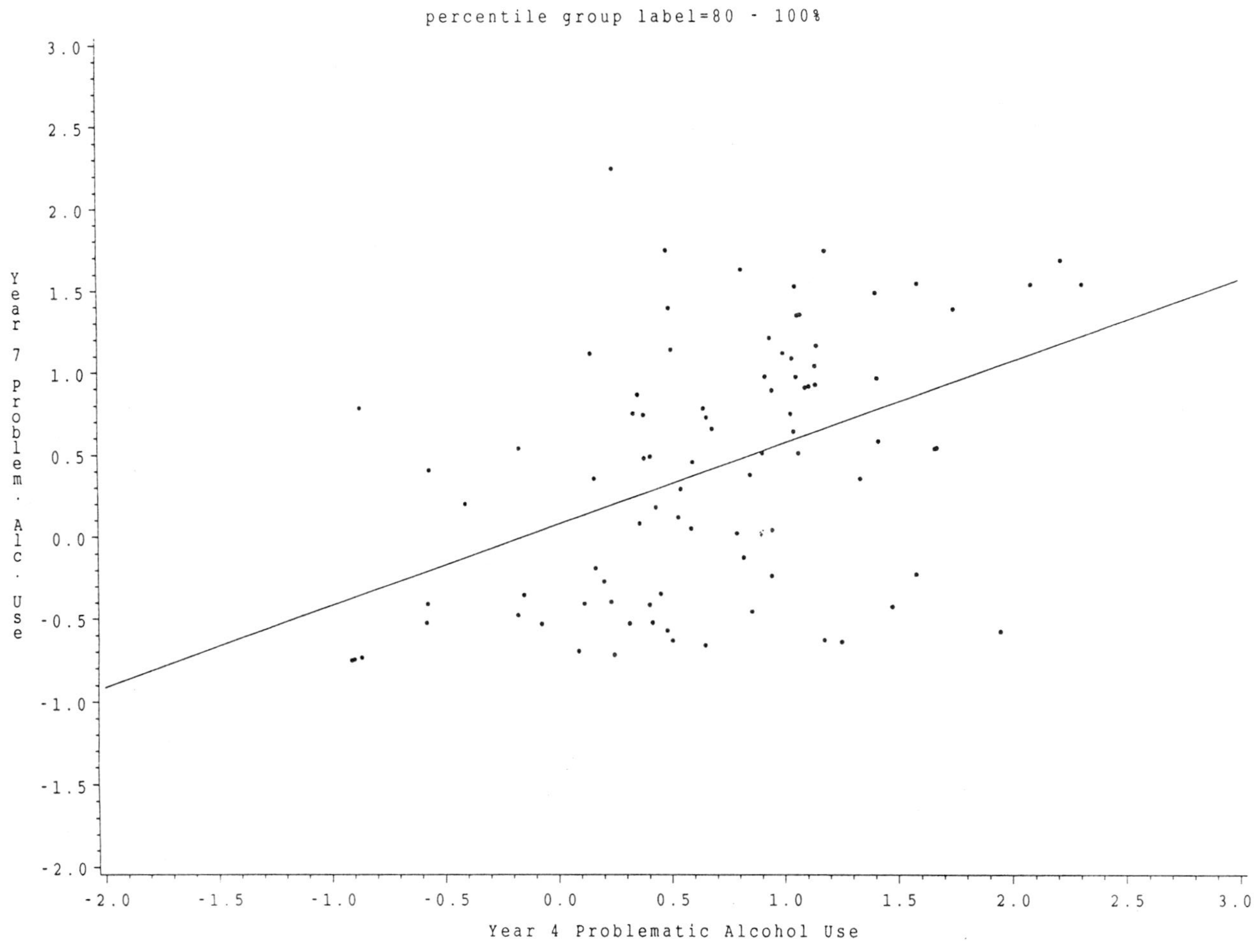
percentile group label=80 - 100%
Year 7 Problem. Alc. Use
3.0
2.5
2.0
1.5
1.0
0.5
0.0
-0.5
-1.0
-1.5
-2.0
-2.0
-1.5
-1.0
-0.5
0.0
0.5
1.0
1.5
2.0
2.5
3.0
Year 4 Problematic Alcohol Use

FIG. 6.3E.

that those individuals who had particularly low alcohol expectations represent two distinct groups. There is a small cluster of six individuals who have high initial alcohol abuse and who persist in this abuse. Individuals, however, who have a very low level of initial alcohol abuse in this group seem to be quite variable in subsequent alcohol abuse patterns. One conjecture regarding this pattern is that the young adults in this study who have low alcohol expectations represent a mix of individuals. Some individuals have not yet formed their opinions concerning alcohol's effects due to inexperience with alcohol. This inexperience leads to a low endorsement rate of how alcohol would affect the person. Presumably, subsequent assessment of alcohol's expectations for this group would reveal a greater variation in alcohol expectations than assessed initially. The pattern of response for the other four groups does not appear to be grossly distorted, although the extreme alcohol consumption of relatively few individuals is evident from the factor scores on the alcohol abuse factors. Although a follow-up analysis of these data excluding individuals in the lowest 20% on alcohol expectancies resulted in a nonsignificant interaction effect for both the regression and factor models, is it difficult to tell whether the failure to detect an interaction is due to the reduced power of the design, or whether the interaction effect detected is merely due to the aberrant pattern shown in the low expectation group.

Specification Errors. It is possible that other polynomial effects exist in the data besides those suggested by theory. To investigate this possibility, quadratic effects associated with Heavy Alcohol Use and Alcohol Expectancies were also estimated. This was done by merely including the targets f1f1 and f2f2 to the target list in the GENINT program and including f1f1 and f2f2 into the structural model for Year 7 Alcohol Use. These analyses also converged, but had generally poorer overall fits and failed to find any statistically significant effect for the quadratic effects under any estimation technique for either the raw or ranked data. Given that a second expectancy factor was modeled in these data, the question as to whether this factor also contained an interaction with the latent variable of heavy alcohol use, the interaction of heavy alcohol use with the second expectancy factor was also calculated and used to predict subsequent alcohol use. This parameter also failed to explain a statistically significant amount of variability in the criterion under all methods of estimation.

Another way to explore whether specification errors unduly influenced the analysis is to first calculate the factor loadings associated with the base measurement model and then, rather than estimating these as free parameters in concert with the interactive model, to fix these numeric values in an effort to see if the estimate of the interactive effect is unduly affected. The GENINT program will handle such analyses, appropriately multiplying numeric values for cross-product effects. In one sense, such an analysis can be thought of as somewhat similar to a two-stage least squares solution for the data. Although not presented here in the interests of space, the magnitude of the effect associated with the interactive effect retained its general magnitude and significance under this model as well.

Effects of Outliers. It is possible that a significant interaction effect could be due merely to the presence of a few atypical but influential data points in a study, regardless of the pattern of the scatterplots sorted by expectancies. To examine this possibility, the factorially complex interaction model was rerun, excluding 10% of the individuals in the study who were highest in multivariate nonnormality (Bollen & Arminger, 1991). These models were based on both raw and ranked version of the restricted sample. These analyses resulted in statistical significance for the interaction parameter across both GLS and ML.

SUMMARY

At this point, it seems particularly appropriate to repeat one of he cautions that Jöreskog and Yang (1996) raised regarding use of interactive effects for latent variables. Specifically, they argued that the procedure not be applied routinely to any data, but that a theoretical justification must first be made for its use. For these real-world data, it is difficult to determine the true state of affairs underlying the data. Nevertheless, some analyses appear more reasonable than others, given the pattern of results. First, in contrast to estimates based on simulated data, large differences were found in the linear measurement model across the estimation techniques. In factor models based on raw data, standardized estimates of the contribution of prior Heavy Alcohol Use were higher under ML and WLS estimates than for GLS estimates. For

Alcohol Expectancies, the reverse pattern obtained, with higher standardized estimates associated with GLS solutions than for ML and WLS. Ranking the data appeared to raise the standardized effect of prior Heavy Alcohol Use for GLS solutions to more comparable values found in ML and WLS. Differences in standardized effects are in some part no doubt due to the different metrics of the manifest variables, but ranking and normalizing the data did not reduce these differences entirely. WLS estimates from the factor models revealed only the autoregressive effect of prior Heavy Alcohol Use and failed to detect any significant effect for Alcohol Expectancies. Although use of the rank transform is no panacea (as discussed later), it appears to be useful in the study of interactions by reducing the degree of correlation between the main effect and the interaction latent variable.

In addition to differences in path coefficients, estimates of covariance/correlation between the two predictor constructs varied as a function of estimation technique. For the raw data, the ML estimate of the association between the two predictors was highest, and ranking the data appeared to produce comparable standardized estimates comparable to this value. Differences in covariances between exogenous latent variables, as well as factor variances, have significant implications for the estimation of interactive effect because these characteristics determine the value of the variance for the interaction latent variable, and thus affect the value of the path coefficient associated with the latent variable interaction.

The central question of interest for these analyses, however, is whether Heavy Alcohol Use is a product of prior heavy use and expectancies. Even though one could make a case that the interaction effect is present across the majority of minimization criteria, it does not seem reasonable to think that obtaining a statistically significant result associated with the interaction vector conclusively proves the interaction's existence. A position of methodological skepticism for these data would require the researcher to actively look for alternative explanations for any proposed effect. For these data, for example, it seems reasonable to believe that the reason for the interaction in these data may well be an artifact of the inexperience of some individuals with alcohol's effect. In a slightly related vein, the extreme alcohol consumption patterns of relatively few individuals could make them very influential observations in the statistical models of interactions. Finally, even though the latent variable interaction models proposed in the literature are elegant, the question remains as to whether the effort used to fit the models is worth it, given that the gains in the magnitude of unique effects associated with the interaction terms seem relatively modest compared to the more familiar general linear model/regression approaches.

Although it seems reasonable to mention the pitfalls that may be present in making the inference from a significant interaction effect to actually demonstrating differential trajectories or true moderator effects, it is necessary to keep in mind that the jump from a significant interaction term to a conclusive demonstration of a moderator effect is also difficult for standard multiple regression models employing interaction terms. For multiple regression models, however, these difficulties are present along with model misspecification and misestimation due to measurement error.

When comparing structural models across minimization techniques, an additional point is evident. Across all analyses, WLS estimates of the latent variable interaction effect were consistently small. Clearly, the divergent pattern of results from the WLS technique reiterate the point that much larger sample sizes are required than were available in this study (Muthén & Kaplan, 1992) and echoes Jöreskog and Yang's (1996) recommendation that only large sample sizes be used for such minimization techniques. Alternatively, it may be that the WLS technique is just poorly suited to detect interaction effects of this nature.

DISCUSSION

Even though the real world analyses presented deal with only one study, it is hoped that the variety and extent of the models considered points to the utility of the GENINT program in fitting interactive and quadratic effects. Although a comparison of computer resources necessary to fit latent variable interaction models was not presented here, use of the restricted model, which employed nonlinear constraints for cross-product effects, results in substantial savings. The flexibility and ease of use of the model will also hopefully encourage researchers to consider more realistic measurement models for their data.

Even though the analysis presented here concerns only one data set, some recommendations regarding practice for fitting such latent variable models with interactive effects also appears appropriate. First, researchers who wish to investigate interactive or quadratic effects at the latent variable level should pay particular attention to the linear measurement model before fitting interactive effects. Given that the GENINT program can model interactive effects successfully with factorially complex effects, it seems reasonable to explore this relatively neglected aspect of model construction involving product or quadratic effects. Second, although rank normalizing the data in order to reduce the skew of manifest variables does not appear to improve or alter fit statistics for estimates of linear effects, it appears that the fit of interactive models was improved substantially. (Recall, though, that the evident convergence across ML and GLS estimates may be due to the rescaling of the manifest variables to similar values under the rank normalization.)

Given that standard remedies for discontinuous scaling are not possible in interactive effects, perhaps normalizing techniques should be explored in future work as either a way to secure more consistent solutions for the data or as ancillary analyses to judge the robustness of a proposed effect. Use of more accurate measurement models is of particular value in use of the Kenny and Judd model for latent variable interaction or quadratic effects. The Kenny and Judd model, because it is a full information estimation technique, meaning that misspecification of part of the model may easily bring about bias in estimates associated with other parts of the model (Bollen, 1995).

As mentioned before, the Kenny and Judd factor model is only one possible model of latent variable interaction effects. The two-stage last squares approach of Bollen (1995) appears particularly promising in that it requires less programming effort, does not require normality, allows the researcher to explore whether other effects than simple products and quadratic effects exist and readily allows the researcher to incorporate intercept information into the model. Similarly, Jöreskog and Yang's (1996) extension of the Kenny and Judd model also allows the researcher to incorporate intercept information into the interactive model. Clearly, additional research is needed to inform researchers as to the relative advantages of these approaches as well as their susceptibility to the effects of nonnormality. Using values from the linear measurement model as fixed values, as opposed to start values, allows the researcher to specify two-step approaches to estimation, which may also prove useful in developing and defending models with polynomial effects.

Finally, given that the exploration of interactive and quadratic effects expands to yet another magnitude the possible models that can be used to fit the data, it is important for a researcher to carefully consider only those effects predicted by theory as a base model, and then to consider alternatives to such a model only to the extent that they constitute a plausible alternative explanation. Given the pernicious effects of non-normality on the proper identification of nonlinear factor models, a researcher must be careful to make the case for an interactive effect based on assumptions that appear reasonable and consider additional models or alternative as assumptions as a way of addressing the concerns that a reasonable skeptic might raise about the claimed effect. The examination of factor score plots is one approach that enables a more critical examination of whether a proposed interaction effect is, in fact the most reasonable interpretation to attach to a significant latent variable interaction term. As such, it is hoped that the GENINT program described here will promote efforts at such methodological skepticism regarding proposed structural relationships.

NOTES

Please address correspondence concerning this manuscript to the author at 210 McAlester Hall, University of Missouri–Columbia, Columbia, MO 65211, or via e-mail at: woodpk@missouri.edu. The program and worked examples described in this chapter can be obtained at the home page at: http://www.missouri.edu/~wood

REFERENCES

Anderson, J. C., & Gerbing, D. W. (1988). Structural equation modeling in practice: A review and recommended two-step approach. *Psychological Bulletin, 103*, 453–460.

Baltes, P. B., Reese, H. W., & Nesselroade, J. R. (1977). *Life-span developmental psychology: Introduction to research methods.* Hillsdale, NJ: Lawrence Erlbaum Associates.

Bollen, K. A. (1989). *Structural equations with latent variables.* New York: Wiley.

Bollen, K. A. (1995). Structural equation models that are nonlinear in latent variables: A least squares estimator. In P. V. Marsden (Ed.), *Sociological Methodology, Vol. 25*, pp. 223–251. Washington, DC: American Sociological Association.

Bollen, K. A., & Arminger, G. (1991). Observational residuals in factor analysis and structural equation models. In P. V. Marsden (Ed.), *Sociological Methodology Vol. 21* (pp. 235–262). Washington, DC: American Sociological Association.

Bohrenstedt, G. W., & Goldberger, A. S. (1969). On the exact covariance of products of random variables. *Journal of the American Statistical Association, 52*, 1439–1442.

Busemeyer, J. R., & Jones, L. E. (1983). Analysis of multiplicative combination rules when the causal variables are measured with error. *Psychological Bulletin, 93*, 549–562.

Cooper, M. L., Russell, M., Skinner, J. B., Frone & Mudar, P. (1992). Stress and alcohol use: Moderating effects of gender, coping, and alcohol expectancies. *Journal of Abnormal Psychology, 101*, 139–152.

Friedrich, R. J. (1982). In defense of multiplicative terms in multiple regression equations. *American Journal of Political Science, 26,* 797–833.

Gotham, H. (1995). Relations between personality traits, alcohol expectancies and alcohol use. Unpublished master's thesis, University of Missouri.

Hayduk, L. A. (1987). *Structural equation modeling with LISREL: Essentials and advances.* Baltimore: Johns Hopkins University Press.

Jaccard, J., & Wan, C. K. (1995). Measurement error in the analysis of interaction effects between continuous predictors using multiple regression: Multiple indicator and structural equation approaches, *Psychological Bulletin, 117,* 348–357.

Jöreskog, K. G., & Yang, F. (1996). Nonlinear structural equation models: The Kenny- Judd model with interaction effects. In G. A. Marcoulides & R. E. Schumacker (Eds.), *Advanced structural equation modeling* (pp. 57–88). Mahwah, NJ: Lawrence Erlbaum Associates.

Kendall, M. H., & Stuart, A. (1958). *The advanced theory of statistics (Vol. 1).* London: Griffin.

Kenny, D. A., & Judd, C. M. (1984). Estimating the nonlinear and interactive effects of latent variables. *Psychological Bulletin, 96,* 201–210.

Kushner, M. G., Sher, K. J., Wood, M. D., & Wood, P. K. (1994). Anxiety and drinking behavior: Moderating effects of tension-reduction alcohol outcome expectancies. *Alcoholism: Clinical and Experimental Research, 18,* 852–860.

Loehlin, J. C. (1992). *Latent variables models: An introduction to factor, path, and structural analysis* (2nd ed.). Hillsdale, NJ: Lawrence Erlbaum Associates.

McDonald, R. P. (1980). A simple comprehensive model for the analysis of covariance structures: Some remarks on applications. *British Journal of Mathematical and Statistical Psychology, 33,* 161–183.

Muthén, B., & Kaplan, D. (1992). A comparison of some methodologies for the factor analysis of nonnormal Likert variables: A note on the size of the model. *British Journal of Mathematical and Statistical Psychology, 45,* 19–30.

Ping, R. A. (1995). A parsimonious estimating technique for interaction and quadratic latent variables. *Journal of Marketing Research, 32,* 336–347,

Ping, R. A. (1996). Latent variable interaction and quadratic effect estimation: A two-step technique using structural equation analysis. *Psychological Bulletin, 119,* 166–175.

Sher, K. J., Walitzer, K. S., Wood, P. K., & Bront, E. B. (1991). Characteristics of children of alcoholics: Putative risk factors, substance use and abuse, and psychopathology. *Journal of Abnormal Psychology, 100,* 427–448.

Wood, P. K., & Erickson, D. J. (1998). Estimating interaction and nonlinear effects with SAS. In R. E. Schumacker & G. A. Marcoulides (Eds.), *Interaction and nonlinear effects in structural equation modeling* (pp. 101–124). Mahwah, NJ: Lawrence Erlbaum Associates.

7

Evaluation of D.A.R.E. (Drug Abuse Resistance Education) With Latent Variables in the Context of a Solomon Four Group Design

Jodie B. Ullman
California State University, San Bernardino

Judith A. Stein
University of California, Los Angeles

Richard L. Dukes
Colorado University, Colorado Springs

This chapter outlines the advantages and demonstrates the use of the Solomon Four Group Design in the context of a multiple groups structural equation model (SEM) with latent means. The data are aggregated at the level of the classroom. Hypotheses that test threats to internal validity such as pretest sensitization, history, and maturation are illustrated and analogs in appropriately constrained models are presented. This analytic approach is then demonstrated with data from an ongoing D.A.R.E. evaluation. Pretest sensitization was found for two of the D.A.R.E. core constructs; acceptance of risky behavior and institutional bonds. Classrooms of students who participated in D.A.R.E. reported higher self-esteem, greater resistance to peer pressure, stronger institutional bonds, and were less accepting of risky behaviors.

Drug use prevention programs are difficult to evaluate because of threats to validity such as history, maturation, pretest sensitization, imperfect measurement, attrition, and faulty statistical analyses. When evaluating the impact of drug education programs, it would be useful to employ an analytic technique that can explicitly account for, model, and measure the rapid growth and developmental changes occurring among youth in both experimental and control groups. Changes in youngsters after a drug education intervention may have nothing to do with the program. Conversely, a lack of change in drug attitudes and behaviors may reflect historical trends in society or maturation of the individual young person. Furthermore, even the act of pre-testing youngsters before they enter a drug program may sensitize them to the material that will be forthcoming.

The focus of this study is D.A.R.E. (Drug Abuse Resistance Education). D.A.R.E. is the most widely used school-based drug use prevention program in the United States (Ennett, Tobler, Ringwalt, & Flewelling, 1994). Beginning in Los Angeles, California in 1983, D.A.R.E has grown into an international program that is presented in approximately 20 countries (Becker, Agopian, & Yeh, 1992; Rogers, 1993). In the D.A.R.E. program, a trained police officer delivers the curriculum once a week for a semester to fifth or sixth graders. The D.A.R.E. curriculum is designed to (a) build self-esteem, (b) increase resistance to peer pressure, (c) delay onset of experimentation with alcohol, tobacco, and other drugs, and (d) increase institutional bonds, that is, bonds to parents, teachers, and police.

Past research on the immediate effects of D.A.R.E. has been inconclusive. Immediately following the program, Dukes, Ullman, and Stein (1995) found that students who participated in D.A.R.E. had higher self-esteem, stronger institutional bonds, and endorsed fewer risky behaviors than students who did not participate in the program. However, a 3-year follow-up found no differences between students who had D.A.R.E. and those who did not (Dukes, Ullman, & Stein, 1996). Faine and Bohlander (1989) found that D.A.R.E students had greater self-esteem, more resistance to peer pressure, stronger bonds to police and stronger antidrug attitudes immediately following D.A.R.E. participation, but 1 year later, there were no significant differences between D.A.R.E. and non-D.A.R.E. students. Ringwalt, Ennett, and Holt (1991) found that D.A.R.E students had less positive attitudes toward drugs and greater resistance to peer pressure than non-D.A.R.E. students, but they found no differences in self-reported drug use. Clayton, Cattarello, and Walden (1991) found that D.A.R.E students had stronger antidrug attitudes than did students who had not participated in D.A.R.E. However, Clayton et al. (1991) found no differences in self-esteem, resistance to peer pressure, or reported drug use in D.A.R.E. and non-D.A.R.E. participants.

When assessing the immediate effects of D.A.R.E., Harmon (1993) found that D.A.R.E. participants had stronger antidrug attitudes, were more assertive, and had a stronger belief in prosocial norms than nonparticipants. No differences between participants and nonparticipants were found for self-esteem or reported drug use. Rosenbaum, Flewelling, Bailey, Ringwalt, and Wilkinson (1994) found evidence that gender moderated reported cigarette usage 1 year after participation in D.A.R.E. Specifically, young women who received D.A.R.E. reported that they quit using cigarettes more often than young men who had not received the program.

These inconsistencies in past evaluations of D.A.R.E. led Ennett et al. (1994) to conduct a meta-analysis of eight D.A.R.E. evaluations. Effect sizes were calculated for each of the domains of interest. The largest mean effect size was for knowledge of alcohol, tobacco, and other drugs (.42). The other mean effect sizes were substantially smaller: self-esteem (.06), social skills (.19), antidrug attitudes (.19), positive attitudes toward police (.13), and self-reported drug use (.06). Thus, although the research on the effectiveness of D.A.R.E. has been inconclusive, it may be that D.A.R.E. is moderately effective in presenting knowledge and building social skills and antidrug attitudes. However, it appears less effective in the specific areas it should be targeting, such as reported drug use.

Past evaluations of the D.A.R.E. program have been flawed by limitations in their design and analysis. For instance, D.A.R.E. is implemented in classrooms of students; however, the analysis generally is performed at the level of the individual student (Clayton et al., 1991; see Dukes et al., 1995 for an aggregated classroom example; Faine & Bohlander, 1989; Rosenbaum et al., 1994). Analyzing classroom level data at the level of the individual is problematic. Because the D.A.R.E program is implemented at the classroom level, the children within a classroom potentially are more related to one another than to children in other classrooms. This nonindependence violates fundamental assumptions that underlie the general linear model (Keppel, 1991; Woodward, Bonett, & Brecht, 1990). Therefore, it is statistically inappropriate to analyze data from classroom implementations at the level of the individual student. The solution to this problem employed in this study is analysis of classroom level data.

One obvious problem with this solution is sample size. Many studies do not have the luxury of having enough classrooms to perform a full multivariate analysis. With a small number of aggregated groups, power can be severely diminished. However, using aggregated classroom scores leads to more stable means with less random noise and smaller error variances (Biglan & Ary, 1985). A second problem with this approach is that generalizations can be made only to classrooms of students and not to students as individuals.

Rather than analyze classroom level data, another solution to the problem of level of analysis is to use hierarchical linear modeling techniques (HLM). HLM is variant of the class of *nested models*, in which a model is developed for both the classroom-level data and the individual data. Further discussion of this type of model is outside the scope of this chapter. We refer the reader to the chapter by Khoo and Muthén in this volume for an example of an HLM analysis of clustered data. In addition, Bryk and Raudenbush (1992) present an excellent introduction to HLM.

A second important issue beyond the unit of analysis is that the appropriate variable to employ in hypothesis testing is a latent variable, not necessarily an observed variable. The D.A.R.E. program focuses on: (a) increasing children's self-esteem, (b) increasing students' institutional bonds, (c) increasing resistance to peer pressure, and (d) reducing endorsement of risky drug behaviors. These four foci are hypothetical constructs, and can be represented best with several interrelated measured variables that are subsumed by one overarching construct. Most past research has analyzed the effectiveness of D.A.R.E. via measured variables instead of the constructs. These measured variables may not fully capture the construct at an appropriate level of abstraction. Furthermore, analyzing data based strictly on measured variables leads to an analysis that contains measurement error. Therefore an analysis that is performed at the level of the construct and is free of measurement error is preferable to other approaches.

Finally, D.A.R.E evaluations often lack controls for threats to internal validity such as maturation, history, and pretest sensitization (Cook & Campbell, 1979). Over the semester in which D.A.R.E. is given students attitudes towards drugs, resistance to peer pressure, institutional bonds, and self-esteem may change due to maturational processes within the students. Additionally, during the time period when students are participating in D.A.R.E., societal events may occur that would change drug behaviors even in the absence of D.A.R.E. Occurrences such as these are called history effects. Without a pretest, it is impossible to assess and estimate the degree of maturation or assess any history effects that might have occurred between the time that D.A.R.E. was given to students and the postprogram evaluation. Unfortunately, use of a pretest introduces the possibility of pretest sensitization (Braver & Braver, 1988; Cook & Campbell, 1979). Pretest sensitization refers to possible reactivity to a pretest. Simply participating in the pretest may alter behavior through heightened awareness of various issues regardless of whether or not a student participated in D.A.R.E. For example, the information given on resisting peer pressure may be enough to alter responses at the posttest even for those students who did not participate in D.A.R.E. The Solomon Four Group Design, diagrammed in Fig. 7.1, addresses these problems explicitly (Braver & Braver, 1988; Cook & Campbell, 1979).

THE SOLOMON FOUR GROUP DESIGN AND ANALYSIS

To rectify the limitations in design and analysis and the problems of past research, one remedy may be use of the Solomon Four Group Design with latent variables. A combination of these two approaches can ameliorate several limitations. The Solomon Four Group Design allows for explicit examination of pretest sensitization, history effects, and maturation, as well as treatment effects. Use of latent variables allows tests of hypotheses at the construct level. Furthermore, these constructs are free of measurement error (Bentler, 1995; Bollen, 1986).

	Pretest		Posttest
Group 1	$O_1\ (\mu_{11})$	D.A.R.E.	$O_2\ (\mu_{12})$
Group 2	$O_3\ (\mu_{21})$	No D.A.R.E.	$O_4\ (\mu_{22})$
Group 3		D.A.R.E.	$O_5\ (\mu_{32})$
Group 4		No D.A.R.E.	$O_6\ (\mu_{42})$

FIG. 7.1. Solomon Four Group Design with observations (O_i) and population means (m_{ii}) indicated.

There is some disagreement regarding the appropriate method of analysis for a Solomon Four Group Design (Braver & Braver, 1988; Cook & Campbell, 1979; Huck & Sandler, 1973). A brief review of studies employing Solomon Four Group designs revealed that most studies employed a series of 2 x 2 analysis of variance (ANOVA) on posttest scores (Duryea, 1983; Gibson & Bogat, 1993; Mungas & Walters, 1979; Ragon, Kittleson, & St. Pierre, 1995; Spirito, Overholser, Ashworth, Morgan, & Benedict-Drew, 1988). Although most of these studies employed multiple dependent variables, only few studies explicitly take into account the correlation between these dependent variables (Spirito et al., 1988).

The pretest data generally were not analyzed even when researchers found evidence of pretest sensitization (Spirito et al., 1988). Gibson and Bogat (1993) performed *t* tests on the pretest variables to check randomization, but they did not include pretest information in their treatment effects analysis. Duryea (1983, p. 252) stated "Pretest scores are not analyzed in this design." No mention was made of pretest data in Mungas and Walters (1979), or Ragon et al. (1995). Additionally, in our review of the literature we found no studies that tested for maturation or history effects (see Dukes et al. 1995 for an exception).

Traditional Solomon Four Group Analysis Approach

Pretest Sensitization. This analysis is performed through a 2 x 2 ANOVA of posttest scores. Refer to Fig. 7.2 for a diagram of this analysis. According to Campbell and Stanley (1963), pretest sensitization is present if there is an interaction between presence of pretest and presence of a significant treatment effect. More specifically, pretest sensitization is present if O_2 is more, or less, different from O_4 and O_5 is equal to O_6. The corresponding null hypothesis for this test of pretest by treatment interaction is:

$$H_0\text{: } (\mu_{12} - \mu_{22}) - (\mu_{32} - \mu_{42}) = 0 \qquad (7.1)$$

where the μ_{ii} represent the population means for dependent variables in each cell of the design. The first subscript indicates group membership, and the second subscript indicates time point. Therefore, μ_{42} is the population cell mean for the fourth group at the second time point, that

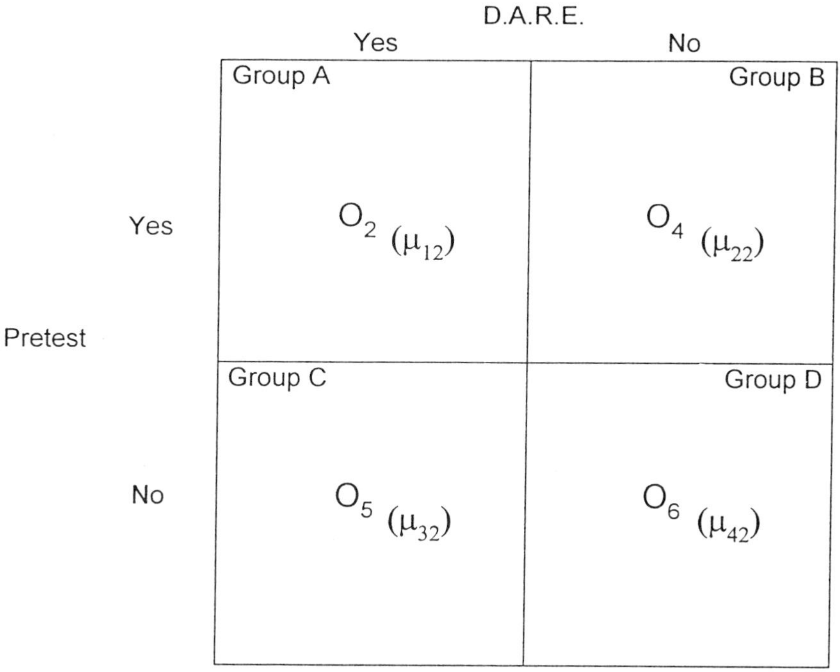

FIG. 7.2. Traditional pretest sensitization analysis with posttest scores.

is, the posttest only, no-treatment group. This subscript notation is used throughout this chapter.

When this pattern of pretest sensitization occurs, the researcher concludes that there is a treatment effect, but it occurs only when the pretest is given (Braver & Braver, 1988; Campbell & Stanley, 1963). Campbell and Stanley (1963) recommended that, under this condition, no further analysis should be performed. This is an extreme test for pretest sensitization, as it does not allow for a treatment effect to be merely enhanced by the pretest. Huck and Sandler (1973) modified the Campbell and Stanley (1963) approach to allow for a treatment effect that is simply enhanced by the pretest. Huck and Sandler (1973) extended Campbell and Stanley (1963) by stating that if the interaction in the 2 x 2 ANOVA on posttest scores is significant, and if O_2 is greater than O_4, and if O_5 is greater than O_6, then there is evidence of pretest sensitization; however, the pretest serves only to enhance the treatment effect.

Treatment Effects. Analysis of treatment effects depends on the presence or absence of pretest sensitization. When pretest sensitization is not present, Campbell and Stanley (1963) suggest looking at the main effect in the 2 x 2 ANOVA on posttest scores. In this analysis, if the main effect for treatment is significant, then there is unequivocal evidence of a treatment effect. Unfortunately, this test is not the most powerful test available because it ignores the pretest information (Braver & Braver, 1988). Therefore, if the result of this test is not significant, the researcher should not infer that there is no treatment effect. An alternative approach would be to perform an ANCOVA on the two pre-post test groups and a *t* test on the posttest only groups. Following these separate analyses, Braver and Braver (1988) suggested combining the results via meta-analytic techniques to achieve maximum power.

When there is evidence of pretest sensitization, the appropriate analysis to evaluate treatment effects is somewhat unclear. Braver and Braver (1988) recommended performing a simple main effect test on the pretested groups (O_2 vs. O_4):

$$H_o\colon \mu_{12} - \mu_{22} = 0 \tag{7.2}$$

and a simple main effect on the post test only groups (O_5 vs. O_6):

$$H_o\colon \mu_{32} - \mu_{42} = 0 \tag{7.3}$$

If each test is significant, then there would be evidence of treatment effects even in the groups not pretested.

Use of the *t* test on scores of the posttest-only groups poses several problems (Cook & Campbell, 1979). Without a pretest as a control, it is impossible to determine if the treatment was associated with any observed differences in posttest scores. Therefore, it is impossible to rule out such threats to validity as history and maturation.

Additionally, the use of an ANCOVA is also potentially problematic. An ANCOVA statistically adjusts groups to make them equivalent (Keppel, 1991). A statistical approach to equate groups is most appropriate in experimental designs in which random assignment has occurred. In many drug use prevention programs, including D.A.R.E., it can be difficult to assign treatments to groups randomly. When nonequivalent groups are statistically equated through ANCOVA methods, the adjusted post test scores may not be interpretable (Tabachnick & Fidell, 1996).

Maturation and History Effects. Specific strategies for assessing history and maturation effects generally are not discussed or tested in traditional Solomon Four Group analyses. However, Campbell and Stanley (1963) suggested that the effects of history could be assessed through a *t* test that compared O_3 vs. O_6.

$$H_o: \mu_{21} - \mu_{42} = 0 \quad (7.4)$$

The effects of maturation could be compared through a repeated measures *t* test comparing O_3 vs. O_4.

$$H_o: \mu_{21} - \mu_{22} = 0 \quad (7.5)$$

AN ALTERNATIVE ANALYSIS APPROACH TO THE SOLOMON FOUR GROUP DESIGN

Our structural equations modeling (SEM) analysis approach differs in several important ways from more traditional approaches. The majority of past methods have been unable to analyze all of the data from the Solomon Four Group Design simultaneously (see Braver & Braver, 1988; Dukes et al., 1995 for an exception). This inability to analyze all of the data simultaneously is a limitation of the traditional approaches to Solomon Four Group analysis. However, with an SEM approach, we are able to estimate multiple group models with appropriate equality constraints, thereby using all of the data and eliminating the problem of wasting valuable data.

In SEM, all four groups of the design are analyzed simultaneously. Therefore, several hypotheses may be tested concurrently. Figure 7.3 presents a simplified diagram for a Solomon Four Group Design specified through SEM with just one construct at each time point. In SEM diagrams, circles represent latent variables (constructs), and rectangles represent measured variables or a constant. A line connecting two variables specifies a direct relationship between the variables. Absence of a line connecting variables signifies that a direct relationship has not been hypothesized. For ease of discussion, the means of latent variables are indicated on the diagram.[1] In our analyses, we examine all four constructs at each time point in each group.

[1]In these models the latent intercepts are also the letent means, and they are labeled μ_{ii} for ease of discussion.

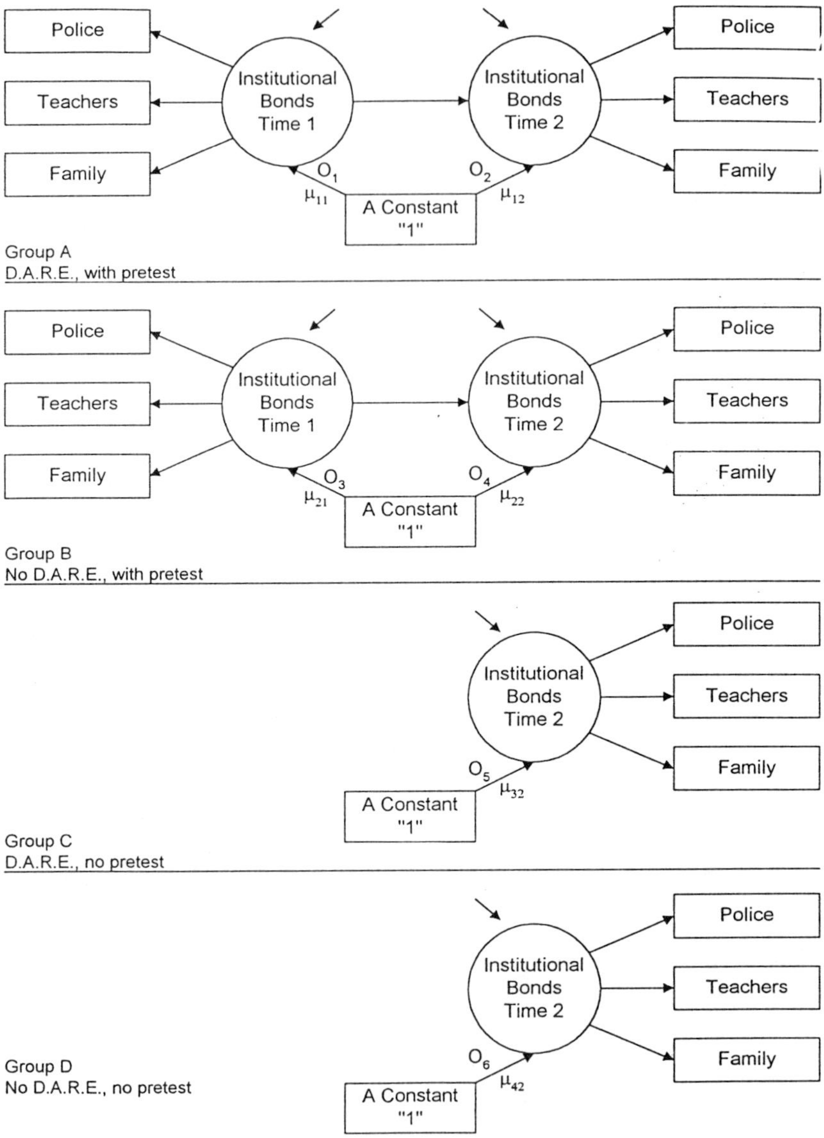

FIG. 7.3. Solomon Four Group specification with one latent variable per time point.

In SEM, hypotheses about latent means can be tested by assessing the significance of the difference between the latent means after the measurement models are shown to be invariant across groups. One or more of the latent means are fixed to a constant, usually 0, and the difference between the fixed mean(s) and other latent means is estimated. SEM computer packages commonly provide the standard errors and appropriate test statistics for the difference between means (Ullman, 1996).

Multivariate and Construct Level Analysis. A second important difference in the SEM method versus the more traditional methods is our use of multiple latent variables in each group and at each time point in the Solomon Four Group Design. Using this approach, we are able to examine the effects of treatment on error-free constructs simultaneously. Additionally, the constructs of interest in the D.A.R.E. program are correlated with one another; therefore, it is an advantage to use a multivariate analysis approach such as SEM that allows explicit estimation of these intercorrelations. Finally, only SEM techniques provide hypothesis testing at the level of the construct.

Additionally, there are at least four other differences in our analytic approach versus the more traditional analytic approaches. These differences described next are not exclusive to SEM techniques. These analyses could be performed using statistical techniques, such as ANOVA/MANOVA.

Randomization Check. Prior to any analyses, we performed a randomization check on pretest scores. If the pretested groups are equivalent at the onset with regard to the target constructs, we are able to combine the pretest data for later analyses. This comparison of O_1 and O_3 translates to the following hypothesis,

$$H_o: \mu_{11} - \mu_{21} = 0 \tag{7.6}$$

Pretest Sensitization. Pretest sensitization is assessed through a test of the interaction between the treatment and the presence of a pretest. As illustrated in Fig. 7.4, pretest sensitization is present if O_2 is significantly greater than O_4 and if O_5 is not significantly different from O_6 :

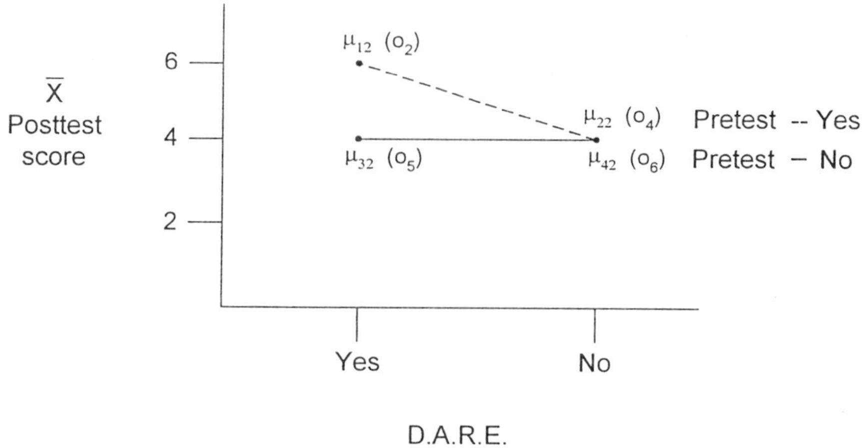

FIG. 7.4. Example of presence of pretest sensitization (Campbell & Stanley, 1963).

$$H_0\text{: } (\mu_{12} - \mu_{22}) - (\mu_{32} - \mu_{42}) = 0 \tag{7.7}$$

History and Maturation. History effects occur when an observed effect may be the result of events unrelated to the treatment having occurred in the same time frame as the treatment (Cook & Campbell, 1979). For example, in drug use prevention studies, history effects could take the form of heightened media or community attention to substance use. If present, these history effects distort the effects of the treatment. History effects are different from maturation. Maturation effects occur when the observed effect may be the result of growth in an individual that is not related to the treatment (Cook & Campbell, 1979). As with history effects, maturation effects may also distort the treatment effect. Referring to Fig. 7.3, both of these threats to validity can be examined simultaneously though examination of the comparison of O_3 and O_4 (maturation) :

$$H_0\text{: } \mu_{21} - \mu_{22} = 0 \tag{7.8}$$

Notice that one difficulty in assessing maturation is that this test is confounded with both history effects and pretest sensitization. If this hypothesis is rejected, it is difficult to make conclusions about maturation effects because of the confounding with both history and pretest sensitization.

History effects can be tested through a comparison of O_3 and O_6, or, if the pretest groups are equivalent, O_1 and O_3 can be combined and compared to O_6,

$$H_0: \mu_{21} - \mu_{42} = 0$$

or (7.9)

$$H_0: \left(\frac{\mu_{11} + \mu_{21}}{2} - \mu_{42}\right) = 0$$

History effects are not confounded with maturation effects because different students are involved in the comparisons. The pretest is given at the beginning of the semester and the posttest at the end of the semester. All of the students are roughly equivalent in age at the pretest and at the posttest.

Treatment Effects. When there is evidence that the pretested groups are equivalent, the data from these groups can be considered together. In this type of SEM model, the combining of the groups is accomplished by fixing both means (μ_{11} and μ_{21}) to 0. The difference between these combined pretest latent means and the posttest latent means then can be estimated. Several hypothesis can be tested simultaneously,

$$H_0: \left(\frac{\mu_{11} + \mu_{21}}{2} - \mu_{12}\right) = 0 \quad (7.10)$$

$$H_0: \left(\frac{\mu_{11} + \mu_{21}}{2} - \mu_{22}\right) = 0 \quad (7.11)$$

$$H_0: \left(\frac{\mu_{11} + \mu_{21}}{2} - \mu_{32}\right) = 0 \quad (7.12)$$

$$H_0: \left(\frac{\mu_{11} + \mu_{21}}{2} - \mu_{42}\right) = 0 \quad (7.13)$$

Equation 7.10 tests the hypothesis that there is no difference between the combined pretested groups and the pre-post test group that had D.A.R.E.. Equation 7.11 tests the hypothesis that there is no difference between the combined pretested groups and the pre-post test group that did not have D.A.R.E. If D.A.R.E. had an effect, given adequate power, we would expect the hypothesis tested in Equation 7.10 to be rejected, and we would expect the hypothesis in Equation 7.11 not to be rejected.

Additionally, we can test the parallel hypotheses using the pretested groups and the posttest-only groups. Equation 7.12 tests the hypothesis that there is no difference between the combined pretested groups and the posttest only group that had D.A.R.E. Equation 7.13 tests the hypothesis that there is no difference between the combined pretested groups and the posttest-only group that did not have D.A.R.E. If D.A.R.E. had an effect, we expect the hypothesis given in Equation 7.12 would be rejected and the hypothesis in Equation 7.13 would not be rejected.[2]

The design implied by testing the hypotheses given in Equations 7.15 and 7.16 is an untreated control group design with separate pretest and posttest samples (Cook & Campbell, 1979). Given comparability of the separate pretest and posttest groups, this design is preferable to the posttest-only design, and it is especially useful when pretest sensitization is present (Cook & Campbell, 1979).

AN EMPIRICAL EXAMPLE

Methods

This study is based on an evaluation of the D.A.R.E. program in Colorado Springs, Colorado. This study spans a 6-year period from 1990 to 1995.[3] As reported in Dukes, et al., (1995, 1996) and Dukes, Stein, and Ullman (1997), the D.A.R.E. program is given one semester per academic year, in which approximately half the schools in Colorado Springs receive D.A.R.E. and half do not. Over 10,000 students from 60 elementary schools in five school districts were included in the evaluation. The entire D.A.R.E. evaluation includes information gathered from students, teachers, parents, and school administrators during the program; however, the study reported in this chapter employed only the self-report information from the students.

[2]Note that Equation 7.13 is identical to Equation 7.9 that tests history effects. Therefore, considered alone this test is confounded and cannot be interpreted unambiguously. This hypothesis given in Equation 7.13 will be rejected when there are history effects but not treatment effects.

[3]Analyses not reported in this chapter show a general cohort effect. Each year, self-esteem of students increased as did their antidrug attitudes and bonds with police, family, and teachers. These effects are not the focus of this chapter, so they are not discussed further.

The D.A.R.E. curriculum lasted one semester. It was delivered to students during the last year of elementary school. As a new cohort began to receive D.A.R.E., students in half of the schools were assigned to the control group. All of the students eventually received D.A.R.E., but only a randomized half of the schools received it in the fall. This situation offered an opportunity to employ the Solomon Four Group Design before the remainder received D.A.R.E. in the spring. Additionally, students in half of the classes at D.A.R.E. schools and half the classes at control schools were randomly assigned to receive a pretest that measured attitudes toward their self, family, teachers, police, peer group, and risky behaviors regarding alcohol, tobacco, and other drugs. At the end of the semester, students in all the classes received a posttest that was identical to the pretest.

Instrument. The questionnaire consisted of a 5-point Likert response scale items that queried the core constructs of D.A.R.E.: (1) opinion of self (Rosenberg, 1965), (2) resistance to peer pressure, (3) bonds with family, police, and teachers, (4) attitude toward use of alcohol, tobacco, or other drugs; and specific aspects of the D.A.R.E. curriculum. The means and standard deviations are presented in Table 7.1.[4] Additionally, the survey queried demographic characteristics, such as academic performance, gender, ethnicity, and occupational prestige of the head of the household.

Classroom Level Data. Data from 597 classes were available for analysis. Seven classrooms in Group A (pre- and posttest group with D.A.R.E.) and 6 classes in Group B (pre- and posttest group without D.A.R.E.) were missing either the pretest or the posttest due to failure of the teacher to administer the tests or loss of the packet in shipment. These 13 (2%) cases were deleted from further analysis. Therefore, the analyses were performed on data from 584 classrooms (Group A, which corresponds to O_1 and O_2 discussed previously [$N = 138$], Group B, O_3 and O_4 [$N = 120$], Group C, O_5 [$N = 162$], Group D, O_6 [$N = 164$]). The average number of students per class was 21.7. If the usable responses

[4]Overall, our sample is not accepting of risky behavior. Although it is possible that our measures of acceptance of risky behaviors are subject to floor effects, we do not think that this is a problem in our analysis. As students enter into the teenage years we would expect an increase in endorsement of risky behaviors in the absence of treatment effects.

in a classroom dropped below nine students, the case was omitted from the analysis.[5]

RESULTS

Model, Estimation, and Assessing Fit. All models are estimated using maximum likelihood estimation in the EQS structural equations program (Bentler, 1995). Goodness of fit for models is assessed on the basis of the chi-square/degrees of freedom ratio and the Comparative Fit Index (CFI; Bentler, 1990, 1995). Chi-square/degrees of freedom ratios equal or less than 2:1 and Comparative Fit Indices of .90 or greater are indicative of good-fitting models. Additionally, we have taken a conservative approach, and interpret coefficients that are statistically different from zero using a two-tailed z test criterion with $\alpha = .05$. We also have chosen not to report effect sizes in this analysis. The issue of appropriate effect size estimation in latent mean structural equation models is outside the scope of this chapter.

Analysis Overview. Confirmatory factor analytic models are tested in each group of the Solomon Four Group Design. After confirmation of the underlying factor structures within each group, a four-group model is estimated to serve as a baseline for later comparisons. After estimation of the baseline model with no equality constraints across groups, the equality of the factor loadings of the four confirmatory models is tested. Next, the equality of the factor regression paths in Groups A and B is tested. These are the regression paths between O_1 and O_2 (Group A) and between O_3 and O_4 (Group B). The significance of the differences between the successively more constrained models is assessed with chi-square difference tests. Following these tests of invariance across groups, latent means are estimated. The hypothesis regarding pretest sensitization was examined through chi-square difference tests and the Lagrange Multiplier test. Finally, hypotheses regarding

[5]The decision to use classrooms with nine or more students was a somewhat arbitrary cutpoint. Our concern was for stability of scores. The distribution of classroom sizes was examined and there was a natural cut-point between classes with at least nine students and those with less than nine students.

TABLE 7.1

Latent variables and measured variables, means, and (standard deviations) employed in analysis of program effects

		Group A D.A.R.E. (N = 138)		*Group B No D.A.R.E. (N = 120)*		*Group C D.A.R.E. (N = 162)*	*Group D No D.A.R.E. (N = 164)*
Latent Variable	*Measured Variable*	*Pretest*	*Posttest*	*Pretest*	*Posttest*	*Posttest Only*	*Posttest Only*
Self-Esteem							
	Rosenberg Self Esteem (items 1–7)	24.74	25.17	24.17	23.80	24.41	23.55
		(2.82)	(3.18)	(2.53)	(2.77)	(3.30)	(3.04)
	Rosenberg Self-Esteem (items 8–10)	8.42	8.84	8.27	8.55	8.49	8.24
		(1.18)	(1.31)	(1.16)	(1.09)	(1.29)	(1.17)
Resistance to Peer Pressure							
	"Don't let other kids talk me into doing things I don't want to do"	3.96	3.99	3.88	3.93	3.98	3.92
		(.32)	(.29)	(.30)	(.39)	(.31)	(.27)
	"Sometimes I do things that are wrong so kids will keep liking me"	3.98	4.04	3.96	4.00	3.97	3.88
		(.33)	(.34)	(.33)	(.35)	(.35)	(.33)
Institutional Bonds							
	"Police prefer to catch you doing wrong/ are not fair dealing with kids"	4.10	4.28	4.06	4.10	4.21	4.01
		(.27)	(.27)	(.25)	(.30)	(.24)	(.32)

Table 7.1 continued

Table 7.1 continued

Latent Variable	Measured Variable	Group A D.A.R.E. (N = 138)		Group B No D.A.R.E. (N = 120)		Group C D.A.R.E. (N = 162)	Group D No D.A.R.E. (N = 164)
		Pretest	Posttest	Pretest	Posttest	Pretest Only	Posttest Only
	"Teachers prefer to catch you doing wrong/are not fair dealing with kids"	4.06	4.09	4.05	3.96	4.02	3.88
		(.26)	(.29)	(.22)	(.35)	(.30)	(.35)
	"Parents are too strict/don't understand/I'm not important to family"	4.07	4.10	4.04	4.00	4.06	3.95
		(.22)	(.27)	(.23)	(.31)	(.25)	(.27)
Acceptance of risky behavior							
	"O.K. to drink alcohol"	1.17	1.16	1.19	1.21	1.15	1.20
		(.14)	(.14)	(.17)	(.17)	(.13)	(.15)
	"O.K. to smoke cigarettes"	1.21	1.20	1.24	1.26	1.19	1.25
		(.16)	(.17)	(.17)	(.19)	(.16)	(.19)
	"O.K. to drink or use drugs at party if all are"	1.18	1.16	1.20	1.22	1.16	1.20
		(.14)	(.14)	(.15)	(.17)	(.13)	(.14)
	"O.K. to smoke pot/try to satisfy curiosity"	1.19	1.18	1.22	1.24	1.17	1.22
		(.13)	(.13)	(.13)	(.14)	(.11)	(.13)
	"Drugs on street safe if know seller"	1.21	1.16	1.19	1.19	1.18	1.20
		(.15)	(.14)	(.15)	(.14)	(.12)	(.13)

maturation, history, and treatments effects are tested based on latent mean differences evaluated with *z* tests.

Confirmatory Single Group Factor Analyses. As an initial step, the factor structure underlying each of the constructs was confirmed in each model. The hypothesized model diagrammed in Fig. 7.5 was estimated in the two pretest and posttest groups (Groups A and B). Figure 7.6 illustrates the hypothesized model that was estimated in the two posttest-only groups (Groups C and D). Table 7.1 presents the latent and measured variable names and the means and standard deviations for the measured variables.

There was strong support of the hypothesized measurement model in each group: Group A, χ^2 (224, N = 138) = 318.09, Comparative Fit Index (CFI) = .95, $p < .05$; Group B, χ^2 (224, N = 120) = 406.95, CFI = .90, $p < .05$; Group C, χ^2 (48, N = 164) = 82.395, CFI = .96, $p < .05$; Group D, χ^2 (48, N = 162) = 119.02, CFI = .92, $p < .05$.

Equality of Factor Models and Regression Coefficients. An initial four-group model was tested, in which all of the factor loadings and regression paths were freely estimated. This unconstrained model provided a baseline for comparison with other more restrictive models in which the factor loadings and regression paths were constrained to equality across groups and time points. There was evidence of an excellent fit in this unconstrained model, χ^2 (544, N = 584) = 926.52, CFI = .93, $p < .05$. Next a model was estimated in which all of the factor loadings were constrained to equality across groups and time points. Again there was evidence of a good fit, χ^2 (584, N = 584) = 980.54, CFI = .93, $p < .05$. Equating the factor loadings across groups and time did not significantly degrade the model, χ_{diff}^2 (40, N = 584) = 53.99, $p > .05$. Given that equating the factor loadings did not degrade the model, we next constrained the factor regression paths in Group A and Group B to equality across groups. This model fit the data, χ^2 (588, N = 584) = 985.64, CFI = .93, $p < .05$. Additionally, these added constraints did not significantly degrade the model fit, χ_{diff}^2 (4, N = 584) = 5.09, $p > .05$.

Randomization Check of Pretested Groups. As a test of the initial equality of the groups at the pretest, the difference between the latent means of each construct at the pretest (Groups A and B) was tested. This procedure corresponds to a test of the following hypothesis:

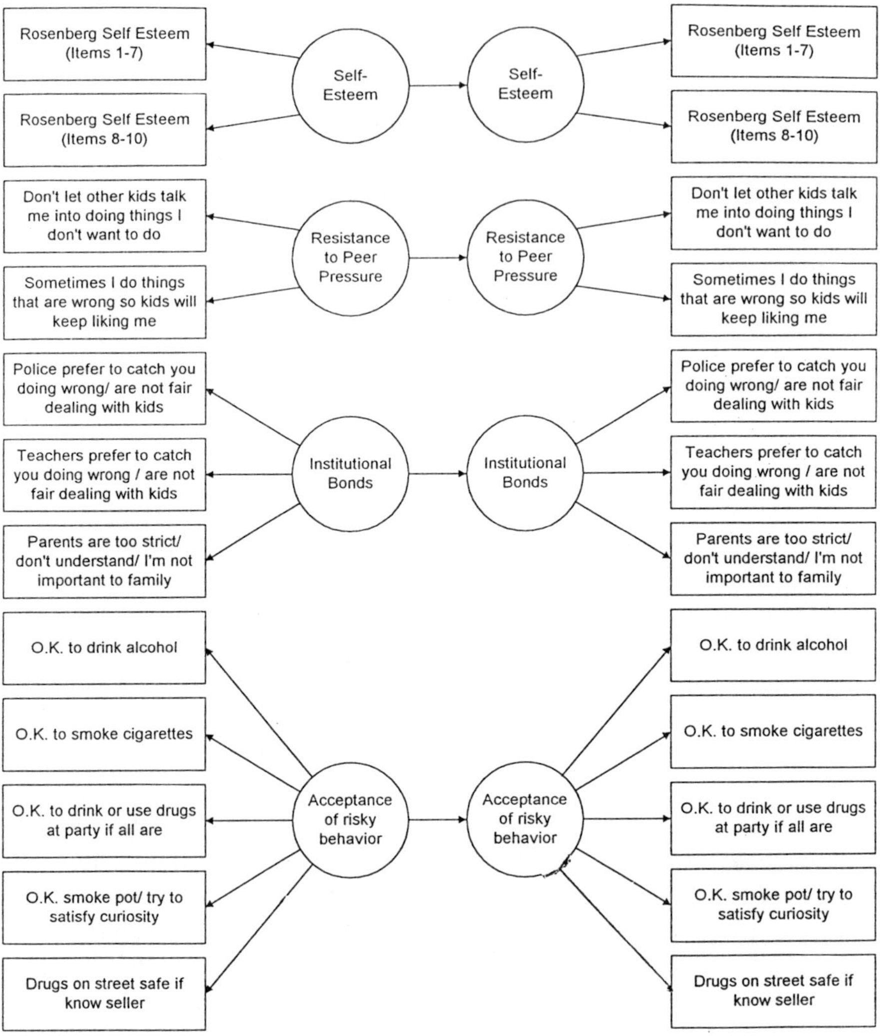

FIG. 7.5. Hypothesized model for pretest and posttest groups (groups A and B). Note: residuals for measured and latent variables and covariances among latent variables were estimated but are not included in the diagram.

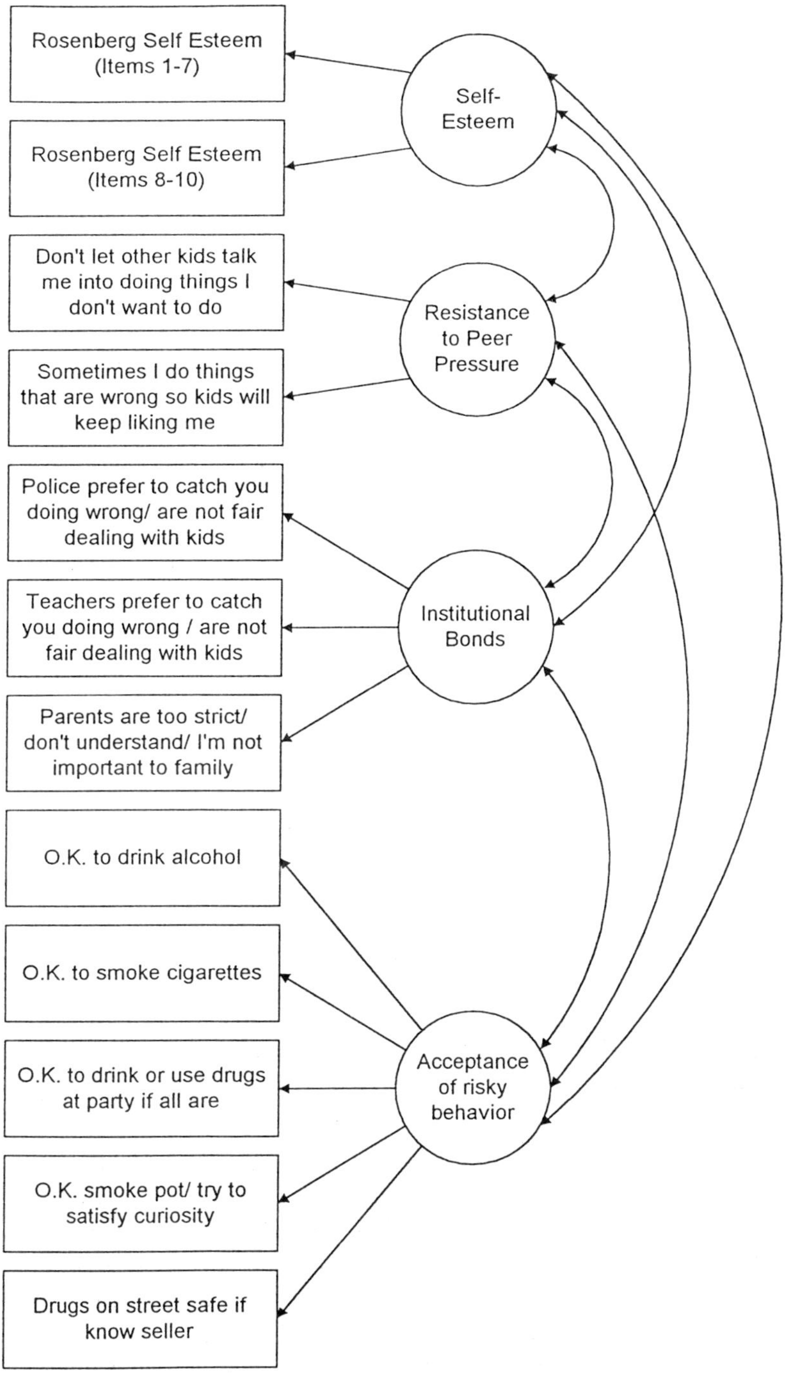

FIG. 7.6. Hypothesized model for posttest-only group (groups C and D). Note: residuals were estimated but are not included in the diagram.

$$H_0: \begin{bmatrix} \mu_{11\ \text{Self-Esteem}} \\ \mu_{11\ \text{Institutional Bonds}} \\ \mu_{11\ \text{Resistance to Peer Pressure}} \\ \mu_{11\ \text{Acceptance of Risky Behavior}} \end{bmatrix} - \begin{bmatrix} \mu_{21\ \text{Self-Esteem}} \\ \mu_{21\ \text{Institutional Bonds}} \\ \mu_{21\ \text{Resistance to Peer Pressure}} \\ \mu_{21\ \text{Acceptance of Risky Behavior}} \end{bmatrix} = 0 \quad (7.14)$$

for each construct simultaneously. In these equations, the first subscript indicates the group and the second subscript indicates the time point. This hypothesis is evaluated with a z test available in EQS. There was no evidence that the pretested groups differed from one another on any constructs at the first time point: Self-Esteem $z = -1.86, p > .05$; Resistance to Peer Pressure $z = -1.21, p > .05$; Institutional Bonds $z = -1.13, p > .05$; Acceptance of Risky Behaviors $z = 1.61, p > .05$. Note that, for the sake of clarity in the text, all latent variables are capitalized.

Pretest Sensitization. Pretest sensitization for the four constructs was tested simultaneously. Therefore, the following hypothesis was tested:

$$H_0: \left(\begin{bmatrix} \mu_{12\ \text{Self-Esteem}} \\ \mu_{12\ \text{Institutional Bonds}} \\ \mu_{12\ \text{Resistance to Peer Pressure}} \\ \mu_{12\ \text{Acceptance of Risky Behavior}} \end{bmatrix} - \begin{bmatrix} \mu_{22\ \text{Self-Esteem}} \\ \mu_{22\ \text{Institutional Bonds}} \\ \mu_{22\ \text{Resistance to Peer Pressure}} \\ \mu_{22\ \text{Acceptance of Risky Behavior}} \end{bmatrix} \right) - \left(\begin{bmatrix} \mu_{32\ \text{Self-Esteem}} \\ \mu_{32\ \text{Institutional Bonds}} \\ \mu_{32\ \text{Resistance to Peer Pressure}} \\ \mu_{32\ \text{Acceptance of Risky Behavior}} \end{bmatrix} - \begin{bmatrix} \mu_{42\ \text{Self-Esteem}} \\ \mu_{42\ \text{Institutional Bonds}} \\ \mu_{42\ \text{Resistance to Peer Pressure}} \\ \mu_{42\ \text{Acceptance of Risky Behavior}} \end{bmatrix} \right) = 0 \quad (7.15)$$

Pretest sensitization was examined through the use of the Lagrange Multiplier (LM) test and chi-square difference tests. A model was estimated that tested Equation 7.15. Then, the LM test was employed to help determine the locus of the pretest sensitization. For the model with the hypothesis given in Equation 7.15, χ^2 $(636, N = 584) = 2450.21, p < .05$. Based on the LM test, constraints implied for Acceptance of Risky Behavior and Institutional Bonds were released and freely estimated. Releasing these constraints led to a significantly improved model, χ_{diff}^2 difference $(2, N = 584) = 14.12, p < .001$. Classrooms of students who received the pretest and participated in D.A.R.E. reported less acceptance of risky behavior and stronger institutional bonds than students who participated in D.A.R.E. but who were not pretested.

Maturation. Maturation was assessed by testing the following hypothesis:

$$H_0: \begin{bmatrix} \mu_{21\,\text{Self-Esteem}} \\ \mu_{21\,\text{Institutional Bonds}} \\ \mu_{21\,\text{Resistance to Peer Pressure}} \\ \mu_{21\,\text{Acceptance of Risky Behavior}} \end{bmatrix} - \begin{bmatrix} \mu_{22\,\text{Self-Esteem}} \\ \mu_{22\,\text{Institutional Bonds}} \\ \mu_{22\,\text{Resistance to Peer Pressure}} \\ \mu_{22\,\text{Acceptance of Risky Behavior}} \end{bmatrix} = 0 \tag{7.16}$$

This hypothesis compares the pretest and posttest scores in Group B and was tested using univariate z tests. No significant effects for maturation were found: Self-Esteem $z = -1.80$, $p > .05$; Institutional Bonds $z = -1.54$, $p > .05$; Resistance to Peer Pressure $z = 1.62$, $p > .05$; Acceptance of Risky Behavior $z = 1.87$ $p > .05$.[6]

History. History effects were examined through a test of the following hypothesis, which compared the combined pretested groups (Groups A and B) with the posttest group that did not receive D.A.R.E. (Group D),

$$H_0: \begin{bmatrix} \dfrac{\mu_{11\,\text{Self-Esteem}} + \mu_{21\,\text{Self-Esteem}}}{2} \\ \dfrac{\mu_{11\,\text{Institutional Bonds}} + \mu_{21\,\text{Institutional Bonds}}}{2} \\ \dfrac{\mu_{11\,\text{Resistance to Peer Pressure}} + \mu_{21\,\text{Resistance to Peer Pressure}}}{2} \\ \dfrac{\mu_{11\,\text{Acceptance of Risky Behavior}} + \mu_{21\,\text{Acceptance of Risky Behavior}}}{2} \end{bmatrix} - \begin{bmatrix} \mu_{42\,\text{Self-Esteem}} \\ \mu_{42\,\text{Institutional Bonds}} \\ \mu_{42\,\text{Resistance to Peer Pressure}} \\ \mu_{42\,\text{Acceptance of Risky Behavior}} \end{bmatrix} = 0 \tag{7.17}$$

This hypothesis was tested using a z test. History effects were found for three of the constructs: Self-Esteem $z = -2.95$, $p < .05$; Resistance to Peer Pressure $z = -2.15$, $p < .05$; Institutional Bonds $z = -5.07$, $p < .05$. These results indicate that, between the time of the pretest and the posttest, self-esteem, resistance to peer pressure, and institutional bonds decreased, but acceptance of risky behaviors by students did not change, $z = 1.35$, $p > .05$.

[6]Note this hypothesis confounds history and pretest sensitization with maturation.

Treatment Effects. Because treatment effects were assessed in both the pretest and the posttest-only groups, two hypotheses assessed these effects. These hypotheses are the same as those given in Equations 7.10 and 7.12.[7] Equations 7.18 and 7.19 make the general equations given in Equations 7.10 and 7.12 specific to the exact D.A.R.E. constructs examined in this analysis. Equation 7.18 is the exact specification of Equation 7.10 and Equation 7.19 is the exact specification of Equation 7.12. The first hypothesis (derived from general Equation 7.10) compared the combined pretested groups (Groups A and B) with the posttest group that received D.A.R.E. (Group A):

$$H_0: \begin{bmatrix} \dfrac{\mu_{11\text{ Self-Esteem}} + \mu_{21\text{ Self-Esteem}}}{2} \\ \dfrac{\mu_{11\text{ Institutional Bonds}} + \mu_{21\text{ Institutional Bonds}}}{2} \\ \dfrac{\mu_{11\text{ Resistance to Peer Pressure}} + \mu_{21\text{ Resistance to Peer Pressure}}}{2} \\ \dfrac{\mu_{11\text{ Acceptance of Risky Behavior}} + \mu_{21\text{ Acceptance of Risky Behavior}}}{2} \end{bmatrix} - \begin{bmatrix} \mu_{12\text{ Self-Esteem}} \\ \mu_{12\text{ Institutional Bonds}} \\ \mu_{12\text{ Resistance to Peer Pressure}} \\ \mu_{12\text{ Acceptance of Risky Behavior}} \end{bmatrix} = 0 \quad (7.18)$$

Again this hypothesis was tested using z tests. Immediately following the D.A.R.E. program, classrooms of students reported higher Self-Esteem $z = 5.33$, $p < .05$; greater Resistance to Peer Pressure $z = 2.76$, $p < .05$; stronger Institutional Bonds $z = 4.96$, $p < .05$; and less Acceptance of Risky Behaviors $z = -2.20$, $p < .05$ than classrooms of children who did not participate in the D.A.R.E. program.

Although Equation 7.18 is a more powerful test than the hypothesis given in Equation 7.19, due to the pretest sensitization, treatment effects

$$H_0: \begin{bmatrix} \dfrac{\mu_{11\text{ Self-Esteem}} + \mu_{21\text{ Self-Esteem}}}{2} \\ \dfrac{\mu_{11\text{ Institutional Bonds}} + \mu_{21\text{ Institutional Bonds}}}{2} \\ \dfrac{\mu_{11\text{ Resistance to Peer Pressure}} + \mu_{21\text{ Resistance to Peer Pressure}}}{2} \\ \dfrac{\mu_{11\text{ Acceptance of Risky Behavior}} + \mu_{21\text{ Acceptance of Risky Behavior}}}{2} \end{bmatrix} - \begin{bmatrix} \mu_{32\text{ Self-Esteem}} \\ \mu_{32\text{ Institutional Bonds}} \\ \mu_{32\text{ Resistance to Peer Pressure}} \\ \mu_{32\text{ Acceptance of Risky Behavior}} \end{bmatrix} = 0 \quad (7.19)$$

[7]We have not included tests of Equations 7.11 and 7.13 in this section.

also were assessed with a different hypothesis (derived from general Equation 7.12):

This hypothesis compares the combined pretested groups (Groups A and B) with the posttest-only group that received D.A.R.E. (Group C). Using z tests, only one significant difference was detected: Students who received D.A.R.E. reported less endorsement of risky behaviors than students on the pretest, Acceptance of Risky Behaviors $z = -2.60$, $p < .05$. No other significant differences were detected: Self-Esteem $z = -.142$, $p > .05$; Resistance to Peer Pressure $z = .737$, $p > .05$; and Institutional Bonds $z = 1.66$, $p > .05$.

DISCUSSION

This chapter illustrates the evaluation of a drug use prevention program, D.A.R.E., using a combination of a Solomon Four Group Design and latent variable models. Using SEM multisample techniques to analyze a Solomon Four Group Design had several advantages. For instance, no data were wasted, that is, all the data were used in the analyses. Data collection in a Solomon Four Group Design can be expensive and time-consuming. Therefore, an analysis technique that makes use of all of the data is a strong advantage. Another advantage of using the Solomon Four Group Design is the ability to test hypotheses regarding various threats to validity. This chapter demonstrates specification of hypotheses about latent means through SEM techniques that allow one to partition and to explicate effects of pretest sensitization, history, and maturation, as well as treatment effects. Furthermore, SEM analyses allow for a multivariate examination of theoretically error-free constructs. The observed variables employed in drug prevention program research are measured with error. This error may distort treatment effects. Another advantage of the SEM approach is the ability to test hypotheses at an appropriate level. Hypotheses generally are made at the level of the construct; therefore, a construct level analysis is highly desirable and may be more meaningful.

Data gathered from a large D.A.R.E. program in Colorado Springs were employed as an empirical example of these techniques. These data were analyzed at the classroom level. Pretest sensitization was found for two of the D.A.R.E. constructs: Acceptance of Risky Behaviors and Institutional Bonds. Following participation in D.A.R.E., the classrooms of students who took the pretest had less acceptance of risky behaviors and stronger institutional bonds than those classrooms of students who received the D.A.R.E. curriculum but did not receive the pretest. No evidence of pretest sensitization was found for Self-Esteem or Resistance to Peer Pressure.

There was also evidence of history effects but no evidence of maturation effects. At first glance, this seems somewhat counterintuitive. History effects, which were assessed by comparing latent mean differences between classrooms of students who took the pretest with classrooms of students who took only a posttest, were found for self-esteem, resistance to peer pressure, and institutional bonds. We found that, over time, students in non-D.A.R.E. classrooms reported less self-esteem, less resistance to peer pressure, and decreased institutional bonds than classrooms of different students prior to D.A.R.E. No maturation effects were found for any of the D.A.R.E. constructs. Maturation was assessed by comparing pre- and posttest scores of classes that did not receive D.A.R.E.

The test of maturation effect is confounded with both history and pretest sensitization. The same three constructs (Self-Esteem, Resistance to Peer Pressure, and Institutional Bonds) were affected by both pretest sensitization and history. Pretest sensitization increased mean levels of each construct, and history effects showed a decrease in mean levels of each construct. Therefore, if pretest sensitization was increasing mean levels of these constructs as history was decreasing the mean levels of the same constructs, any differences in maturation could be obscured. This confounding is a limitation of analysis of the Solomon Four Group Design. However, our findings point out the complexity of evaluating program effects among young people who are experiencing rapid development and changes in their attitudes and lives. Therefore, it may be particularly important to test for pretest sensitization effects when studying this age group.

The presence of the history effects may also signal presence of a randomization failure. Perhaps the classrooms in the posttest-only group are systematically different from the other groups. Specifically, perhaps these classrooms, had they been pretested, would have scored lower than the others at the pretest.

Two different hypotheses were employed to examine treatment effects. Past Solomon Four Group analyses often have employed combinations of ANCOVA and posttest-only *t* tests to assess treatment effects. As discussed in the introduction, both of these approaches are problematic. In this study, we assessed treatment effects with two smaller research designs derived from the Solomon Four Group Design. First, using a pre-post untreated control group design, we found that classrooms of students who received D.A.R.E. reported higher self-esteem, greater resistance to peer pressure, stronger institutional bonds, and were less accepting of risky behaviors. Unfortunately, conclusions regarding self-esteem, resistance to peer pressure, and institutional bonds were ambiguous due to the demonstrated presence of both pretest sensitization and history effects. Given this ambiguity, treatment effects also were assessed through the use of an untreated control group, with separate pre- and posttest samples derived from the Solomon Four Group Design. Using this design, treatment effects were detected only for Acceptance of Risky Behaviors. Specifically, following participation in D.A.R.E., classrooms of students were less accepting of risky behaviors. Because this effect was detected in both treatment effects analyses, we can conclude more confidently that immediately following D.A.R.E., acceptance of risky behaviors is reduced.

This chapter demonstrates a combination of analytic techniques that may be particularly beneficial to evaluators of drug use prevention programs geared toward young people. During the time in young people's lives when they are participating in the D.A.R.E. program, they are also experiencing major life changes. These children are maturing into young adults and are becoming more aware of both themselves and their surroundings. During this time period, students may be particularly influenced by changes in themselves and their environments. These multiple forces of change need to be considered in evaluations that are targeted to this age group, especially because many changes may be in the direction of greater rebelliousness or antisocial attitudes and behaviors. The Solomon Four Group Design is particularly useful with this age group because of the inclusions of specific tests for maturation, history, and pretest sensitization, as well as multiple tests of treatment effects.

There are limitations, however, to the use of the Solomon Four Group Design. Maturation and history threats to validity are confounded, and therefore, are difficult to interpret unambiguously. Other designs, such as the cohort sequential design, may be better choices for evaluating programs in which these threats to validity are strong and there is a need to disentangle the effects of separate threats. A second limitation in this chapter is the use of classroom-level data. By using the classroom as the unit of analysis, we were unable to examine potential moderators at the individual level. It might be reasonable to suggest that such variables as gender and ethnicity may moderate the effectiveness of D.A.R.E. In principle, the approach demonstrated here could be combined with a HLM approach in an SEM context. Combining these two techniques would allow individual moderators to be examined. Demonstration of this approach was outside the scope of this chapter.

In conclusion, through the use of multiple group SEM in the context of a Solomon Four Group Design, we have demonstrated a technique to reduce several analytic and methodological problems. We would encourage applied researchers to employ these techniques to partially disentangle the complex results of assessment of drug prevention programs in populations experiencing many competing sources of change.

ACKNOWLEDGMENTS

This research was supported in part by funds from the El Pomar Foundation, the Colorado Springs Public Schools, the Colorado Springs Police Department, and Grant DA01070-24 from the National Institute on Drug Abuse. Views expressed in the paper are those of the authors. They do not necessarily reflect official positions of the organizations mentioned above. The authors gratefully acknowledge the administrative and technical assistance of Wendy Sallin, Sandra Matthews, Cynthia Dukes, Julio Angulo, and Gisele Pham.

REFERENCES

Becker, H. K., Agopian, M. W., & Yeh S. (1992). Impact evaluation of Drug Abuse Resistance Education (DARE). *Journal of Drug Education, 22*, 283–291.

Bentler, P. M. (1990). Comparative fit indexes in structural models. *Psychological Bulletin, 107*, 238–246.

Bentler, P. M. (1995). *EQS: Structural equations program manual*. Encino, CA: Multivariate Software Inc.

Biglan, A. & Ary, D. V. (1985). *Methodological issues in research on smoking prevention*. National Institute on Drug Abuse Research Monograph Series Monograph 63:170–195.

Bollen, K. A. (1986). *Structural equations with latent variables*. New York: Wiley.

Braver, M. C., & Braver, S. L. (1988). Statistical treatment of the Solomon Four-Group design: A meta-analytic approach. *Psychological Bulletin, 104*, 150–154.

Bryk, A. S., & Raudenbush, S. W., (1992). *Hierarchical linear models: Applications and data analysis methods*. Newbury Park, CA: Sage.

Campbell, D. T., & Stanley, S. L. (1963). *Experimental and quasi-experimental designs for research*. Chicago: Rand McNally.

Clayton, R. R., Cattarello, A., & Walden, K. P. (1991). Sensation seeking as a potential mediating variables for school-based prevention interventions: A two year follow-up of DARE. *Journal of Health Communications, 3*, 229–239.

Cook, T. D., & Campbell, D. T. (1979). *Quasi-experimentation: Design and analysis issues for field settings*. Boston: Houghton Mifflin.

Dukes, R. L., Stein, J. A., & Ullman, J. B. (1997) Long-term impact of Drug Abuse Resistance Education (D.A.R.E.): Results of a 6-year follow-up. *Evaluation Review, 21*, 483–500.

Dukes, R. L., Ullman, J. B., & Stein, J. A. (1995). An evaluation of D.A.R.E. (Drug Abuse Resistance Education) using a Solomon Four Group design. *Evaluation Review, 19*, 409–435.

Dukes, R. L., Ullman, J. B., & Stein, J. A. (1996). Three-year follow-up of drug abuse resistance education (D.A.R.E.). *Evaluation Review, 20*, 49–66.

Duryea, E. J. (1983). Utilizing tenets of inoculation theory to develop and evaluate a prevention alcohol education intervention. *Journal of School Health, 53*, 250–256.

Ennett, S. T., Tobler, N. S., Ringwalt, C. L., & Flewelling, R. (1994). How effective is Drug Abusive Resistance Education? A meta-analysis of Project DARE outcome evaluations. *American Journal of Public Health, 84*, 1394–1401.

Faine, J. R., & Bohlander, Jr., E. (1989). *DARE in Kentucky schools 1988–89: An evaluation of the Drug Abuse Resistance Education Program.* Bowling Green, KY: Western Kentucky University Social Research Laboratory.

Gibson, G., & Bogat, G. A. (1993). Pretesting effects in evaluation of a sexual abuse education program for preschool children. *Journal of Child Sexual Abuse, 2*, 15–31.

Harmon, M. A. (1993). Reducing the risk of drug involvement among early adolescents. *Evaluation Review, 17*, 221–239.

Huck, S. W., & Sandler, H. M. (1973). A note on the Solomon 4-group design: Appropriate statistical analyses. *Journal of Experimental Education, 42*, 54–55.

Keppel, G. (1991). *Design & analysis: A researcher's handbook* (3rd ed.) Englewood Cliffs, NJ: Prentice-Hall.

Mungas, D. M., & Walters, H. A. (1979). Pretesting effects in evaluation of social skills training. *Journal of Consulting and Clinical Psychology, 47*, 216–218.

Ragon, B. M., Kittleson, M. J., & St. Pierre, R. W. (1995). The effect of a single affective HIV/AIDS education program on college students' knowledge and attitudes. *AIDS Education and Prevention, 7*, 221–231.

Ringwalt, C., Ennett, S. T., & Holt, K. D. (1991). An outcome evaluation of project DARE (Drug Abuse Resistance Education). *Health Education Research: Theory and Practice, 6*, 327-337.

Rogers, E. M. (1993). Diffusion and re-invention of project DARE. In T. E. Backer & E. M. Rogers (Eds.)., *Organizational aspects of health communication campaigns: What works?* (pp. 139-162). Newbury Park, CA: Sage

Rosenberg, M. (1965). *Society and the adolescent self-image*. Princeton, NJ: Princeton University Press.

Rosenbaum, D. P., Flewelling, S. L., Bailey, C. L., Ringwalt, C. L., & Wilkinson, D. L. (1994). Cops in the classroom: A longitudinal evaluation of Drug Abuse Resistance Education (DARE). *Journal of Research in Crime and Delinquency, 31*, 3–31.

Spirito, A., Overholser, J., Ashworth, S., Morgan, J., & Benedict-Drew, C. (1988). Eva;iatopm pf a siocode awaremess curriculum for high school students. *Journal of the American Academy of Child & Adolescent Psychiatry, 27*, 705–711.

Tabachnick, B. G., & Fidell, L. S. (1996). *Using multivariate statistics*, (3rd ed.). New York: HarperCollins College Publishers.

Ullman, J. B. (1996) Structural Equations Modeling. In B. G. Tabachnick & L. S. Fidell (Eds.), *Using multivariate statistics* (3rd ed.); (pp. 709–812). New York: HarperCollins College Publishers.

Woodward, J. A., Bonett, D. G., & Brecht, M-L. (1990). *Introduction to linear models and experimental design*. San Diego: Harcourt Brace.

8

Epidemics of Smoking: Modeling Tobacco Use Among Adolescents

David C. Rowe
University of Arizona

Joseph L Rodgers
University of Oklahoma

Michael Gilson
University of Arizona

Smoking behavior can be modeled using an Epidemic Modeling of Social Activities (EMOSA) approach. The approach borrows from both the mathematics of epidemic disease and that of the diffusion of social innovations to describe how naive nonsmokers become first experimental smokers and then regular smokers. In some models, the possibilities of quitting and relapse are also included. The models consist of difference equations that predict the prevalence of individuals at each smoking stage. In their application to data on smoking, two main discoveries have emerged. The first is that the transition from naive nonsmoker to experimental smoker requires social contacts; in the model's terminology, it is a prevalence-driven transition. In contrast, social influence had little effect on the transition from experimental smoker to regular smoker. EMOSA models were also applied to the spread of cigarette smoking among Hispanic and Black females. The models particularly suggested that Black girls are less susceptible to psychological or physiological addiction to nicotine.

Beyond the predictions of tribal rainmakers, the weather can be forecast in two distinct ways. One is to look at the averages over the last couple of years and to predict, for any day, cloudiness and rain on the basis of the past history. Another method is to use a large computer that contains a mathematical model of the processes generating weather. The model advances from starting conditions, the situation in the world today, to the weather a few days or weeks hence. As simulation models have become better, forecasters have become bolder: Weather can now be forecast 5 days in advance.

These two methods–prediction and simulation–have their analogues in the social sciences. The closest equivalence to prediction is to use linear models. A linear model can be completely atheoretical, like looking up a past weather history. Consider the application of linear models to the prediction of adolescent smoking. If a model works, it gives useful information about who will smoke among adolescents. Thus, we might predict smoking among adolescents from parental smoking and social class, undoubtedly with some success. Nonetheless, a linear model is not a simulation because it does not reproduce the processes of social interaction and their effects; it cannot incorporate the many processes that create the observed relationships directly. We discuss this important distinction in more detail in Rodgers, Rowe, and Buster (1998).

In this chapter, we discuss our Epidemic Models of the onset of social activities (EMOSA). Like linear models, EMOSA models can yield predictions about behavior. Unlike linear models, they are also simulations of the processes of social interaction that lead to behavioral outcomes. EMOSA models borrow their mathematics from that used in models of both epidemic disease and the diffusion of social innovations (Bailey, 1975; Mahajan & Peterson, 1985; Rogers, 1983). The models consist of a set of difference equations. The equations are used to predict the proportion of individuals of all ages who display a particular social activity (e.g., sexual intercourse, drinking alcohol, or smoking cigarettes) from data on the proportion of the youngest members of a cohort who have adopted the activity. The predictions include a social influence process because the model includes the effect of a behavior being transmitted in a face-to-face, interacting population. These models capture the idea that the opportunity for many behaviors depends on the mere prevalence of the behavior in a population; social diffusion of a behavior is more rapid when many people have already acquired a behavior than when few have acquired it. Parameters in the model represent the rate of diffusion of the social activity. Many processes are implicit in these parameters. To capture how a social process works, separate parameters can be estimated for subdivisions of the population, and the change in the parameters can be compared with theoretical predictions. In our applications, we have examined the effect of such moderators as parental smoking. However, our technique can be extended to a variety of social and individual difference variables, for example, the effect of religiosity on rates of diffusion.

Some scholars might raise the issue of whether these models are truly multivariate. To determine the multivariate status of linear models, it is relatively straightforward to count the number of dependent and independent variables. This process is not so straightforward in nonlinear dynamic systems models because the parameters in these models imply many interacting processes. For example, a transition parameter might implicitly reflect personality traits and parental influences. As noted previously, these processes can be made explicit when samples are subdivided and a separate parameter is estimated for each subgroup (e.g., separate transition parameters for high religious and low religious subgroups). Thus, in our smoking models, there is one obvious dependent variable (smoking status), several potential independent variables (sex, parental smoking, age, race), and many interrelated processes at work.

First, we discuss how an epidemic model can capture some social processes and detail their assumptions. Second, we present some findings from our previous research on smoking. Third, we illustrate the EMOSA approach with new data on the smoking of Black and Hispanic girls. Finally, we conclude with some thoughts on improving and extending the EMOSA modeling approach.

THE CASE FOR SMOKING AS AN EPIDEMIC PROCESS

A short history of smoking illustrates the *epidemic* properties of smoking (Corti, 1932; Goodman, 1993). The smoking of tobacco was invented by the Native Americans. In 1492, Christopher Columbus's crew first observed smoking among the Arawak people of the Caribbean. Columbus was offered dried tobacco leaves as a present. From these early explorations, smoking was brought to Europe in the 1550s. Sir Walter Raleigh, who was later to spend many years in a Tower of London prison (but for other offenses), helped to popularize pipe smoking among the English aristocracy. He even convinced the Queen to smoke, and soon smoking had spread among all classes of English society. As more people smoked, its hazards became apparent: Careless smokers could burn down whole neighborhoods. A fire in Constantinople in 1633 led to outlawing smoking in a campaign that included the death penalty. Yet, it was not eliminated; public concern failed to halt the spread of smoking. In an era in which communication across oceans was no faster than a ship under sail, smoking made its global spread in under 100 years. Each culture that encountered tobacco adopted it and accommodated to it. The Persians most likely invented the water pipe for cooling tobacco smoke; the Arabs popularized it. In the modern era, China produces 38% of the world's tobacco crop, and the United States produces 10%. Between 1985 and 1988, Americans smoked about 2,900 cigarettes per capita.

In its speed and extent of spread, smoking rivaled the spread of true epidemic diseases. In each population it encountered, smokers brought tobacco products to nonsmokers, and induced the latter to try it. Even in places without the benefit of newspapers, radio, and television, smoking could be spread by the face-to-face contacts among smokers and nonsmokers. Its strong appeal is also apparent from its history. Once smoking had obtained a foothold in a minority of a country's people, it was nearly impossible to eliminate (woe to California today). Smokers had a strong craving for tobacco, once they were hooked, and they demanded more. In the terminology of infectious disease, smoking became an *endemic* condition, self-sustaining in any sizable human population.

Nonetheless, no one is born a smoker. To be maintained, new smokers must be constantly recruited from among young people, mainly adolescents because for the most part smoking is initiated only by the young. One exception may be conditions of war; when Sarajevo was under sniper and artillery fire, newspapers reported that adults started smoking. Under peacetime conditions, however, almost all newly initiated smokers are teenagers. In the United States, where smoking prevalence in the big state of California has fallen to a mere 20%, older smokers are mainly trying to quit.

Because new children are constantly initiated into smoking, EMOSA models are a good way to think about the process involved. As noted previously, we coined the acronym Epidemic Models of Onset of Social Activities (EMOSA) for models that use epidemic-like principles to explain the spread of smoking behaviors that are socially influenced, including drinking of alcoholic beverages, and sexual intercourse (Rodgers & Rowe, 1993; Rowe & Linver, 1995).

In this chapter, we focus on the spread of cigarette smoking, first looking at findings from our three previous studies of smoking and then presenting some new findings from the National Longitudinal Study on Adolescent Health (Add Health). The first study fit smoking data from cities in North Carolina and Florida (Rowe & Rodgers, 1991). The model was fit to just a single stage of smoking, the transition from *nonsmoker* to *ever smoked*. The next two studies fit EMOSA models to smoking of a largely white, middle class sample in Bloomington, Indiana (Rowe, Chassin, Presson, Edwards, & Sherman, 1992; Rowe, Chassin, Presson, & Sherman, 1996). The sample was a longitudinal cohort, so extensive data existed for each respondent. The first of the Bloomington studies fit a staged sequence of smoking transitions, from nonsmoker through quitting and relapse. The second added another dimension to the model, namely, whether the parents of the adolescents were smokers. In each case, the model predicts the proportion of individuals at the different stages (i.e., the proportion of smokers), and the goodness-of-fit of the model was evaluated by comparing the predicted proportions to the observed ones.

When we use EMOSA models in this chapter, we intend neither a moral nor a medical judgment about smoking. Smoking is like an epidemic infectious disease insofar as the dynamics of its spread and maintenance in a population may be *modeled* by the same set of equations as infectious diseases. To distinguish *social* from *biological* infection, in the remainder of this article we use the nondisease vocabulary for aspects of the model (i.e., *experienced*, not *infecteds*).

The Basic Dynamics of Epidemic Model of Smoking

In our first study to examine smoking, the data were prevalence curves from Raleigh, North Carolina, Tallahassee, Florida, and Baltimore, Maryland (Rowe & Rodgers, 1991). The data were prevalences of having *ever smoked* a cigarette. For example, in Baltimore, 7% of Black males had ever smoked at age 12, but 33% had ever smoked by age 17. The prevalence curves varied by location and race. Blacks ages 12 to 17 years smoked less than Whites 12 to 17 years old. Smoking was greater in the south than in Baltimore. An EMOSA model of smoking fit the data well. One weakness in this study was that these data could not distinguish among stages of smoking.

In our EMOSA models on data from Bloomington, Indiana, we distinguished between experimental smokers and regular smokers (Rowe et al., 1992; Rowe et al., 1996). In a typical Bloomington prevalence curve, experimental smoking increased from about 20% in sixth grade to 40% in 11th grade; regular smoking increases from about 0 to 8% over the same period. These figures, of course, depend on the exact definition of experimental and regular smoking, as well as on the location and period of the sample. The general pattern, however, of increasing prevalence with grade-level from junior high through high school is universal in the United States, and probably in most countries. EMOSA modeling attempts to reproduce, in part, the social influence processes that generate these prevalence changes.

Although EMOSA models can be mathematically complex in an effort to capture many subtleties of social processes (Huckfeldt, Kohfeld, & Likens, 1982; Mahajan & Peterson, 1985; Rodgers & Rowe, 1993), an idea that is common to many such models is the distinction between prevalence-driven processes and constant rate processes. In the real world, prevalence-driven processes mean that people meet and influence one another. In these meetings, some people can be designated as *naive* and others as *experienced*. In most EMOSA models, the mixing of people occurs randomly. This assumption is a strong one, and in some ways unrealistic; we will return to discuss its strengths and limitations below.

In the smoking application, the prevalence-driven processes refer to the contact between an experienced cigarette user and the naive, nonsmoker (to date, we have not modeled the use of chewing tobacco). Constant rate processes, on the other hand, involve no individual-level social contagion or social influence. Exposure to media advertisements or to smoking in movies and TV programs may act as a constant rate influence, because it would not be crucial for a naive person to meet directly with an experienced smoker for the naive person to adopt smoking. In contrast, we assume that, in many cases, the first use of smoking occurs in the former type of face-to-face meeting, where an experienced user makes cigarettes available to a nonsmoker in a setting conducive to first use, away from watchful parental eyes.

There are data to support this view of smoking initiation (Friedman, Lichtenstein, & Biglan, 1985). Teenagers were asked about the situations in which they had first used cigarettes. In the typical scenario, Bob or Sue is at home in the afternoon with a friend. The friend offers a cigarette, most often purchased from a store or vending machine, with the friend either saying nothing or offering some encouragement, like "why don't you try this?" Interestingly, the recipients of the cigarettes also seem to differ initially in their intentions. Eighteen percent of them refused to smoke. Among those who smoked, some inhaled deeply yet some did not inhale at all. Those experimenters who were more enthusiastic in their initiation to cigarettes subsequently used them more often. From the beginning, some individuals appear to be primed to become smokers more than others: an issue of individual differences, or population "heterogeneity," to which we return later. Friedman et al.'s (1985) data also support the idea that peers and face-to-face contacts are usually responsible for the spread of smoking behavior among teenagers.

In our EMOSA models, prevalence-driven change had a different mathematical representation from constant rate change. Specifically, the proportion of experienced was given by an equation of the general form of Equation 8.1:

$$P_{(a+1)} = TP_a(1\text{-}P_a) + P_a, \tag{8.1}$$

where *a* is year, age, or grade level, P is the proportion of experienced, and T is a transition rate parameter. The processes of random mixing in the population are captured by the first product term on the right hand side of Equation 8.1. In general, if objects are allowed to mix, with the proportion of one type of object, p, equal to P_a, and the other, q, equal to $1\text{-}P_a$, then the mixing rates under random contact would be:

$$p^2 + 2pq + q^2. \tag{8.2}$$

The first term, the probability of experienced meeting experienced, and last term, that of naive meeting naive, cannot create new experienced individuals, and so they can be ignored. The center term–representing the probability of contacts between naive individuals and adopters–is the same as the first, right-hand term in Equation 8.1, except for the constant 2. As this constant can be absorbed into the value of the rate parameter, T, the two equations make the same assumption of random mixing.

A constant rate process takes a different mathematical form. This equation is:

$$P_{a+1} = T(1-P_a), \tag{8.3}$$

where the terms have been defined previously. In a constant rate process, a constant proportion of naive individuals become adopters in each time period. For example, this model is often used to describe media effects. The first time a new advertisement appears on TV, the majority of viewers are naive and may be influenced. After successive exposures, the size of the naive group in the population would diminish.

The processes of face-to-face, prevalence-driven conversion versus constant rate conversion create, for similar parameter values, quite different growth curves. In the constant rate process, the greatest number of conversions occur early, and then the rate of new converts declines. In the face-to-face process, the growth curve is *S*-shaped, with early conversions limited by the small number of the experienced users and late conversions limited by the small number of the naive. The rate of conversion is thus most rapid midway through the process. The two curves are illustrated in Fig. 8.1 for a starting proportion of adopters of .05 and a transition rate of .6.

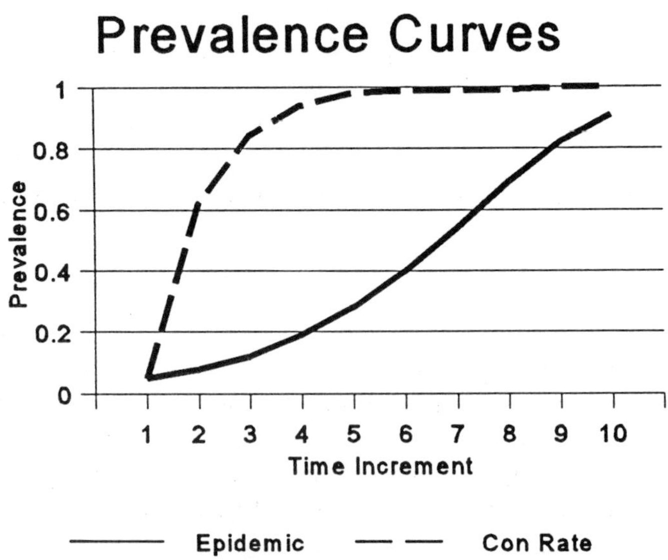

FIG. 8.1.

Assumptions of Epidemic Models of Smoking

Basic epidemic models of smoking, and other epidemic models more generally, make several assumptions: (a) that mixing in a population occurs at random, (b) that transition rate parameters remain constant over the course of the epidemic process, and (c) that mixing is within a generation.

The random mixing assumption commonly leads to objections about the unrealistic nature of epidemic models. Children, after all, are not blue and white balls in urns, so favored in illustrations of probability theory. Children mix nonrandomly; for example, friendship networks in schools take the form of cliques, where college-bound students have more frequent than random contact with one another, as do athletes and rebellious drug users (Rowe, Woulbroun, & Gulley, 1994). On the other hand, there is also considerable mixing among students. Many friendship groups in school are short-lived; they dissolve between the start and end of a school year. Berndt and Hoyle (1985) found that 30% to 50% of friendships were unstable within one academic year. According to Cohen (1977), a stable clique retains half its membership over one academic year. Twenty-six percent of Cohen's cliques were unstable and lost more than half their membership in an academic year. Although these friendship changes are not random, they do create new pathways for the flow of social influence.

Some nonrandomness of mixing does not undermine an EMOSA model. Consider the two examples in Table 8.1.

In the hypothetical example, the frequency of adopters was .6 and that of naive individuals, .4. In the upper example, random mixing occurred, in which the cell frequencies followed a binomial distribution. In the lower example, population composition was the same, but mixing was nonrandom. Nonrandom mixing increased the sizes of the off-diagonal cells, where the proportion of like-status individuals in mutual contact increased 25% (i.e., from 52% to 65%). The example assumes that nonrandomness increased contacts between similar individuals, because of the general similarity of friends for physical and behavioral traits (Kandel, 1978). Because this type of nonrandomness decreased the prevalence of Naive/Experienced contacts, the overall result would be a reduction of recruitment. Thus, non-random mixing tends to slow the "epidemic" spread of new behaviors or social innovations. The general form of spread, however, remains unchanged.

TABLE 8.1

Hypothetical Example of Rates of Random and Nonrandom Mixing

		Random Mixing	
		Person A	
		N	A
Person B	A	.24	.36
	N	.16	.24
		Nonrandom Mixing	
		Person A	
		N	A
Person B	A	.20	.40
	N	.25	.15

The idea that random mixing is closer to reality than most people think is developed theoretically in other settings (Dodd, 1995). In Dodd's experiments on the diffusion of knowledge, messages were given via air-dropped leaflets or via interviewers to people at random, and the diffusion of the messages was studied. Models that assumed randomness of mixing fit the observed spread of the messages well. Dodd makes the important observation that randomness does not imply that a behavior is purposeless to each individual involved. As he commented, "Acts that seem highly purposeful *to the individual*, such as the appointments he makes to see people in the course of a day, may also be random-like *in the aggregate* in a large population" (italics in original, p. 397). Randomness is created by the large number of influences on any particular act, which are not highly correlated among one another, and which are also not highly correlated across different members of a population. His conclusion is that:

> "Random causation, in short, should be thought of not as irregular and chaotic but as *multiple and diverse* causation where every cause is but a small share of the total causation" (italics in original, p. 397).

In the case of smoking, many influences would exist for a particular occasion where a nonsmoker might try a cigarette. They involve any temporary motivations of participants, the availability of a cigarette, the absence of adults who might interfere, the particular activities planned, and so on. Friendships and acquaintanceship groups change throughout adolescence. The particular individuals present might seem purposive to the individuals involved, but in the aggregate, many such groupings will have a component of randomness because many influences have decided whether one person joins a group on one particular occasion.

We also explored the random mating assumption in data on sexual intercourse from the Adolescent Sexuality Study (Rowe, Rodgers, & Meseck-Bushey, 1989). We were able to correlate the sexual intercourse status of best friends (coding nonvirgin = 1, virgins = 0). Under a random mating assumption, these correlations would be zero; if there was positive assortment, they would be greater than zero. The actual correlation (f coefficient) for White male with best female friend was .15. White females with best male friend was .11. Hence, some assortment took place, but it was relatively small in comparison to a perfect correlation (1.0), or in comparison to adolescent friends' similarity for age, race, and other traits (Kandel, 1978). In a simulation study, we imposed different degrees of nonrandom mixing on the epidemic model of sexual intercourse. In this model, T represented the rate at which opposite-sex virgins contacting a male nonvirgin (T_f) or female nonvirgin (T_m) had sexual intercourse. Under random mixing, the transition parameter, T_m, had a value of .899 for White males and T_f of .221 for White females. With a f coefficient of nonrandom mixing of .10, T values became .985 and .205, respectively. The chi-square fit worsened from 5.7 to 6.0. The general pattern was that both the T and chi-square values changed slowly with greater deviation from random mixing. Thus, our epidemic model had similar properties when relatively low and empirically realistic levels of nonrandom mixing exist as it did when mixing was assumed to be truly random.

On the other hand, some specific types of nonrandom mixing would violate an epidemic model. The mixing process assumes that everyone in the group is eventually reachable from any other individual. In school networks, some students are *isolates* who have almost no contact with other students. The EMOSA approach predicts that such social isolates would not become adopters of smoking or other peer-influenced behaviors. With school network data, this is a testable theory. As well, if some isolates adopt the behavior, it may be through their nonschool or family social networks and thus be consistent with an epidemic model. To achieve a more realistic model of mixing, one must move to either quite complex sets of equations or to computer simulations.

Another assumption of the model is that the transition rate parameter, T, stays the same for an individual regardless of the prevalence of smoking in a population and the individual's age, but that exposure is less likely at a younger age. The T parameter can be defined as, "the conditional probability of adoption, given a naive person contacts an experienced person." One test of whether this assumption is reasonable is to test model fit. A single value of T may fail to fit, indicating failure of the model.

We also accept the likely population heterogeneity of T. Individuals vary in their susceptibility, once they are exposed to the behavior. In the smoking application, we know that fewer than one third of adolescents currently become adult smokers. Even in societies where smoking is more strongly favored, a large minority remain nonsmokers.

Regular smoking is a strongly heritable behavior (Heath & Madden, 1995). Both the initiation and persistence of smoking are heritable (in most estimates, 40% to 60% of total variation is genetic). According to Heath and Madden, personality traits mediate some of the heritability of smoking initiation (e.g., the greater impulsivity of smoker-initiates). In contrast, although persistence was substantially genetically influenced, commonly assessed personality traits failed to explain this genetic effect. Heath and Madden speculated that persistence may depend more on the physiological response to nicotine than on general personality traits. With regard to the heterogeneity of T, the genetic findings indicate that T is heterogeneous from person to person. However, as personality traits and drug-response are probably stable over time, the dependence of T on these sources of heterogeneity would suggest that temporal changes may not be a major source of heterogeneity. The earlier initiators may be susceptible because of their genetic dispositions, but their T value would not necessarily change if they somehow missed opportunities for an early initiation.

The last major assumption is within-generation mixing. This assumption is problematic because, if taken strictly, the epidemic process could never start. That is, all children arrive as naive teenagers, so no child exists to start the epidemic diffusion of smoking among peers. Conceptually, we can avoid this problem if we allow that some children find other routes to adopting smoking than peer contagion, and thus they are present in small numbers to seed the process. This might occur if some young teens were encouraged to smoke by their smoking parents, or some by older teens, a violation of the within-generation mixing assumption. In support of the assumption, we know that the majority of teenagers have friends of similar age. The age-segregation created by grade levels in junior and senior high schools helps to impose an age-segregation of friendship and acquaintanceship networks. If the assumption works for the majority of transmissions, it is reasonable and useful, although not strictly true. Epidemic models can be written to permit inter-generational mixing (Rowe & Rodgers, 1994). Relaxing this assumption, however, increases greatly the number and complexity of the model's equations. There is also some decrease in model accuracy without intergenerational mixing, but it is not large. At present, we have not made a systematic comparison of models with and without intergenerational mixing. Nonetheless, for the most part, we favor imposing this assumption to reduce the complexity of epidemic models.

Estimation of Epidemic Model Parameters

Although in some work we have written specialized programs to estimate epidemic models, in more recent work, we have used the SAS's PROC NLIN program. It offers a general means for estimation of model parameters for epidemic models. Examples of programming code can be found in Rodgers and Rowe (1993).

The purpose of parameter estimation is simple to understand. For a fixed number of initial *experienced* at the first age (represented as a_1 or a_0), each value of T can generate a different predicted prevalence curve. The object of estimation is to find the value of T that generates predicted prevalences that are as close as possible to the observed prevalence. With just one unknown, T, as in Fig. 8.1, a good estimate of its value could be found by trial and error. However, such guess work becomes daunting if two or more parameters are estimated. To deal with this problem, the SAS PROC NLIN searches for the best values of parameters through the multidimensional space of possible values.

A mathematical function called a *loss function* describes how close the predicted value is to the observed one. One loss function is the least squares:

$$L = \text{sum of } (Po_a - Pe_a)^2, \tag{8.4}$$

where L is the function's value, Po_a is the observed proportion at age *a*, Pe_a is the model's estimated proportion at age *a*, and the sum is taken over all ages. If the observed and estimated proportions always agree, the value of L is zero. The larger L, the greater the discrepancy. A modification of Equation 8.4 yields another loss function, the chi-square function. Using the chi-square as the loss function has another advantage. Because the theoretical chi-square test sampling distribution is available, it is possible to use it to statistically evaluate goodness-of-fit. As with the least-squares loss, a good fit is indicated by a small value, and a perfect fit by a zero value.

A model's chi-square must be evaluated against its degrees of freedom. The degrees of freedom is usually one less than the number of prevalences. For example, in Fig. 8.1, the values at a_1 were used to seed the model, so the number of observations was 9, not 10. One parameter was estimated, T. So, for this equation, the degrees of freedom available to test the fit would be df=8. According to a chi-square table, any chi-square greater than 15.5 is significant, with df=8. The model hopefully would produce a lower chi-square, but a larger one does not necessarily mean that the model is poor. With a large *N*, chi-squares are often significant, even in a good model. The size of the predicted–observed discrepancy, the pattern of under- and overestimation, and the sample size all must be considered in evaluating fit.

Applying Epidemic Models to Smoking Cigarettes

In our epidemic work on smoking, we had two central research questions in mind. First, would an epidemic model fit? With its simplifying assumptions, the model might well fail. Second, we wanted to know whether particular transitions involved in the adoption of cigarette smoking were constant rate versus face-to-face, that is, prevalence-driven.

Stages of smoking and parental smoking status added several layers of complexity to our epidemic model (Rowe et al., 1996). These stages were: nonsmoker, triers ("smoked once or twice, or one cigarette a month"), regular smokers ("smoked weekly or more"), and quitters ("no longer smoke, but in the past was a regular smoker"). Because studies consistently find that the children of smokers are more likely to smoke than the children of nonsmokers (e.g., Eysenck, 1980), we also separated the sample by whether the parents smoked. We were interested in the process underlying this family effect. For example, if parental smoking increased susceptibility to peer influence, we would expect a larger prevalence-driven parameter for the children of smokers than nonsmokers. Our idea was to see which model parameters were most affected by the smoking status of the parents. With this background, we can present the set of epidemic equations used in the model.

The model has eight equations: four equations for the offspring of smoking parents and four for those of nonsmoking parents. Except for the subscript on parameters T and j indicating group membership, the equations were identical. For this reason, only the four equations for the children of smoking parents are presented next.

Because the equation models the proportion of nonsmokers, the subscript of the first proportion (NS) refers to nonsmoking status. The proportion of nonsmoking children (P_{NS}) is described by the following equation:

$$P_{NSa+1} = P_{NSa} - T_1(P_{NSa} - W_1)(\text{CONTACTING GROUP}), \quad (8.5)$$

where P_{NSa+1} is the proportion of nonsmokers at age $a+1$, P_{NSa} is the proportion of nonsmokers at age a, W_1 is the proportion immune, T_1 is the rate parameter, and the CONTACTING GROUP is the proportion of everyone who smokes (i.e., triers and regular smokers, regardless of parental smoking status). Notice that the equation adjusts the contact groups by the proportion of *immune* children of smoking parents, W_1. By definition, immune individuals are those who were still nonsmokers by adulthood. We assume that these individuals never had an encounter with another person that could lead effectively to smoking behavior. By analogy to individuals immune to infectious disease, they cannot "catch" smoking, regardless of their exposure to other adolescents who smoke.

The difference equations iterate over a period of one grade level at a time (e.g., a may represent sixth grade and $a+1$, seventh grade). Notice that the equation's right-hand term (i.e., new triers) is to be subtracted from the proportion of nonsmokers (age a) in this equation and then added to the next equation, which represents the proportion of triers (age $a+1$):

$$P_{TRa+1} = P_{TRa} + T_1(P_{NSa} - W_1)(\text{CONTACTING GROUP}) - j_1 P_{TRa}, \quad (8.6)$$

where P_{TR} is the proportion of triers (of smoking parents), j_1 is the constant rate parameter, and the other symbols have the same interpretation as before. Notice that in the term beginning with j_1 no social contacts would be involved. This term represents individuals who were becoming regular smokers, at a constant rate.

This term for new regular smokers is added to the following equation:

$$P_{RSa+1} = P_{RSa} + j_1 * P_{TRa} - U * P_{RSa} + V * P_{ESa}, \tag{8.7}$$

where P_{RS} is the proportion of regular smokers, P_{ES} is the proportion of ex-smokers, U is the quitting rate, and V is the relapse rate. The last equation completes the cycle, and describes the proportion of ex-smokers (in the children of smoking parents group):

$$P_{ESa+1} = LNS - P_{NSa+1} - P_{TRa+1} - P_{ESa+1}, \tag{8.8}$$

where LNS is the proportion of children who are of smoking parents in sixth grade. Four parallel equations appear for the children of nonsmoking parents. In these parallel equations, a second immune group constant is used, W_2, which represents the proportion of *immune* children of nonsmoking parents. Together, all eight proportions estimated must add to 1.0 (i.e., 100% = the full population).

Rowe et al. (1996) explored the equations and parameters that best describe the social transmission of smoking behavior. These equations can be modified in several ways that are substantively meaningful. First, the *prevalence-driven* element of equations 8.5 and 8.6 and can be removed (i.e., the CONTACTING GROUP is dropped). In this modification, both the nonsmoker → trier and trier → regular smoker transitions are, by mathematical definition, constant rate ones. A different modification is to make the second transition prevalence-driven. This change is made by adding a contacting group to the two terms involving the parameter j as follows:

$$j_1 * P_{TRa} * (\text{CONTACTING GROUP}), \tag{8.9}$$

where j_1 and P_{TRa} are as defined previously and the CONTACTING GROUP consists of all regular smokers, regardless of parental smoking status.

Findings From EMOSA Models of Smoking

Our results consistently favored one model of the epidemic process that placed the social influences early in the process (Rowe et al., 1992, 1996).

Specifically in the best models, the transition (T) from nonsmoker to regular smoker was prevalence-driven. In other words, your friends get you involved in smoking. In contrast, the transition from experimental smoker to regular smoker (*j*) was constant rate, as though it depended on nonsocial processes, such as the physiological response to nicotine.

The estimated T parameters were similar for males and females (about .50, Rowe et al., 1996). Although we treated the sexes as noninteracting in the epidemic processes, the similarity of their parameters and prevalence curves suggests that some transmission of smoking occurred in mixed-sex groups. The children of smoking parents were more peer-influenced: The T parameter was about .15 greater for the children of parents who smoked.

The *j* parameter was the constant rate parameter for the transition between experimental smoker and regular smoker. The value of *j* was not significantly larger in the children of smokers than nonsmokers. Nonetheless, this parameter was *242%* greater in magnitude in the sons of smokers vs. the sons of nonsmokers (*j*=.068 vs. .028). It was 11% greater in the daughters of smokers than nonsmokers (.109 vs. .09). These findings are suggestive that being a child of a smoking parent, especially a son, carries a greater risk of addiction to cigarettes. A larger sample size than in the Bloomington study may be necessary for the statistical power to demonstrate a statistical difference in this parameter.

The epidemic model developed for smoking can be applied to other questions about smoking. In our work, the division of the school sample was based on the smoking status of parents, but other sample subdivisions can be explored using this epidemic model. Given the differences in smoking rates among ethnic groups, they represent an interesting sample subdivision for epidemic modeling. We give an example of this type of modeling next. Any characteristic of a child is a potential source of sample subdivision for epidemic modeling. For example, personality traits, such as sociability and impulsivity, might have a strong effect on the T transition but none on *j*. On the other hand, if a measure of addictive potential were developed and available, it might be used to partition a sample into low- and high-risk groups that demonstrate large differences in *j*. These examples illustrate a range of potential issues that are approachable using epidemic modeling.

An Epidemic Model of Black and Hispanic Girls Prevalence of Smoking

In many samples, adolescents of African heritage (Blacks) smoke less than most other ethnic and racial groups in the United States (Adlaf, Smart, & Tan, 1989; Rogers, 1991; Warheit, Biafora, Zimmerman, Gil, Vega, & Apospori, 1995; but see Bettes, Dusenbury, Kerner, James-Ortiz, & Botvin, 1990 for a negative finding). There are no compelling theoretical or physiological explanations of this racial/ethnic difference. Our EMOSA approach allows us to investigate several hypotheses about ethnic/racial differences in smoking rates. One explanation is that Black individuals resist social pressures more than those in other ethnic/racial groups, implying differences in the T parameter of the EMOSA model. Although not mutually exclusive, another explanation is that Blacks are less susceptible to the addictive properties of the psychoactive drug in cigarettes, nicotine. This explanation suggests greater group differences in the j parameter.

To evaluate these hypotheses, we modeled the prevalences of experimental and regular smoking in two groups: Black girls and Hispanic girls. The data for our investigation come from the National Longitudinal Study on Adolescent Health (Add Health, see details on the methods and sample reported in Resnick et al., 1997). In the Add Health Study, data were collected on adolescents attending 80 high schools and their feeder junior high schools from around the country. Although the average number of children per school was about 200, several schools were sampled exhaustively. The completely sampled school on which we focus is located in a large metropolitan area and has within it about equal-sized groups of Filipino, Black, and Hispanic students and a small number of White students. To simplify our model, we compare here just the Black and Hispanic girls from this one school. Each respondent reported the age at which they first had smoked a whole cigarette and the age at which they began regular smoking. This information yielded a prevalence curve, the proportion of experimental and regular smokers at each age. Figure 8.2 presents prevalence curves for Black and Hispanic girls.

The form of these curves is similar to that found in Bloomington, with the onset of regular smoking lagging after the onset of experimental smoking and rising much more slowly. The Black girls experimented less, and were less likely to become regular smokers than Hispanic girls. At age 17, three times as many Hispanic girls (14%) than Black girls (4%)

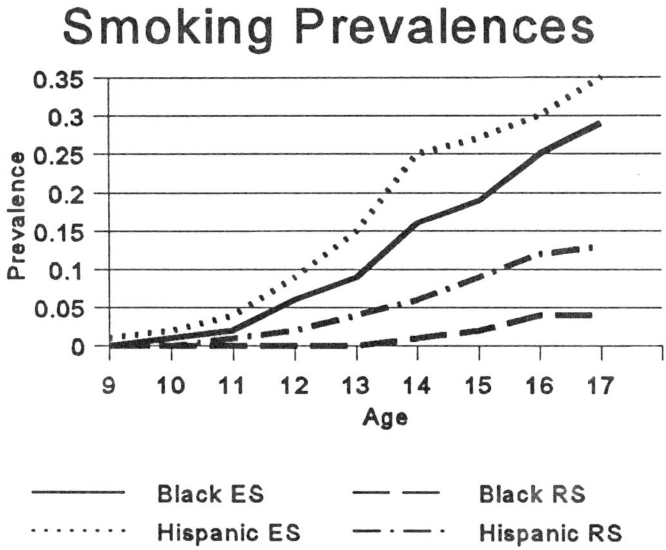

FIG. 8.2.

became regular smokers. Although not shown in Fig. 8.2, the pattern for Black and Hispanic boys was similar.

Equations for the Epidemic Model of Girls' Smoking

The model of girls' smoking required six equations. Table 8.2 shows the six equations. They were based on the best-fit epidemic model of the Bloomington data. Six, instead of eight, equations are used because no information was available on ex-smokers (nor would we expect many in an adolescent population). Thus, the third and sixth equations accumulated the number of regular smokers.

As before, the proportion of nonsmokers was reduced by those who become experimental smokers via prevalence-driven social contacts. The parameters TB and TH set the rate of this process for Black and Hispanic girls, respectively. Girls who become experimenters move from Equation 1 to 3 or Equation 4 to 6 in Table 8.2 if they become regular smokers. The latter is a constant rate process, with rate parameters j_B and j_H, respectively. Based on our previous experience with rates of adults who have never tried smoking, 10% of girls of both ethnic/racial groups were

TABLE 8.2

Equations for Epidemic Model of Black and Hispanic Girls' Smoking

Group	Equation
Black Females	
1. Nonsmoker	$NSB_{a+1} = NSB_a - T_B(NSB_a - .08)(\text{Contacting Group})$
2. Experimental Smoker	$ESB_{a+1} = ESB_a + T_B(NSB_a - .08)(\text{Contacting Group}) - K_B ESB_a$
3. Regular Smoker	$RSB_{a+1} = RSB_a + j_B(ESB_a)$
Hispanic Females	
4. Nonsmoker	$NSH_{a+1} = NSH_a - T_H(NSH_a - .12)(\text{Contacting Group})$
5. Experimental Smoker	$ESH_{a+1} = ESH_a + T_H(NSH_a - .12)(\text{Contacting Group}) - K_H ESH_a$
6. Regular Smoker	$RSH_{a+1} = RSH_a + j_H(ESH_a)$

Note: *N* = 260 Hispanic females; *N* = 174 Black females.

assumed to be immune. The immune group is represented by the proportions .08 and .12, in Equations 1 and 3 in Table 8.2, respectively.

The rate of the prevalence-driven process depends on the size of the contacting group. There are two extreme hypotheses: (1) that the contact group consists only of individuals from the same group, and (2) that both groups mix freely. Unhappily, a free mixing-rate parameter for the Black and Hispanic girls could not be estimated, because of its high correlation with the parameters T_W and T_B.[1]

Table 8.3 presents the parameter estimates and chi-square for the epidemic models.

In the model in which the contact group was of the same ethnic/racial origin, the overall chi-square was 108.3. The fit for the model of groups mixing freely was 104.7 Neither fit equaled those obtained in a larger sample of Bloomington students (chi-squares 24–26, df=25, Rowe et al., 1996). Our less satisfactory chi-squares were about three times their degrees of freedom. One explanation of the fit quality is a difference in some detail of the models; for instance, the present model also did not include the possibility of quitting and relapse, as did the Bloomington models. A second, more likely explanation for relatively poor fit is that one assumption of the epidemic model was violated: Namely, that the majority of mixing between nonsmokers and smokers took place in this

[1]A mixing parameter, *W*(Contacting Group), was tried that represented the proportion of a group that mixed both with their own group and with the other group.

TABLE 8.3

Parameter Estimates From Model, Within-Group Mixing Only

Parameter	*Estimate*	*SE*
Within-Group Mixing Only[a]		
T_B	2.11	.18
j_B	.07	.09
T_H	1.41	.05
j_H	.22	.05
$X^2(32) = 108.3$		
Random Mixing of All Groups[b]		
T_B	.69	.06
j_B	.07	.14
T_H	.96	.06
j_H	.20	.05
$X^2(32) = 104.7$		

[a]Contacting Group, ESB + RSB or ESH + RSH
[b]Contacting Group, ESB + RSB + ESH + RSH

one large school. We know from other data on the sample that 90% of students' sexual partners were drawn from outside this school's student body. The racial and ethnic diversity of the school may have led some students to find their friendship groups in nonschoolmates from their home, urban communities.

As shown in Table 8.3, when we assumed within-group mixing, experimental smoking actually spread faster among Blacks than Whites (T parameters = 2.11 and 1.41, respectively). Note that these T values were greater than 1.0. As T can be understood as the number of contacts between smokers and nonsmokers times the probability of adoption per contact, the parameter may legitimately exceed 1.0. However, T values greater than 1.0 can lead to badly behaved equations; they must be viewed with some caution. The j parameter that sets the rate of transition from experimental to regular smoker was three times greater in Hispanics than in Blacks (.22 vs. .07). There was some imprecision to the Black's j_B parameter (SE = .09), however, because few made the transition from experimental to regular smoker.

In the panmixing model, Hispanics had a greater adoption rate for experimental smoking than Blacks (T_H = .96 vs. T_B = .69). These T parameters were also in the 0 to 1 range. Blacks again had the slower rate of transition from experimental smoker to regular smoker (j_B = .07 vs. j_H = .20). Thus, in both models, Blacks seem less susceptible to the physical and psychological rewarding properties of tobacco, because once they are exposed, they have a very low rate of transition to regular smoking.

We can, of course, only speculate on the reasons for the slower transitions in Blacks than Hispanics. One interesting possibility would be some kind of physiological resistance to nicotine. This idea is not so farfetched as it may seem. A gene confers on some Asians a strong resistance to becoming addicted to alcohol (Tu & Israel, 1995). Perhaps some Blacks are resistant to nicotine because of a similar genetic difference; but, no proof of this exists today. Perhaps some Blacks are resistant because stronger social norms may exist in the Black community that discourage cigarette smoking beyond experimentation. Whatever the cause, it should be something specific to smoking because the use of many drugs is as prevalent among Blacks as among other ethnic and racial groups, if not moreso.

Depending on the mixing model, we have quite different conclusions about the T parameter. Under panmixing, the T parameter was greater for Hispanics than Blacks, whereas the reverse held under the within-group mixing model. The within-group model is favored by data on schools' friendship networks, showing that friendship choices are largely within ethnic group in many senior high schools in Add Health. The panmixing model is favored by its more reasonable parameter estimates (less than 1.0) and by the idea that the group with the lower smoking rate would have a lower T, that is, the lower likelihood of a naive smoker trying cigarettes on an encounter with a smoker. We expect that actual mixing patterns are closer to the within-group mixing model than the panmixing model, but that some inter-group mixing also occurs. It would be interesting to discover schools in which Black smoking rates have increased to be closer to those of Hispanics. Fitting an epidemic model to these data might reveal whether the change is primarily in the socially driven T transition between nonsmoker and experimental smoker (e.g., a change in social norms), or in the j parameter, implying that Blacks and Hispanics are equally susceptible to the addictive properties of tobacco, thus contradicting the implications of the present findings.

EMOSA modeling may itself be a stimulus for deeper theoretical thinking. Group phenomena are often regarded as relatively independent of individual-level processes. However, group-level phenomena may emerge from individual-level actions through non-linear processes. Changes in age at marriage and age at first parity might reflect EMOSA-like social processes within cohorts. Social norms, usually thought of as determining individual-level behavior, may arise from the frequency and mixing patterns of individual interactions, and be consequences, not causes. Delinquent acts, which are usually done in groups, may be susceptible to an EMOSA spread (Jones, 1998). The EMOSA concept also focuses greater attention on initial conditions, because in many nonlinear processes, small changes in initial conditions may lead to large changes in final outcome. A population composition might be looked at for its ability to sustain the spread of certain behaviors. For instance, the stochastic processes associated with true disease epidemics mean that they tend never to arise, or to spontaneously die out, in small populations. As always, a tension exists between models that provide for greater realism, but become unwieldly in their complexity, and those that are simple and aid understanding, but fail to capture a process in its full complexity. An important goal is to combine individual differences and social dynamics into models of behavior.

FUTURE DIRECTIONS OF EPIDEMIC MODELS

Our examples gave an introduction to the EMOSA modeling approach. The approach is most appropriate when group-level data exist on the prevalence of smoking or other adolescent onset behaviors. It is most interesting when population subdivisions are available that suggest something about either the process of adoption and spread of experimental smoking, or something about the transition from experimental to regular smoking. More detail about the SAS computer code used to fit EMOSA models can be found in our review article on same (Rodgers & Rowe, 1993).

A major limitation of EMOSA models is that they do not handle individual-level data well. In this way, the epidemic model differs markedly from the correlational and regression analyses that are the hallmark of most research on smoking. Regression models handle many predictors simultaneously (although the number of truly independent predictors may be small). In regression models, many hypotheses about smoking are framed in terms of personality and attitudinal variables at the individual level (e.g., Chassin, Presson, & Sherman, 1984; Chassin, Presson, Sherman, Corty, & Olshavsky, 1984). Although an EMOSA model can handle a single individual-level variable, it must give up information as individuals are grouped, for example, smoking behavior in high- versus low-IQ individuals, whereas the original variable was continuous. Thus, the EMOSA model can handle only a few variables at once, because each variable requires a further subdivision of a population. Prevalence of a behavior is not itself a variable in a regression model; it is in the EMOSA model. The regression model predicts for each individual, the EMOSA model only for groups of similar individuals.

One approach to moving the EMOSA model to the individual-level is computer simulation. Simulation work should probably begin with a deterministic model, such as those illustrated in this chapter. The model's assumptions about groups and their mixing can be incorporated into a computer simulation. In a computer simulation, each individual in a population can be represented. The person can be given their trait values, expressed as standard deviation scores. In turn, trait scores can be translated into a probability of adopting smoking, or any other behavior, by a simple mathematical function such as the logistic curve. Random number generators can handle the processes of random mixing. If anything, simulation work offers too many choices, because a great variety of simulations can be created for an epidemic process, even in a single school. However, in other contexts, the computer simulation approach has led to important insights.

In summary, the Epidemic Models of Onset of Social Activities (EMOSA) approach to modeling social behaviors in adolescence, including smoking, has been barely tapped. The ideas and methods here are ready themselves for adoption, if not epidemically, at least with a noticeable rate of growth in the scientific community.

ACKNOWLEDGMENTS

Some data were provided by the National Longitudinal Study on Adolescent Health (Add Health), a program project designed by J. Richard Udry (PI) and Peter Bearman, and funded by grant P01-HD31921 from the National Institute of Child Health and Human Development to the Carolina Population Center, University of North Carolina at Chapel Hill, with cooperative funding participation by the National Cancer Institute;

the National Institute of Alcohol Abuse and Alcoholism; the National Institute on Deafness and Other Communication Disorders; the National Institute of Drug Abuse; the National Institute of General Medical Sciences; the National Institute of Mental Health; the National Institute of Nursing Research; the Office of AIDS Research, NIH; the Office of Behavior and Social Science Research, NIH; the Office of the Director, NIH; the Office of Research on Women's Health, NIH; the Office of Population Affairs, HHS; the National Center for Health Statistics, Centers for Disease Control and Prevention, HHS; the Office of Minority Health, Centers for Disease Control and Prevention, HHS; the Office of Minority Health, Office of the Assistant Secretary for Health, HHS; the Office of the Assistant Secretary for Planning and Evaluation, HHS; and the National Science Foundation.

REFERENCES

Adlaf, E. M., Smart, R. G., & Tan, S. H. (1989). Ethnicity and drug use: A critical look. *International Journal of Addictions, 24,* 1–18.

Bailey, N. T. J. (1975). *The mathematical theory of infectious diseases.* New York: Hafner Press.

Berndt, T. J., & Hoyle, S. G. (1985). Stability and change in childhood and adolescent friendships. *Developmental Psychology, 21,* 1007–1015.

Bettes, B. A., Dusenbury, L., Kerner, J., James-Ortiz, S., & Botvin, G. J. (1990). Ethnicity and psychosocial factors in alcohol and tobacco use in adolescence. *Child Development, 61,* 557–565.

Chassin, L., Presson, C. C., & Sherman, S. J. (1984). Cigarette smoking and adolescent psychosocial development. *Basic and Applied Social Psychology, 5,* 295–315.

Chassin, L., Presson, C. C., Sherman, S. J., Corty, E., & Olshavsky, R. W. (1984). Predicting the onset of cigarette smoking in adolescents: A longitudinal study. *Journal of Applied Social Psychology, 14,* 224–243.

Cohen, J. (1977). Sources of peer group homogeneity. *Sociology of Education, 50,* 227–241.

Corti, C. (1932). *A history of smoking.* New York: Harcourt Brace.

Dodd, S. C. (1995). Diffusion is predictable: Testing probability models for laws of interaction. *American Sociological Review, 20,* 392–401.

Eysenck, H. (1980). *The causes and effects of smoking.* London: M. T. Smith.

Friedman, L. S., Lichtenstein, E., & Biglan, A. (1985). Smoking onset among teens: An empirical analysis of initial situations. *Addictive Behaviors, 10,* 1–13.

Goodman, J. (1993). *Tobacco in history: The culture of dependence.* New York: Routledge.

Heath, A. C., & Madden, P. A. F. (1995). Smoking and addictive behaviors: Epidemiological, individual, and family factors. In J. R. Turner, L. R. Cardon, & J. K. Hewitt (Eds.), *Behavior genetic approaches in behavioral medicine* (pp. 45–66). New York: Plenum.

Huckfeldt, R., Kohfeld, C. W., & Likens, T. W. (1982). *Dynamic modeling: An introduction.* Newbury Park, CA: Sage.

Jones, M. B. (1998). Behavioral contagion and official delinquency: Epidemic course in adolescence. *Social Biology, 45,* 134–142.

Kandel, D. B. (1978). Homophily, selection, and socialization in adolescent friendship pairs. *American Journal of Sociology, 36,* 427–436.

Mahajan, V., & Peterson, R. A. (1985). *Models for innovation diffusion.* Beverly Hills, CA: Sage.

Resnick, M. D., Bearman, P. S., Blum, R. W., Bauman, K. E., Harris, K. M., Jones, J., Tabor, J., Beuhring, T., Sieving, R. E., Shew, M., Ireland, M, Beaninger, L. H., & Udry, J. R. (1997). Protecting adolescents from harm: Findings from the National Longitudinal Study on Adolescent Health. *Journal of the American Medical Association, 278,* 823–832.

Rodgers, J. L., & Rowe, D. C. (1993). Social contagion and adolescent sexual behavior: A developmental EMOSA model. *Psychological Review, 100,* 479–510.

Rodgers, J. L., Rowe, D. C., & Buster, M. (1998). Nonlinear dynamic modeling and social contagion: A response to Stoolmiller's comments on EMOSA modeling. *Developmental Psychology, 34,* 1096–1113.

Rogers, E. M. (1983). *Diffusion of innovations.* New York: Free Press.

Rogers, R. G. (1991). Demographic characteristics of smokers in the United States. *Social Biology, 38,* 1–12.

Rowe, D. C., Chassin, L., Presson, C., Edwards, D., & Sherman, S. J. (1992). An "epidemic" model of adolescent cigarette smoking. *Journal of Applied Social Psychology, 22,* 261–285.

Rowe, D. C., Chassin, L., Presson, C., & Sherman, S. J. (1996). Parental smoking and the "epidemic" spread of cigarette smoking. *Journal of Applied Social Psychology, 26,* 437–454.

Rowe, D. C., & Linver, M. R. (1995). Smoking and addictive behaviors: Epidemiological, individual, and family factors. In J. R. Turner, L. R. Cardon, & J. K. Hewitt (Eds.), *Behavior genetic approaches in behavioral medicine* (pp. 67–84). New York: Plenum.

Rowe, D. C., & Rodgers, J. L. (1991). Adolescent smoking and drinking: Are they 'epidemics'? *Journal of Studies on Alcohol, 52,* 110–117.

Rowe, D. C., & Rodgers, J. L. (1994). A social contagion model of adolescent sexual behavior: Explaining race differences. *Social Biology, 41,* 1–18.

Rowe, D. C., Rodgers, J. L., & Meseck-Bushey, S. (1989). An "epidemic" model of sexual intercourse prevalences for Black and White adolescents. *Social Biology, 36,* 127–145.

Rowe, D. C., Woulbroun, E. J., & Gulley, B. L. (1994). Peers and friends as nonshared environmental influences. In E. M. Hetherington, D. Reiss, & R. Plomin (Eds.), *Separate social worlds of siblings: The impact of nonshared environment on development* (pp. 159–173). Hillsdale, NJ: Lawrence Erlbaum Associates.

Tu, G., & Israel, Y. (1995). Alcohol consumption by Orientals in North America is predicted largely by a single gene. *Behavior Genetics, 25,* 59–65.

Warheit, G. J., Biafora, F. A., Zimmerman, R. S., Gil, A. G., Vega, W. A., & Apospori, E. (1995). Self-rejection/derogation, peer factors, and alcohol, drug, and cigarette use among a sample of Hispanic, African-American, and White Non-Hispanic adolescents. *International Journal of the Addictions, 30,* 97–116.

9

Using Latent Transition Analysis to Examine the Relationship Between Perceived Parental Permissiveness and the Onset of Substance Use

Stephanie L. Hyatt
Linda M. Collins
The Pennsylvania State University, The Methodology Center

This chapter describes Latent Transition Analysis (LTA) and demonstrates the application of this relatively new methodology to the understanding of substance use research questions. LTA is a methodology for estimating and testing stage-sequential models of individual growth in longitudinal data. One example of stage-sequential development is the onset of substance use in adolescence. This sequence usually begins by experimentation with either alcohol or cigarettes, and can progress to higher levels of substance use experimentation. It is interesting to consider what factors might influence an adolescent's progression through the stages of substance use. In this study, a stage-sequential model of substance use onset for 9th and 10th graders is examined in relation to perceived parental permissiveness toward alcohol use by their children. Adolescents who perceive their parents as highly permissive are expected to be at-risk for higher rates of experimentation with various substances, and higher probabilities of advancing in their level of substance use. It was found that adolescents with strict parents (low perceived parental permissiveness) were less likely to have begun experimentation with any substance by 9th grade, whereas adolescents with parents seen as highly permissive were more likely to have tried alcohol, tried cigarettes, been drunk, tried marijuana, and tried cocaine by 9th grade. Additionally, these at-risk adolescents were 17 times more likely than the low permissiveness group to transition from the "No Use" level in 9th grade to the most advanced level of substance use by 10th grade. LTA provides an interesting way to examine this stage-sequential process.

This chapter describes Latent Transition Analysis (LTA; Collins & Wugalter, 1992) and demonstrates the application of this relatively new methodology to the understanding of substance use research questions. Latent Transition Analysis (LTA) is a methodology for estimating and testing stage-sequential models of individual growth in longitudinal data. Stage-sequential processes are characterized by change expressed as movement between stages. Usually the stages are qualitatively different. LTA has been used to model substance-use onset for seventh and eighth graders (Graham, Collins, Wugalter, Chung, & Hansen, 1991; Collins, Graham, Long, & Hansen, 1994), the mastery of math skills during high school (Collins & Wugalter, 1992), cigarette quitting and relapse behavior (Martin, Velicer, & Fava, 1996), and HIV-risk behaviors (Posner, Collins, Longshore, & Anglin, 1996).

There have been numerous studies providing evidence for the usefulness of conceptualizing substance-use onset as a stage-sequential process. Stage-sequential models enable researchers to identify pathways to higher levels of substance use and to illuminate group differences in these pathways. Results of studies examining substance-use onset as a stage-sequential process show the key role that alcohol and tobacco play early in the onset process. Based on a sample in seventh grade at the first occasion of measurement and in eighth grade at the second occasion of measurement, Collins, Graham, Long, and Hansen (1994) defined eight drug-use stages based on experimentation with alcohol, cigarettes, drunkenness, and more advanced use (recent alcohol or cigarette use, or any marijuana use). The most common pathway to advanced use began with alcohol use, then progressed to cigarette use, drunkenness, and finally advanced use. It was found that although adolescents usually begin substance use by experimenting first with alcohol, then cigarettes, a small subgroup begins by experimenting first with cigarettes then alcohol. Interestingly, adolescents who began the substance-use onset process by trying cigarettes first tended to be at greater risk for accelerated onset trajectories. In a sample of high school students followed into young adulthood, Yamaguchi and Kandel (1984) defined five drug use classes: (1) alcohol, (2) cigarettes, (3) marijuana, (4) other illicit drugs, and (5) prescribed psychotropic drugs. Clear temporal sequencing of stages of drug use were found. For males, the most common temporal sequencing included using alcohol before marijuana, and using both alcohol and marijuana before trying illicit drugs. For females, marijuana use was typically preceded by either alcohol use or cigarette use (not specifically alcohol use as for males). Additionally, illicit drug use by women was most often preceded by experimentation with alcohol, cigarettes, and marijuana (whereas for males, cigarettes were not necessarily tried before illicit drugs).

Although peer influence has often been emphasized as the single most important predictor of substance-use onset, parental influence also plays an important role. Kandel, Kessler, and Margulies (1978) examined various factors related to initiation into three substance-use stages: hard liquor use, marijuana use, and other illicit drug use (all substances other than alcohol, cigarettes, and marijuana). Even though most parents in their sample discouraged adolescent use of liquor, many of those parents reported requent use of liquor themselves. Thus, the relationship between parental factors and adolescent substance use is complete, involving both the effect of parental behavior and the effect of parental values. Additional factors such as strength of parent–child bonds may also play a role in parental effects on substance-use onset. Kandel and colleagues found that adolescents who felt their parents discouraged or forbid marijuana use were less likely to start using marijuana than those who reported more permissive parental attitudes. Hawkins, Catalano, and Miller (1992) suggested that family attitudes toward substance use may be a risk factor in the onset of adolescent substance use. McDermott (1984) found evidence that permissive parental attitudes toward drug use may actually have a larger effect on adolescent substance use than exposure to parental drug use. Research suggests that parental permissiveness toward alcohol use predicts adolescent alcohol use (Barnes & Welte, 1986), and parental permissiveness toward cigarette and marijuana use is related positively to adolescent cigarette and marijuana use (Brook, Gordon, Whiteman, & Cohen, 1986). These studies suggest that parental permissiveness about alcohol use could influence the onset of substance use at various stages. If a high level of parental permissiveness is, in fact, a risk factor for substance-use onset, then youths who have highly permissive parents will be expected to have higher substance-use prevalence rates and/or higher probabilities of transitioning to more advanced levels of substance use.

The study in this chapter expands on previous work in two ways. First, we analyze prospective longitudinal data gathered on adolescents and asking about alcohol, cigarettes, marijuana, and cocaine so that subjects can be followed through levels of various combinations of substance use. Using LTA, we will examine substance-use stage prevalence, and incidence of transitions between substance-use stages in order to examine pathways to adolescents' first experimentation with advanced substances. We can also identify stages that indicate higher risk for more advanced use in the future. Second, this study examines whether the incidence of substance-use stages and prevalence of stage transitions varies according to degree of perceived parental permissiveness about alcohol use.

LATENT TRANSITION ANALYSIS

Latent Transition Analysis (LTA; Collins & Wugalter, 1992), an extension of latent class theory (Goodman, 1974), is a method for estimating and testing stage-sequential models of development in longitudinal data. It is helpful to draw a distinction between the dynamic and the static part of LTA models. The dynamic part, which changes over time, is the movement through the stage sequence. In the empirical example presented here, this is adolescent substance use. The static part, which does not change over time, is an exogenous predictor that can be used to divide a sample into groups, in order to compare the probability of stage membership and the incidence of stage transitions across groups. The static variable used in this study is perceived parental permissiveness. Both the static and dynamic parts of the model can be treated as latent (unobserved) and measured by multiple categorical manifest (observed) items. In the case of the dynamic part of the model, these are measured repeatedly over time, as in this study where the manifest indicators are six items measured in 9th and 10th grade. These items each measure a characteristic related to substance use that helps to indicate the subject's current stage of substance use. The use of multiple indicators increases the reliability of the measures by estimating random measurement error and removing this error from subsequent analyses. For technical reasons, the operational definitions of stages used in LTA are called *latent statuses* (see Collins & Wugalter, 1992). The static latent variable in these analyses, perceived parental permissiveness, was measured by two questionnaire items that asked adolescents how angry their parents would be if they found out the adolescent had tried drinking alcohol or had been drunk.

Latent Transition Analysis (LTA) was used to fit a proposed model of substance use onset to the data. (In this case, the model was based on past empirical research on related data sets involving similar stage-sequential models of substance-use onset.) These analyses produced a goodness-of-fit statistic and parameter estimates of the probability of membership in each latent status of substance use and of transitioning to a different latent status between 9th and 10th grade. Also, introducing the level of perceived parental permissiveness as an independent variable allows the estimation of the probability of membership in each of the two classes, as well as how rates of stage membership and transition probabilities differ for the two groups.

The Mathematical Model

To aid explanation, the LTA mathematical model is presented in terms of two times of measurement, two manifest indicators of the static latent variable, and six manifest indicators of the latent status at each time. The model can be directly extended to other latent class or transition problems. Let the first time of measurement be Time t, and the second be Time $t + 1$. Also, let the static latent variable divide the sample into latent classes $c = 1, \dots, C$ and be measured by a first manifest indicator with $a = 1, \dots, A$ response categories and a second manifest indicator with $b = 1, \dots, B$ response categories. Suppose also that the six manifest indicators of latent status, measured at two time points, are Item 1 with $i, i' = 1, \dots, I$ response categories, Item 2 with $j, j' = 1, \dots, J$ response categories, and so on, where $i, j, k, l, m,$ and n refer to item responses at Time t and i', j', k', l', m', and n' refer to item responses at Time t+1. Let $p = 1, \dots, S$ denote the latent status at Time t and $q = 1, \dots, S$ the latent status at Time t+1.

In any data set, different individuals will respond to a set of items in different ways. A response pattern, which is a set of possible item responses for an individual, can then be represented by $Y = \{a, b, i, j, k, l, m, n, i', j', k', l', m', n'\}$, the vector of responses to: the two latent class indicators, the six latent status indicators at Time t, and the six latent status indicators at Time $t + 1$. The number of people who respond in a particular way can be tabulated, and used to determine the observed proportion of people giving each possible response pattern. In other words, this is the proportion in each cell of the contingency table formed by cross-tabulating all the items. An LTA model attempts to reproduce these observed proportions, in a manner conceptually similar to how a structural equation model attempts to reproduce an observed covariance matrix.

Parameters Estimated in the Model

There are five sets of parameters estimated in this model: γ (gamma) parameters, δ (delta) parameters, τ (tau) parameters, and two sets of ρ (rho) parameters, one set for the static latent classes and one set for the dynamic latent statuses. The estimated probability of having a particular response pattern is:

$$P(Y)=\sum_{c=1}^{C}\sum_{p=1}^{S}\sum_{q=1}^{S}\gamma_{c}\rho_{a|c}\rho_{b|c}\delta_{p|c}\rho_{i|p,c}\rho_{j|p,c}\rho_{k|p,c}\rho_{l|p,c}\rho_{m|p,c}\rho_{n|p,c}\tau_{q|p,c}\rho_{i'|q,c}\rho_{j'|q,c}\rho_{k'|q,c}\rho_{l'|q,c}\rho_{m'|q,c}\rho_{n'|q,c} \qquad (9.1)$$

where

γ_c is the probability of being in latent class c;

$\rho_{a|c}$ is the probability of having the value a on the first latent class indicator given membership in latent class c, and $\rho_{b|c}$ is the probability of having the value b on the second latent class indicator given membership in latent class c;

$\delta_{p|c}$ is the probability of being in latent status p given membership in latent class c;

$\rho_{i|p,c}$ is the probability of response i to Item 1 at Time t, given membership in latent status p at Time t and membership in latent class c (similarly, $\rho_{i'|q,c}$ represents the probability of response i' to Item 1 at Time $t + 1$, given membership in latent status q at Time $t + 1$ and membership in latent class c); and,

$\tau_{q|p,c}$ is the probability of membership in latent status q at Time $t + 1$ given membership in latent status p at Time t and membership in latent class c.

All parameters are estimated using the EM algorithm (Dempster, Laird, & Rubin, 1977), which iterates until a prespecified convergence criterion is reached.

New Features of LTA

The analyses presented here were performed using WinLTA, a new version of the LTA program (Brunner & Schafer, 1997; Collins, Schafer, Hyatt, Flaherty, & Hames, 1999). This program has two important improvements over the previous version, LTA (Collins, Wugalter, & Rousculp, 1992). The first improvement in the program is the added capability to handle missing data in the manifest indicators. The EM algorithm uses information from the model and incomplete data to estimate parameters. LTA's missing-data algorithm has been tested on data missing completely at random (MCAR), where items have a random probability of being missing, and on data missing at random (MAR), where items have a random probability of being missing conditional on the response to a measured variable (Hyatt & Collins, 1998). For example, MCAR would occur if missingness does not depend on any variables relevant to the study at hand. MAR, on the other hand, would occur if males and females had a different probability of having missing information on cocaine use, and if gender was measured and available for use in the missing data analysis. In a simulation study, Hyatt and Collins (1998) demonstrated that when ρ parameters are strong ($\rho = .9$) and the sample size is relatively large ($N = 1{,}000$), LTA's missing-data algorithm can successfully recover parameter estimates using data sets with up to 50% missing data. The amount of missing data in this study falls well within these bounds, thus it was reasonable to apply LTA's missing data routine.

Although the EM algorithm is useful for producing maximum likelihood parameter estimates for latent transition models, it does not provide a measure of uncertainty of those estimates. The second feature added to LTA is the estimation of standard errors for each parameter by means of data augmentation (Schafer, 1997). A variability estimate for computing confidence intervals is useful in making statistical comparisons across groups.

A SUBSTANCE-USE ONSET EXAMPLE

In this study, sequential stages of involvement in alcohol, tobacco, marijuana, and cocaine and the effects of perceived parental permissiveness are investigated by following adolescents from 9th to 10th grade. LTA, a methodology for testing stage-sequential models of individual growth, is used to test the goodness-of-fit of a model of substance-use onset. In this study, the phrase "substance-use onset" refers to the process of experimenting with new substances over time. This type of analysis allows us to estimate the probability of an adolescent having a high level of perceived parental permissiveness, the probability of membership in each stage of substance use given whether or not the adolescent perceives their parents as highly permissive, and the probability of transitioning to a higher level of substance use (experimentation with additional substances) in 10th grade given the individual's 9th grade substance-use level and whether or not the adolescent perceives his or her parents as highly permissive. Statistically significant differences in substance use stage prevalence and transition probabilities will be examined for adolescents with low and high levels of perceived parental permissiveness. The results of the analyses will provide insight as to which stages of growth position adolescents for higher risk of marijuana and cocaine use, and whether there is an effect of the level of perceived parental permissiveness on substance use onset.

METHODS

Subjects

The *Adolescent Alcohol Prevention Trial* (AAPT; Graham, Rohrbach, Hansen, Flay, & Johnson, 1989) was a prospective longitudinal panel study of adolescent substance-use onset in the American youth population. The survey employed a sample of 12 junior high schools in Los Angeles and Orange Counties, California stratified by size, test scores, and ethnic composition. Annual waves of data were collected in four panels, with the first panel beginning in the fall of 1986 and all four panels ending in the fall of 1991.

The subjects included in this study were $N = 3{,}906$ students who were 9th graders in Fall 1989 or Fall 1990 and 10th graders 1 year later and were administered the form of questionnaire that included indicators of parental permissiveness. This includes approximately one third of the total subjects included in the AAPT data. The sample used in this study was approximately 9.8% Asian, 3.5% Black, 30.5% Hispanic, 42.2% White, 7.7% other ethnic groups, and 4.7% unknown.

Any missing items for the subjects were dealt with using LTA's missing-data routine described before so that the analyses could be conducted on all subjects. The overall percentage of missing pieces of data for this study is 12.0%. The amount of attrition, or students dropping out between the 9th-grade wave and the 10th-grade wave, is 20.2%. Almost all of the missing information in this study is due to attrition, a problem common in longitudinal studies.

Measures

Substance Use. Six items measuring substance use were used in this study to define a subject's level of substance use at each year. Two items were used to measure lifetime drunkenness. First, drunkenness (How many times have you ever been drunk?) was coded 1 for "never been drunk" and 2 for "been drunk once or more." Second, subjects were asked the number of alcoholic drinks consumed on the day when they had the most alcohol, and this was coded 1 for "two or fewer drinks" and 2 for "three or more drinks." The remaining items measured lifetime alcohol use (How many alcoholic drinks have you had in your whole life?), lifetime tobacco use (How many cigarettes have you smoked in your whole life?), lifetime marijuana use (Have you ever used marijuana?), and lifetime cocaine use (Have you ever used cocaine?). Responses on these items were coded as 1 for "never used" and 2 for "used once or more," with the alcohol use item coded as 1 if the subject has only had sips for religious purposes. The simple dichotomization of all six measures is appropriate because the research question involves adolescents' first experimentation with alcohol, drunkenness, cigarettes, marijuana, and cocaine. In this onset model, advancement in the level of substance use is defined by trying a new substance even once.

Parental Permissiveness. Two items were used in this study to define a subject's perception of parental permissiveness in 9th grade. Adolescents were asked, "How angry would your parents be if they found out you tried drinking alcohol?" and "How angry would your parents be if they found out you got drunk?" Responses were coded 1 for "not angry at all," "a little angry," or "pretty angry," representing high parental permissiveness, and 2 for "very angry," representing low parental permissiveness (a more authoritative, or possibly authoritarian, style of parenting). Missing data were coded zero for all items measuring substance use and perceived parental permissiveness.

The Predicted LTA Model

Previously, Collins, Graham, Rousculp, & Hansen (1997) examined a stage-sequential model of substance-use onset developed in this data set. This model was characterized by eight drug-use latent statuses: (1) no drug use, (2) experimentation with alcohol only, (3) experimentation with tobacco only, (4) experimentation with alcohol and tobacco, (5) experimentation with alcohol to the point of drunkenness but no tobacco use, (6) alcohol use, tobacco use, and drunkenness, (7) alcohol use, drunkenness, and advanced use, and (8) alcohol use, tobacco use, drunkenness, and advanced use (advanced use was defined as more regular use of tobacco or alcohol, or any use of marijuana). Based on this previous model used by Collins and colleagues to examine substance use in seventh and eighth grade students, the researchers specified a modified model of onset that was expected to be more appropriate for 9th- and 10th-grade students. The current proposed model of substance-use onset is depicted in Fig. 9.1.

It is characterized by eight drug use latent statuses, each represented by a circle. (Readers familiar with path diagrams for structural equation models should note that the arrows here represent stage transitions rather than regressions.) The current model incorporates the adolescents' first experience with marijuana as a unique stage (as opposed to combining this with advanced use of alcohol and tobacco as was done in previous analyses), and the first experience with cocaine as the most advanced stage.

This onset model proposes that adolescents who first try tobacco go on to try alcohol, but those who begin the process with alcohol use might progress to experimentation with tobacco or proceed directly to experiencing drunkenness. Adolescents who progress from having already used tobacco and alcohol to the point of drunkenness are expected, if they continue with the onset process, to experiment next with marijuana, and then possibly with cocaine. It is important to note the one-way arrows in Fig. 9.1. This implies that backwards motion is impossible according to this model, which is appropriate when all six substance-use items reflect lifetime use.

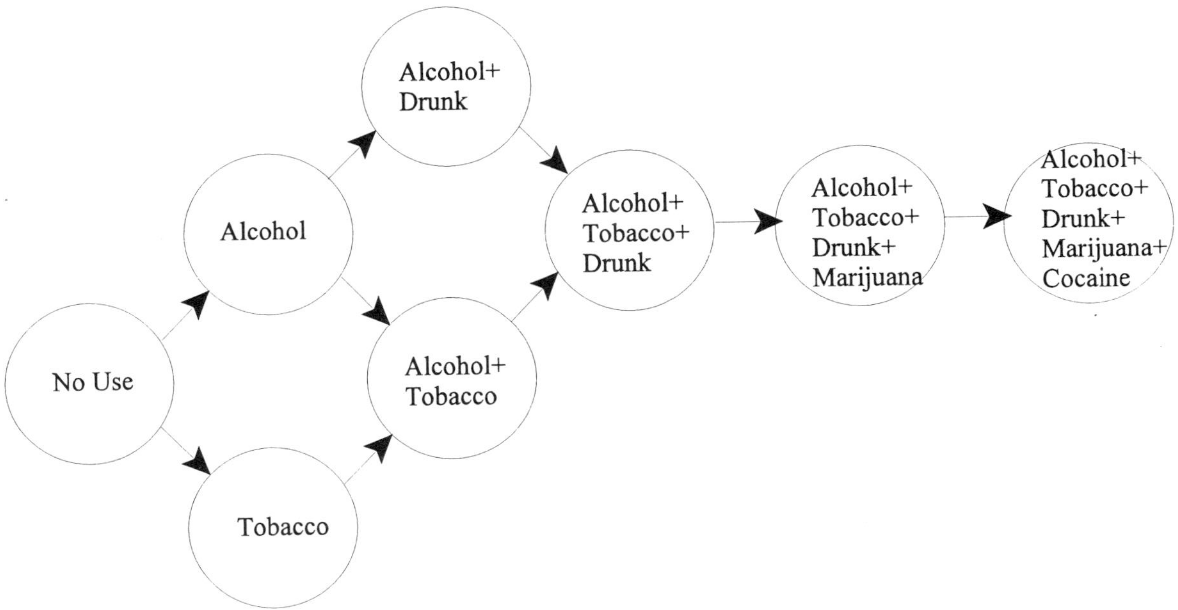

FIG. 9.1. The predicted stage-sequential model of substance-use onset.

It is possible that not all individuals will adhere to the specified model of onset. One type of error that can occur involves individuals giving inconsistent answers over time. For example, subjects who report alcohol use in 9th grade but report no alcohol use in 10th grade do not adhere to the model and therefore contribute to measurement error. Another type of error that may occur involves subjects who report paths that are not consistent with the model. For example, an individual may report having used alcohol, tobacco, and marijuana but may not report having experienced drunkenness. The model proposed here contains no latent status for alcohol, tobacco, and marijuana use without drunkenness, therefore this response would contribute to measurement error and decrease the overall fit of the model.

Identification and Constraints

Identification of a model is imperative in estimating LTA parameters. Underidentification occurs when there are too many parameters to estimate given the amount of information in the data. Positive degrees of freedom is a necessary but not sufficient condition for identification in LTA models. Possible signs of underidentification are parameters that remain at their starting value when they are not fixed to that value, parameter estimates that are very sensitive to different subsets of data or slight changes in the specified model, and parameter estimates that vary according to start values.

Problems with identification can often be eliminated by reducing the amount of estimation performed by the model. To achieve this, parameters can either be fixed or constrained. For a parameter that is fixed to a specified value, no estimation is required. For a set of two or more parameters constrained to be equal to each other, the estimation of only one parameter per equivalence set is required. Besides identification problems, another reason to fix or constrain parameters is to increase the specification and interpretability of the model.

In the model tested here, the γ and δ parameters were estimated freely. The τ parameters, which make up the transition probability matrix, had the same pattern of constraints for each latent class. These parameters were constrained so that illogical transitions between latent statuses could not occur (see Table 9.1).

For example, no backward movement was allowed—a subject who had tried cigarettes by 9th grade could not return to the "No use" latent

TABLE 9.1

τ Parameters Constraints

	10th Grade Latent Status							
9th Grade Latent Status	1	2	3	4	5	6	7	8
1. No Use	1	1	1	1	1	1	1	1
2. Alcohol	0	1	0	1	1	1	1	1
3. Tobacco	0	0	1	1	0	1	1	1
4. Alcohol + Tobacco	0	0	0	1	0	1	1	1
5. Alcohol + Drunk	0	0	0	0	1	1	1	1
6. Alcohol + Tobacco + Drunk	0	0	0	0	0	1	1	1
7. Alcohol + Tobacco + Drunk + Marijuana	0	0	0	0	0	0	1	1
8. Alcohol + Tobacco + Drunk + Marijuana + Cocaine	0	0	0	0	0	0	0	1

status in 10th grade. Three additional illogical τ parameters were set to 0. They are: The transition from "Alcohol only" to "Tobacco only"; the transition from "Tobacco only" to "Alcohol + Drunk"; and the transition from "Alcohol + Tobacco" to "Alcohol + Drunk." All other τ parameters (i.e., all logical transitions) were freely estimated.

The two manifest indicators of perceived parental permissiveness are assumed to be parallel items; therefore, only one ρ parameter for the static latent variable was estimated. Both items are constrained to have the same probability of responding "yes" when a "yes" was expected and "no" when a "no" was expected.

The ρ parameters for the dynamic latent variable were constrained to be equal across time and across latent classes, so that the meaning of the latent statuses did not change over time or across latent classes. It is imperative that the meaning of latent statuses remains the same across times in order to have meaningful transition probabilities. In addition, constraints were imposed so that only two ρ parameters were estimated for each of the six latent status items: one for the probability of responding "yes" when a "yes" was expected, and one for the probability of responding "no" when a "no" was expected. Imposing these restrictions greatly reduces the number of parameters to be estimated in the model and aids in interpreting the results.

RESULTS

Model Fit

Model fit is assessed by comparing the response pattern proportions predicted by the model with the observed response proportions appearing in the data. A goodness-of-fit statistic, G^2, expresses the degree of agreement between the predicted and observed proportions. The G^2 is in theory distributed as a chi-square, with degrees of freedom equal to *number of possible response patterns* minus *number of independent parameters estimated* minus 1. However, LTA models tend to involve large contingency tables and large degrees of freedom. Under these conditions, the G^2 is usually not distributed as a chi-square, and its distribution is unknown, so overall hypothesis testing can be done in only a rough way (Collins, Fidler, Wugalter, & Long, 1993). In the model tested here, the goodness-of-fit statistic was 4,925.26 with 16,305 degrees of freedom. Having the value of the fit statistic substantially less than the degrees of freedom indicates that the model fits the data reasonably well. If desired, it is possible to examine the impact of estimating a particular parameter on the model fit. This can be achieved by comparing the fit of one model, which has a particular parameter fixed to some value, to the fit of a second model, which is identical except that the parameter of interest is freely estimated. This comparison of nested models would allow the user to examine the impact of estimating that parameter on the overall model fit, and could possibly aid in determining the most appropriate model.

Parameter Estimates

ρ Parameters for the Static Part of the Model. These ρ parameters represent the amount of measurement error in the items that measure the static latent variable. Constraints were imposed such that only one ρ parameter was estimated for the static part of the model. Subjects in the low perceived parental permissiveness class had a probability of .828 of answering "very angry" to each of the permissiveness items, and subjects in the high perceived parental permissiveness class had an identical probability of answering something other than "very angry" to each of the permissiveness items. Values close to 0 and 1 indicate good measurement of the latent class; a value above .8 suggests that the two indicators of perceived parental permissiveness in this study are good measures.

ρ Parameters for the Dynamic Part of the Model. These ρ parameters represent the amount of measurement error in the items that measure the dynamic latent variable. Table 9.2 displays these parameter estimates, giving the probability of responding "yes" to each item.

The values were constrained to be equal across time and latent class, therefore the ρ parameters were the same for Time 1 and Time 2, and for high perceived parental permissiveness and low perceived parental permissiveness. Additionally, in order to achieve identification, constraints were imposed such that only two r parameters were estimated for each item. Because each of the six items was dichotomous (responses were coded either 1 or 2), the probability of responding "no" to each item can be calculated by subtracting the shown parameter value from 1.00. Most of these values are near 0 and 1, indicating that the latent variable "substance use" is being measured accurately. The probability of responding "yes" at Time 1 given membership in the "No Use" latent status and the high perceived parental permissiveness latent class, for example, is less than .18 for all six items, which we expect because subjects in this latent status have not used any substance.

TABLE 9.2

ρ Parameters for Response "Yes"

Latent Status	*Ever Tried Alcohol*	*Ever Tried Tobacco*	*Ever Been Drunk*	*3 or More Drinks*	*Ever Tried Marijuana*	*Ever Tried Cocaine*
No use	.177	.057	.037	.044	.020	.005
Alcohol	.983	.057	.037	.044	.020	.005
Tobacco	.177	.951	.037	.044	.020	.005
Alcohol + Tobacco	.983	.951	.037	.044	.020	.005
Alcohol + Drunk	.983	.057	.863	.854	.020	.005
Alcohol + Tobacco + Drunk	.983	.951	.863	.854	.020	.005
Acohol + Tobacco + Drunk + Marijuana	.983	.951	.863	.854	.880	.005
Alcohol + Tobacco + Drunk + Marijuana + Cocaine	.983	.951	.863	.854	.880	.590

γ Parameters. The γ parameters represent the probability of membership in each latent class. Based on these results, a 9th grader has a 25.3% chance of perceiving his or her parents as highly permissive and a 74.7% chance of perceiving his or her parents as having low permissiveness. Clearly, the majority of 9th graders believe that their parents would be angry to find out that they have tried drinking alcohol or been drunk.

δ parameters. In this study the δ parameters represent the probability of being in a certain latent status during the first wave, given membership in the perceived parental permissiveness latent class. Table 9.3 shows the proportion of adolescents in each of the eight substance-use statuses for 9th-grade adolescents with high and low levels of perceived parental permissiveness, along with 95% confidence intervals.

TABLE 9.3

δ Parameters: Probability of Latent Status Membership

Latent Status	*Level of Permissiveness*	*9th Grade*
No use	High	*.065 [.042,.097]
	Low	*.321 [.298,.344]
Alcohol	High	*.140 [.110,.176]
	Low	*.243 [.220,.268]
Tobacco	High	.017 [.008,.034]
	Low	.030 [.023,.039]
Alcohol + Tobacco	High	*.124 [.086,.172]
	Low	*.200 [.179,.222]
Alcohol + Drunk	High	*.089 [.067,.116]
	Low	*.036 [.028,.047]
Alcohol + Tobacco + Drunk	High	*.253 [.199,.315]
	Low	*.112 [.097,.129]
Alcohol + Tobacco + Drunk + Marijuana	High	*.220 [.191,.251]
	Low	*.032 [.022,.046]
Alcohol + Tobacco + Drunk + Marijuana + Cocaine	High	*.092 [.068,.122]
	Low	*.026 [.017,.037]

Note: Point estimate and 95% confidence interval based on data augmentation.

* Indicates significant difference between high and low parental permissiveness.

In 9th grade, the "No use" latent status has the highest expected membership for the low permissiveness group, with 32.1% expected to be abstainers, but the "Alcohol, Tobacco, and Drunkenness" latent status has the highest expected membership for the high permissiveness group, with 25.3% expected to be in this status. It is interesting to note the small percentage (1.7% and 3.0%) of 9th graders who are expected to have tried only tobacco. This indicates that, of the students who are expected to have experimented with tobacco by 9th grade, most are expected to have also experimented with alcohol by 9th grade.

Significant differences between students with low and high levels of perceived parental permissiveness show up in seven of the eight latent statuses. Results suggest that students who perceive low parental permissiveness are almost five times more likely than students who perceive high parental permissiveness to be in the "No Use" latent status (32.1% vs. 6.5%). Students with low perceived parental permissiveness are also more likely to be in the "Alcohol Only" status (24.3% for low permissiveness vs. 14.0% for high permissiveness) and the "Alcohol and Tobacco" status (20.0% for low permissiveness vs. 12.4% for high permissiveness). However, 9th graders who perceive their parents as highly permissive are more likely to fall into all latent statuses involving drunkenness, marijuana use, or cocaine use than those who perceive their parents as less permissive. Students who perceive their parents to be highly permissive are significantly more likely to be in the "Alcohol and Drunkenness" status (8.9% vs. 3.6%) and the "Alcohol, Tobacco, and Drunkenness" status (25.3% vs. 11.2%). By 9th grade, over six times more students with parents perceived as highly permissive are expected to have experimented with marijuana but not cocaine (22.0% vs. 3.2%), and more than three times as many are expected to have experimented with cocaine (9.2% vs. 2.6%). It appears that having parents with high levels of perceived permissiveness may act as a risk factor in adolescent substance use. Parents who are unsuccessful at conveying to their children that they should not try drinking alcohol are more likely to have adolescents who are advanced in substance use.

Prevalence rates can be calculated by combining across statuses with common substance use characteristics. Table 9.4 shows, for adolescents with high and low levels of perceived parental permissiveness, the proportion of students in 9th grade who have experimented with each substance. By 9th grade, substantial differences between these groups have appeared. More than five times as many students with high perceived

TABLE 9.4

Prevalence Rates

Substance	*Level of Permissiveness*	*9th Grade*
None	High	6.5%
	Low	32.1%
Ever Tried Alcohol	High	91.8%
	Low	64.9%
Ever Tried Tobacco	High	70.6%
	Low	40.0%
Ever Been Drunk	High	65.4%
	Low	20.6%
Ever Tried Marijuana	High	31.2%
	Low	5.8%
Ever Tried Cocaine	High	9.2%
	Low	2.6%

parental permissiveness have tried marijuana (31.2% vs. 5.8%). Three times as many students with parents perceived as highly permissive have been drunk (65.4% vs. 20.6%), and more students from this group have tried alcohol and tried tobacco. In fact, 91.8% of the 9th graders who perceive their parents to be highly permissive of alcohol use have already experimented with alcohol. It is clear from this table that 9th graders who perceive their parents as highly permissive have substantially higher prevalence rates for every drug surveyed in this study.

τ Parameters. The τ parameters make up the transition probability matrix for each latent class. Each parameter represents the conditional probability of an individual's transitioning to a certain status at Time 2 given latent status at Time 1 and latent class membership. Table 9.5 displays a matrix of the τ parameters for adolescents with high levels of perceived parental permissiveness and Table 9.6 displays the same matrix for adolescents with low levels of perceived parental permissiveness.

These matrices show the conditional probability of an adolescent transitioning to each substance use status in 10th grade given their 9th-grade substance-use status. Overall, about 25% of these 9th graders

TABLE 9.5

τ Parameters (Transition Probability Matrix): High Parental Permissiveness

	10th Grade Latent Status							
9th Grade Latent Status	*1*	*2*	*3*	*4*	*5*	*6*	*7*	*8*
1. No use	.561	.044	.013	.097	.020	.083	.041	*.142
	[.253,	[.001,	[.001,	[.016,	[.001,	[.013,	[.001,	[.054,
	.831]	.457]	.140]	.343]	.191]	.314]	.448]	.306]
2. Alcohol		.622	—	.028	*.232	.071	.019	.028
		[.477,		[.001,	[.123,	[.019,	[.000,	[.008,
		.750]		.328]	.385]	.202]	.221]	.086]
3. Tobacco			.535	.311	—	.042	.076	.036
			[.157,	[.080,		[.003,	[.002,	[.002,
			.878]	.676]		.362]	.589]	.364]
4. Alcohol + Tobacco				.574	—	.236	.150	.040
				[.406,		[.113,	[.074,	[.001,
				.729]		.416]	.270]	.384]
5. Alcohol + Drunk					.642	.205	.035	.117
					[.322,	[.064,	[.000,	[.010,
					.882]	.461]	.379]	.500]

Table 9.5 continued

Table 9.5 continued

9th Grade Latent Status	*10th Grade L* 1	2	3	4	5	6	7	8
6. Alcohol + Tobacco + Drunk						.661	.285	.054
						[.574,	[.199,	[.002,
						.739]	.387]	.464]
7. Alcohol + Tobacco + Drunk + Marijuana							.891	.109
							[.576,	[.013,
							.987]	.424]
8. Alcohol + Tobacco + Drunk + Marijuana + Cocaine								1,000

Note: Point estimate and 95% confidence interval based on data augmentation.
Rows may not sum exactly to one due to rounding.
* Indicates significant difference between high and low parental permissiveness.

TABLE 9.6

τ Parameters (Transition Probability Matrix): Low Parental Permissiveness

	10th Grade Latent Status							
9th Grade Latent Status	*1*	*2*	*3*	*4*	*5*	*6*	*7*	*8*
1. No use	.766	.097	.011	.053	.032	.031	.002	*.008
	[.717,	[.064,	[.000,	[.024,	[.016,	[.018,	[.000,	[.001,
	.809]	.143]	.224]	.106]	.060]	.052]	.020]	.031]
2. Alcohol		.775	—	.074	*.071	.052	.013	.015
		[.704,		[.030,	[.041,	[.021,	[.002,	[.004,
		.835]		.157]	.117]	.112]	.055]	.052]
3. Tobacco			.374	.406	—	.144	.014	.062
			[.214,	[.222,		[.051,	[.001,	[.002,
			.562]	.617]		.321]	.151]	.442]
4. Alcohol + Tobacco				.715	—	.229	.024	.033
				[.649,		[.172,	[.003,	[.017,
				.773]		.295]	.111]	.059]
5. Alcohol + Drunk					.715	.033	.212	.041
					[.573,	[.001,	[.080,	[.000,
					.827]	.273]	.431]	.417]

Table 9.6 continued

Table 9.6 continued

	10th Grade L							
9th Grade Latent Status	*1*	*2*	*3*	*4*	*5*	*6*	*7*	*8*
6. Alcohol + Tobacco + Drunk						.747	.207	.046
						[.666,	[.115,	[.001,
						.815]	.335]	.487]
7. Alcohol + Tobacco + Drunk + Marijuana							.967	.033
							[.650,	[.000,
							1.000]	.350]
8. Alcohol + Tobacco + Drunk + Marijuana + Cocaine								1.000

Note: Point estimate and 95% confidence interval based on data augmentation.
Rows may not sum exactly to one due to rounding.
* Indicates significant difference between high and low parental permissiveness.

are expected to advance to higher stages of substance use by 10th grade, regardless of the level of perceived parental permissiveness. This indicates that, in general, about the same number of adolescents with high and low levels of perceived parental permissiveness are expected to advance in their status during this year. One possible explanation is that any differences between these two groups has already occurred by 9th grade. A closer inspection of the transition probability matrices, however, reveals more information about group differences in *how* students are advancing in substance use. For example, the probability of staying in the "Alcohol and Drunkenness" latent status from 9th to 10th grade is similar for the two groups (64.2% and 71.5%). But, of the students who did advance in substance use, adolescents who perceive high parental permissiveness are more than twice as likely to experiment with tobacco, marijuana, and cocaine during that year than adolescents who perceive low parental permissiveness. Although these differences are not statistically significant, it is clear that the two groups of adolescents may not be changing in the same way over time.

One important transition probability did show a statistically significant group difference. For adolescents with parents perceived as highly permissive, someone in the "No Use" status at Grade 9 has a 14.2% chance of transitioning to the most advanced status (alcohol use to the point of drunkenness, tobacco use, marijuana use, and cocaine use) by Grade 10. This chance is only 0.8% for adolescents exposed to low levels of perceived parental permissiveness. This finding indicates that parenting style may serve as a risk factor for advancement in substance use over time. Additionally, a second transition probability revealed a significant group difference in the same direction. Adolescents with high perceived parental permissiveness have a 23.2% chance of advancing from the "Alcohol Only" status to the "Alcohol and Drunkenness" status, compared to only 7.1% of adolescents with low perceived parental permissiveness. These two statistically significant group differences provide additional evidence consistent with our prediction that high levels of perceived parental permissiveness may act as a risk factor in adolescent drug use.

It is interesting to note that, for both groups of perceived parental permissiveness, students who have only experimented with tobacco by 9th grade seem to be more likely to advance to the latent status including cocaine use by 10th grade than students who have only experimented with alcohol by 9th grade (although the differences are not significant).

This suggests that adolescents who begin their substance use onset trajectory with tobacco use may be at higher risk for advancing in the future. This is also apparent by examining the probability of staying in the "Tobacco only" latent status for both grades. Adolescents are less likely to remain in the "Tobacco only" latent status than to remain in the "Alcohol only" latent status (53.5 % vs. 62.2% for the high permissiveness group, and 37.4 % vs. 77.5% for the low permissiveness group; the latter difference is significant). In fact, 9th graders who have experimented only with tobacco are more likely to advance in substance use by 10th grade than 9th graders at any other level of substance use. This can be seen by examining the diagonal elements in Tables 9.5 and 9.6. This may be due, however, to the high prevalence of alcohol use at this age; it is common for adolescents who begin their substance-use trajectories with tobacco to progress quickly to experimentation with alcohol.

DISCUSSION

In general, exposure to a high level of perceived parental permissiveness appears to act as a risk factor in adolescent substance-use onset. In the LTA analyses presented here, adolescents with parents who showed high levels of permissiveness were significantly more likely to be in latent statuses involving more advanced use, such as drunkenness, marijuana use, and cocaine use, and less likely to be in latent statuses involving no use or experimentation with alcohol only. The prevalence rate for every type of drug in this study is higher for 9th graders with high perceived parental permissiveness. It is important to keep in mind that these statistics reflect substance use in students approximately 14- and 15-years-old. By 9th grade, more than 90% of the subjects exposed to high parental permissiveness have tried alcohol, 71% have tried cigarettes, two thirds have been drunk, and one third have tried marijuana. The onset process is fairly advanced for these adolescents, suggesting that a substantial number of 9th-grade students are at risk for increased substance use in the upcoming year. The high prevalence rates also indicate that attempts to prevent advanced substance use need to begin well before 9th grade.

Understanding the temporal sequencing of substance use onset is critical in the development of prevention programs for American youth. By using Latent Transition Analysis to examine the effect of perceived parental permissiveness on substance-use onset, we can identify prospectively when high levels of permissiveness act as a risk factor for adolescents. By analyzing longitudinal data on substance-use onset using a stage-sequential model, we can target specific groups of adolescents who are at high risk for using marijuana and cocaine. This study uses substance-use characteristics along with exposure to parental opinions on adolescent alcohol use to predict future use. One noteworthy finding of this study involves the significant effect of high perceived parental permissiveness as a risk factor for future cocaine use. Specifically, these results indicate that adolescents in the "No Use" status exposed to high levels of perceived parental permissiveness in 9th grade have a 14.2% chance of using cocaine in the next year, compared to only 0.8% of those exposed to low levels of perceived parental permissiveness. This is a striking finding for two reasons. First, going from no use of any substance to cocaine use in 1 year is an extremely rapid change. Second, the difference between the groups is large, with nonusing individuals in the high permissiveness group 17 times more likely to make this rapid transition than nonusing individuals in the low permissiveness group.

However, there are good reasons to be very cautious about these findings. One reason is that the numbers of individuals involved are small in absolute terms. A second reason is that in LTA analyses there are many nonindependent parameters, which leads to the possibility of capitalizing on chance when doing many hypothesis tests on individual parameters. It is interesting and informative to look at confidence intervals for the parameter estimates, but when looking at this many confidence intervals there is the possibility that a few significant differences will emerge spuriously. Although a Bonferroni correction could be used, this would be extremely conservative. In this case, we felt we had good reason to predict a priori that differences would be in the direction of more substance use for the adolescents whose parents were perceived as more permissive. It is important to replicate these findings in another data set. If the findings turns out to be valid, it will be important to identify other risk factors that these individuals may be sharing.

Implications for Prevention

We have found evidence for group differences in adolescent substance-use prevalence rates based on the level of perceived parental permissiveness, and we have also established group differences in change over time.

Ninth-grade students who perceive their parents as permissive tend to advance faster along the pathway to substance-use onset, and their profile includes experimentation with more types of drugs than students with less permissive parents. It is likely that the effects of perceived parental permissiveness begin long before 9th grade and endure throughout adolescence. An important follow-up to this study would be to examine the relationship between perceived parental permissiveness of adolescent alcohol use and the onset of adolescent substance use at younger ages, when adolescents frequently experiment with alcohol and tobacco for the first time.

These results suggest that parents should clearly convey their disapproval of substance use in their children. Because parents often find it difficult to discuss substance use with their children, it is important for prevention programs with family or parental components to help parents to gain the skills needed to have such conversations with their children.

Limitations of This Study

One limitation of this study has to do with the small number of indicators of parental permissiveness. Parents of the adolescents were not surveyed, therefore only the adolescents' reports on perceived parental permissiveness could be used. Because the focus of this study was alcohol, not illicit drugs, both of the questions were asked only in relation to alcohol. Additionally, because parental monitoring was not a focus of the AAPT study, few indicators of this important and highly related construct were collected from the adolescents. In general, poor measurement tends to work against finding significant results. Given that we were able to find a significant effect of perceived parental permissiveness on substance-use onset using only two indicators, we feel that perceptions of parental permissiveness play an important role in the effect parents can have on their children's substance-use choices. However, it is undoubtedly true that other related constructs play an equally important role. For example, it is likely that the inclusion of other aspects of parental monitoring in the model would shed more light on the effect parents have on their children's pathways to substance use.

It is possible that our measure of alcohol use could be problematic for subjects who drink for religious purposes only. These subjects would be considered nondrinkers, but might perceive their parents as permissive toward alcohol use due to their religious practices even though religious parents might actually be less permissive. However, three things weaken this potential threat to the results found in this analysis. First, one of the measures of parental permissiveness refers to the adolescent getting drunk, which is very different from having a sip of alcohol for religious purposes. Second, the presence of adolescents who drink only for religious purposes and perceive their parents as permissive toward alcohol use would actually weaken the statistical results, yielding more conservative conclusions. Third, perceived parental permissiveness appears to be a risk factor for several other substances besides alcohol, suggesting that the interpretation of the permissiveness question was clear.

Only two significant differences between the high and low perceived permissiveness groups emerged in the transition probability matrices. Little is known at this point regarding statistical power in latent transition models. One reason for the low number of significant transition probabilities may be that the model includes only two indicators of permissiveness. However, ρ parameters of .83 indicate that these two indicators were measured fairly reliably. Another reason may be the τ parameters themselves. These parameters tend to have the largest standard errors of any of the parameters in the LTA model, and are conditional probabilities based on often quite small sample sizes. Because latent transition models depend on complex contingency tables, extreme cell sparseness can yield low power in detecting statistically significant findings. These two conditions result in wide confidence intervals and make it more difficult to establish statistical significance of group differences.

General Limitations and Strengths of the LTA Approach

LTA combines some of the features of contingency table models and confirmatory factor analysis. Like contingency table models, LTA takes as input cell frequencies, which in this case are response pattern frequencies. LTA models attempt to reproduce these response pattern frequencies, just as log-linear or logistic regression models attempt to reproduce cell frequencies. LTA shares with other contingency table models a limitation on the number of variables that can reasonably be involved in any particular analysis. When the number of cells becomes large, contingency table models can become unwieldy. The model tested here involved six dichotomous items measured at two times, and two more dichotomous items measured once, resulting in a contingency table of 16,384 cells. Although the LTA procedure can handle variables with more than two outcomes (e.g., variables that are not dichotomous), this is often not practical because of the resulting size of the contingency table. The researcher is then faced with having to dichotomize variables, which usually leads to a loss of information. In some situations, such as highly skewed distributions, the loss may be minor; in other situations, dichotomizing variables may be unpalatable.

A related problem arises when it is desired to include exogenous variables, such as the perceived parental permissiveness variable in the model tested here. Unlike continuous variable models, such as regression analyses, where it is possible to build very elaborate models of prediction with numerous exogenous variables, most contingency table models are limited with respect to the number of predictors and/or subgroups it is practical to include. Thus, both the number of exogenous variables and the number of latent classes are limited in LTA models for practical reasons. Researchers who are accustomed to being able to "control for" the effects of a large set of variables while they examine the effects of others may find this limitation frustrating in LTA models. Some interesting work has been done recently by Pfeffermann, Skinner, and Humphreys (1998) on incorporating predictors into LTA models using a logistic regression perspective. Eventually this approach may allow the use of several simultaneous continuous predictors in LTA models.

Despite these limitations, LTA has a number of notable strengths. LTA gives researchers a method for estimating and testing hypotheses about stage-sequential development, yielding a uniquely detailed look at stage-sequential change over time. LTA can reveal instances where change may be occurring at a different rate or in a qualitatively different manner for different subgroups. In the analyses presented here, for example, we were able to see that in the low perceived permissiveness group, it appears that individuals who started the onset process with tobacco are advancing more quickly than those who started with alcohol. Such subgroup differences can be very difficult to detect using continuous variable approaches.

An important feature of LTA is that the estimates of stage incidence and prevalence it provides are adjusted for measurement error. Although adjusting for measurement error is a procedure associated in many people's minds with continuous variables, it is just as important for discrete variables. The effect of measurement error in discrete variables is to cause individuals to be randomly misclassified, in this case misclassified into the wrong stage. In a transition probability matrix involving three or more stages, there are more off-diagonal elements than diagonal elements. Thus, a random misclassification is more likely to move an individual from the diagonal to the off-diagonal than it is to move individuals onto the diagonal. The effect of this is to create a transition probability matrix that has the appearance of more movement between stages than actually occurred. Adjusting for measurement error, as LTA does, gives a more realistic appraisal of the amount of change over time.

ACKNOWLEDGMENTS

The authors are Stephanie L. Hyatt and Linda M. Collins of The Methodology Center and Department of Human Development and Family Studies, The Pennsylvania State University. This research was supported by grant P50 DA 10075, *Center for the Study of Prevention through Innovative Methodology,* from the National Institute on Drug Abuse. The authors would like to thank John W. Graham for allowing them access to data from the Adolescent Alcohol Prevention Trials.

REFERENCES

Barnes, G. M., & Welte, J. W. (1986). Patterns and predictors of alcohol use among 7–12th grade students in New York state. *Journal of Studies on Alcohol, 47,* 53–62.

Brook, J. S., Gordon, A. S., Whiteman, M., & Cohen, P. (1986). Some models and mechanisms for explaining the impact of maternal and adolescent characteristics on adolescent stage of drug use. *Developmental Psychology, 22,* 460–467.

Brunner, R. M., & Schafer, J. L. (1997). *Inference for latent transition models* (Technical Report Number 97-11). University Park, PA: The Pennsylvania State University, The Methodology Center.

Collins, L. M., Fidler, P. L., Wugalter, S. E., & Long, J. L. (1993). Goodness-of-fit testing for latent class models. *Multivariate Behavioral Research, 28,* 375–389.

Collins, L. M., Graham, J. W., Long, J. D., & Hansen, W. B. (1994). Crossvalidation of latent class models of early substance use onset. *Multivariate Behavioral Research, 29,* 165–183.

Collins, L. M., Graham, J. W., Rousculp, S. S., & Hansen, W. B. (1997). Heavy caffeine use and the beginning of the substance use onset process: An illustration of latent transition analysis. In K. Bryant, M. Windle, & S. West (Eds.), *The Science of Prevention: Methodological Advances from Alcohol and Substance Abuse Research,* pp. 79–99. Washington, DC: American Psychological Association.

Collins, L. M., Schafer, J. L., Hyatt, S. L., Flaherty, B. P., & Hames, K. E. (1999). *WinLTA 2.0 User's Manual.* Pennsylvania State University: The Methodology Center.

Collins, L. M., & Wugalter, S. E. (1992). Latent class models for stage-sequential dynamic latent variables. *Multivariate Behavioral Research, 27,* 131–157.

Collins, L. M., Wugalter, S. E., & Rousculp, S. S. (1992). *Latent Transition Analysis (LTA) Program Manual (Version 1.0).* Los Angeles: University of Southern California.

Dempster, A. P., Laird, N. M., & Rubin, D. B. (1977). Maximum likelihood estimation from incomplete data via the EM algorithm. *Journal of the Royal Statistical Society Series B, 39,* 1–38.

Goodman, L. A. (1974). Exploratory latent structure analysis using both identifiable and unidentifiable models. *Biometrika, 61,* 215–231.

Graham, J. W., Collins, L. M., Wugalter, S. E., Chung, N. K., & Hansen, W. B. (1991). Modeling transitions in latent stage-sequential processes: A substance use prevention example. *Journal of Consulting and Clinical Psychology, 59,* 48–57.

Graham, J. W., Rohrbach, L., Hansen, W. B., Flay, B. R., & Johnson, C. A. (1989). Convergent and discriminant validity for assessment of skill in resisting a role play alcohol offer. *Behavioral Assessment, 11,* 353–379.

Hawkins, J. D., Catalano, R. F., & Miller, J. Y. (1992). Risk and protective factors for alcohol and other drug problems in adolescence and early adulthood: Implications for substance abuse prevention. *Psychological Bulletin, 112,* 64–105.

Hyatt, S. L., & Collins, L. M. (1998). *Estimation in latent transition models with missing data* (Technical Report Number 98-22). University Park, PA: The Pennsylvania State University, The Methodology Center.

Kandel, D. B., Kessler, R. C., & Margulies, R. Z. (1978). Antecedents of adolescent initiation into stages of drug use: A developmental analysis. *Journal of Youth and Adolescence, 7,* 13–40.

Martin, R. A., Velicer, W. F., & Fava, J. L. (1996). Latent transition analysis to the stages of change for smoking cessation. *Addictive Behaviors, 21,* 67–80.

McDermott, D. (1984). The relationship of parental drug use and parent's attitude concerning adolescent drug use to adolescent drug use. *Adolescence, 19,* 89–97.

Pfeffermann, D., Skinner, C., & Humphreys, K. (1998). The estimation of gross flows in the presence of measurement error using auxiliary variables. *Journal of the Royal Statistical Society, Series A, 161,* 13–32.

Posner, S. F., Collins, L. M., Longshore, D., & Anglin, D. (1996). The acquisition and maintenance of safer sexual behaviors among injection drug users. *Substance Use and Misuse, 31,* 1995–2015.

Schafer, J. L. (1997). *Analysis of incomplete multivariate data.* London: Chapman & Hall.

Yamaguchi, K., & Kandel, D. B. (1984). Patterns of drug use from adolescence to young adulthood: II. Sequences of progression. *American Journal of Public Health, 74,* 668–672.

10

Prospective Predictors of Smoking Cessation: A Logistic Regression Application

Jennifer S. Rose
Indiana University

Laurie Chassin
Clark C. Presson
Arizona State University

Steven J. Sherman
Indiana University

Dichotomous outcome variables are common in substance use research (e.g., smoker vs. nonsmoker). This chapter describes the use of logistic regression analysis to model such outcomes. A latent variable approach is used to describe the logistic regression model. An observed dichotomous outcome variable can be thought of as having a continuous variable underlying it (Long, 1997). The continuous latent variable is linearly related to the predictors, and is linked to the observed dichotomous outcome by a single cutpoint or threshold value for which larger values have a value of 1 on the observed dichotomous outcome and smaller values have a an observed value of 0. The purpose of logistic regression is to find a linear combination of predictors that maximizes the likelihood of obtaining the observed probability of the sample outcome. Logistic regression is illustrated using an example from a study investigating prospective predictors of successful smoking cessation in young adults.

In many areas of substance use research we encounter questions about whether or not an individual is engaging in some type of substance use and what might predict whether or not a person uses a substance. For example, we might want to know whether or not a person is currently dependent on alcohol, or we might want to know what might predict whether or not an individual will use cocaine in the future. In the area of smoking behavior research, this kind of question arises quite frequently. For instance, we might want to predict whether or not regular smokers will be successful in their attempts to quit smoking.

This type of outcome variable is known as a categorical or discrete outcome. Because it has two categories (e.g., yes/no for whether or not smokers are successful in their quit attempts), it is also known as a dichotomous outcome variable. Only certain statistical approaches can be taken with dichotomous outcomes, and logistic regression is one appropriate method for investigating predictors of this type of outcome variable. In logistic regression, the predictors can be continuous and/or categorical. Logistic regression employs a maximum likelihood estimation method to estimate the probability of being in either one or the other outcome group (e.g., unsuccessful quitter vs. successful quitter) based on values of the predictor variables. The values of the outcome are typically coded as 0 if the outcome is not present (e.g., unsuccessful quitter), or 1 if the outcome is present (e.g., successful quitter). The categories of the outcome variable must be exhaustive and mutually exclusive. This means that all individuals in the chosen sample must be classified into either one or the other category, and an individual cannot belong to both categories. For instance, people may be considered either successful quitters or unsuccessful quitters, but they cannot be considered both.

THE LINEAR PROBABILITY MODEL

Past researchers have often used standard linear regression (with ordinary least squares (OLS) estimation), which is typically used when the outcome variable is continuous, to examine dichotomous outcome variables. For example, in the linear probability model, OLS estimation is used to predict the probability that an outcome will occur. However, there are several problems with this method that make logistic regression a more suitable method of analysis, especially given the recent wide availability of software programs that can estimate logistic regression models. The main problem lies with the types of assumptions that are required in linear regression in order to obtain unbiased parameter estimates and accurate significance tests for the predictor variables.

As may be recalled, there are several major assumptions that you make about the regression model before you begin analysis. The first is that the model is correctly specified, that is, all relevant predictor variables have been included and all irrelevant ones excluded. Although this assumption is often difficult to meet in real life, it still remains an underlying assumption of the regression model. The second assumption is that none of the predictors are collinear, meaning that they are not highly correlated with other predictors. The third is that the disturbance (also known as the error term or residual) has a mean of zero and is not correlated with any of the predictor variables. Furthermore, it is assumed that the disturbance has the same variance across all values of a predictor variable and that it is not correlated across individuals. That is, the disturbance term for each observation is assumed to be homoscedastic and to be serially independent. Another important assumption is that the disturbances are normally distributed, which allows the assumption that the parameter estimates for the predictors (e.g., the regression coefficients) will also be normally distributed. Finally, it is assumed that the relations of the predictor variables to the outcome variable are linear, although higher order relations also can be modeled by including the appropriate term in the analysis (e.g., the square of a predictor to model a quadratic effect, or a multiplicative term to model an interactive effect).

When trying to use linear regression to examine a dichotomous outcome variable, we tend to violate some of these assumptions, and as a result, parameter estimates and significance tests can become suspect. In the linear regression model, the outcome variable is assumed to be continuous, and theoretically, both the predictors and the outcome are free to take on any value ranging from negative to positive infinity. However, unlike a continuous outcome variable, a dichotomous outcome variable can take on only two observed values, and the dichotomous outcome, modeled as a probability, can range from only 0 to 1. This limited range for the outcome variable causes problems because it means that, if the outcome variable only has two values, then the disturbance term can take on only two values for any level of the predictor variables.

As a result, the disturbance is not likely to be normally distributed, and the variance of the disturbance is no longer homoscedastic because it varies systematically with the levels of the predictor variables. In addition, with restrictions on the range of the outcome variable modeled as a probability limited to between 0 and 1, the relations between the predictors and the outcome are not likely to be linear. This is most likely to occur at extreme values of the predictors because the value of the outcome variable can never fall below 0 or go above 1. Thus, the increase in the probability of an outcome occurring with a one-unit change in the predictor at probabilities closer to 0 or 1 will be smaller than it would be if the probability was at .50. The result is an overall S-shaped relation of the predictors to the probability of being in one or the other category of the outcome variable, as opposed to a linear relation. This violation of the assumption of linearity is probably the most serious problem associated with using linear regression for dichotomous outcome variables.

THE LOGISTIC REGRESSION MODEL

As with OLS regression, the logistic regression model requires that the data are normally distributed, homoscedastic, independent, and that there is no collinearity among the predictors. However, one way the logistic regression model can overcome the problems presented by using linear regression with a dichotomous outcome variable is by making the assumption that there is actually a continuous variable underlying the dichotomous outcome variable (Long, 1997). Consider the case of a dichotomous outcome variable that is coded 0 for current nonsmoker and 1 for current smoker. All smokers have an observed value of 1, but they may not all have the same propensity to smoke.

For example, some current smokers may be thinking about quitting soon, whereas other current smokers may have no interest in quitting. Similarly, all current nonsmokers have an observed value of 0, but some current nonsmokers may be adamantly against smoking, whereas others may be thinking about taking up smoking. This continuous latent variable can range from negative infinity to positive infinity and is linearly related to the predictors. It is linked to the observed outcome variable by a cutpoint, or threshold, which has an assumed value of 0. When the value of the continuous latent outcome variable is less than or equal to 0, then our observed dichotomous outcome variable is equal to 0. When the value of the continuous latent variable is greater than zero, then our observed dichotomous outcome variable is equal to 1.

However, because this outcome variable is latent, we can no longer use OLS estimation to derive our parameter estimates. Instead we have to use maximum likelihood estimation.

The drawback to maximum likelihood estimation is that our distributional assumptions are based on large sample, or asymptotic, theory, meaning that the assumptions are more likely to be satisfied as our sample size approaches infinity. As a result, logistic regression tends to require larger sample sizes than are required for linear regression with OLS estimation. The required sample size depends on many things, including (but not limited to) the number of predictors in the logistic regression model, the distribution of these variables, effect sizes, and how rarely the outcome of interest occurs in the sample. However, Long (1997) suggested as a basic rule of thumb that, when using maximum likelihood estimation, the ratio of subjects to parameters should be no less than 10:1, and that the sample size should never be less than 100, regardless of the number of parameters. In addition, using Monte Carlo simulations, Peduzzi, Concato, Kemper, Holford, and Feinstein (1996) found that, in general, 10 outcome events per predictor (with outcome events defined by whichever group has the smallest number) were sufficient to obtain unbiased and accurate parameter estimates, as well as valid tests of significance. For example, if our outcome variable was successful ($N = 120$) versus unsuccessful ($N = 380$) smoking cessation, we would say that we have 120 outcome events. Ideally then, we would want no more than 12 predictors (120/10=12) in our final logistic regression model. If the prevalence was reversed with $N = 380$ successful quitters, and $N = 120$ unsuccessful quitters, the number of outcome events would still be 120 because it is the smaller number, and we would still want a maximum of 12 predictors.

In addition, because the outcome variable is assumed to be latent in the logistic regression model, we know nothing about its distribution, and cannot determine its distribution from the observed dichotomous outcome variable. Thus, the latent outcome variable is unidentified, so we need to identify it by making assumptions about its distribution. That is, we need to give it an origin (i.e., mean) and metric (i.e., variance). In the logistic regression model, the error distribution for the latent continuous outcome variable is assumed to have a mean of 0 and a variance that is equal to $\pi^2/3$ (which is approximately equal to 3.29). This value for the variance of the disturbance term in logistic regression is chosen primarily for its mathematical simplicity (Long, 1997).

The logistic regression model determines the likelihood (or probability) that an outcome *(Y)* will occur, based on a vector of the values of the predictor variables *(X)*. The probability of the outcome occurring (e.g., successful quitter: $Y = 1$) given the values of the predictors is P_1, and is written as:

$$Pr(Y = 1 \mid X). \tag{10.1}$$

The probability for the outcome not occurring (e.g., unsuccessful quitter: $Y = 0$) given the values of the predictors is P_0, which is written as:

$$Pr(Y = 0 \mid X) \tag{10.2}$$

and is equal to $1\text{-}P_1$. Thus, the likelihood of having a value of 1 on the outcome *Y*, given the levels of the predictor variables, is a ratio equal to:

$$P_1/(1\text{-}P_1). \tag{10.3}$$

We use maximum likelihood estimation to solve for the likelihood of the outcome occurring.

The purpose of maximum likelihood estimation in the logistic regression model is to obtain estimates of the regression coefficients that maximize the likelihood (probability) of obtaining the observed value of Y for the sample. Using an example from cigarette smoking research, we might be interested in obtaining a linear combination of predictors that maximizes the likelihood of being a successful quitter. Like linear regression, the probability of the outcome occurring is a function of a linear combination of the predictor variables, which can be written as:

$$P_1 = F(\alpha + \beta X). \tag{10.4}$$

This is known as the *likelihood function*. This linear combination of the predictors includes an estimate for the intercept (α) and parameter estimates for each of the *k* predictor variables multiplied by the values of the predictors ($b_1x_1 \ldots b_kx_k$) for each individual. However, in order to achieve a linear relation between the outcome and the linear combination of the predictors, we need to transform the probability of *Y* so that it is no longer constrained to be between 0 or 1. A natural log transformation of the probability of *Y* allows one to assume a linear relation between the predictors and the outcome variable. This transformed outcome variable is expressed as:

$$ln[P_1/(1\text{-}P_1)]$$

and is called the *log likelihood* or *logit*. With this transformation, the *logit* can now take on any value ranging from positive to negative infinity, is assumed to have a mean of 0 and a variance of $\pi^2/3$, and is linear in its relation to the predictors (Aldrich & Nelson, 1984). One might wonder why we cannot simply transform the outcome variable in the manner described previously, and then conduct an OLS analysis using the transformed outcome variable. Doing so would still result in inefficient parameter estimates resulting from heteroscedastic and non-normal errors.

The outcome can also be interpreted as a predicted probability of being in a particular group by raising the base number for the natural log to the power of the linear combination of the predictors for each individual. This is written as:

$$P_i = exp(\Sigma\beta_k X_{ik})/(1 + exp(\Sigma\beta_k X_{ik})).$$

Because it is a probability, P_i ranges from 0 to 1. When the linear combination of the predictors approaches negative infinity, then P_i approaches a value of 0. When the linear combination of the predictors approaches positive infinity, then P_i approaches a value of 1. It is important to remember that, in the estimated logistic regression model, the form of the relation between the linear combination of predictors and the *probability* of an outcome is nonlinear. That is, the change in the probability of an outcome occurring given a particular predictor depends on the level of that predictor. For example, let us assume that having children is related to a decreased probability of being able to successfully quit smoking. It is quite likely, then, that the difference between having no children and having one child is related to a greater decrease in the likelihood of successful cessation compared to the difference between having one child and having two children. Thus, the rate of change in the probability is not constant across all levels of the independent variable. In the logistic regression model, the relation between the predictors and the probability of an outcome occurring is S-shaped (see Fig. 10.1). This means that the change in the probability of an outcome occurring is likely to be slower at extreme levels of the predictors.

Most applications of logistic regression solve for the probability of having a value of 1, or the presence (as opposed to absence) of the outcome, and most statistical software packages are designed to solve for this.[1] For example, in our smoking cessation example, we are interested in solving for the probability that an individual is a successful quitter ($Y = 1$), as opposed to an unsuccessful quitter ($Y = 0$). Thus, based on the linear combination of predictors for each member of the sample, those with a predicted probability of successfully quitting of .50 or less are classified in the group of unsuccessful quitters, whereas those with a predicted probability greater than .50 are classified as successful quitters.

Parsimony is an important element of any statistical model. With logistic regression, the goal is to find the most parsimonious linear combination of predictors that maximizes the observed probability of the sample outcome. In order to achieve this goal, researchers must take care in building their logistic regression models. There should be a careful balance between theoretical and statistical relevance in evaluating predictors for inclusion in the logistic regression model. Clearly, it is important to include theoretically relevant predictors, but the selection process should be supplemented by univariate analyses in order to determine which measures also appear statistically relevant. This process can help to identify a more parsimonious model, especially when sample size restricts the number of predictors that can be included in any one model. For categorical predictors, one should examine the chi-square associated with a contingency table relating the levels of the categorical variable to the outcome. Contingency tables also help to identify empty cells, which can result in a lack of convergence or in unusually large

[1]One notable exception occurs with SAS PROC LOGISTIC, which solves for the probability that the outcome equals zero. Thus, regression coefficients will have opposite signs than those obtained in software packages that solve for Y= 1, unless the outcome variable is recoded or an option to solve for Y= 1 is chosen. This highlights the importance of knowing how the selected statistical software package codes the outcome variable. Being unaware of this can lead to incorrect interpretation of the relations between predictors and the outcome.

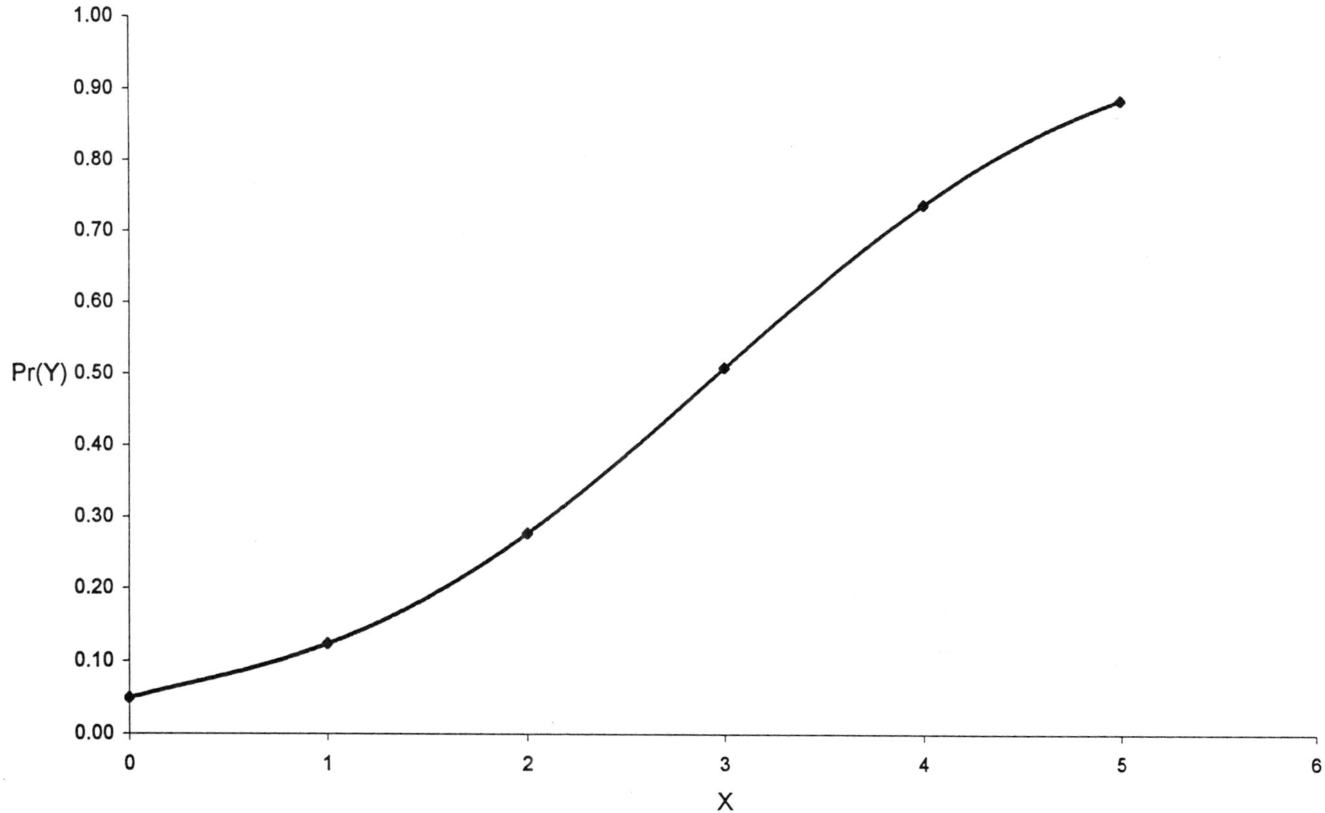

FIG. 10.1. Relation of Predictor X to the Outcome Probability Pr(Y)

estimates of coefficients and/or standard errors. If an empty cell is found, one solution is to collapse the categories in a meaningful way to eliminate the empty cell. For continuous variables, univariate logistic regression analyses or t tests can be conducted. Hosmer and Lemeshow (1989), recommended that variables with a significance level of $p < .25$ and variables that are theoretically important be retained for testing in the multivariate logistic regression model. It is important to keep in mind, however, it is best to avoid using a strictly data-driven approach to building a logistic regression model. Using such an approach, it is entirely possible to come up with a well-fitting, but theoretically incorrect, model. The model-building process should be primarily theoretically guided, but supplemented by statistical analysis.

INTERPRETING THE LOGISTIC REGRESSION MODEL

Similar to the linear regression model, the logistic regression model provides an omnibus test of the null hypothesis that all regression coefficients (excluding the intercept) are equal to zero, and tests for the significance of the regression coefficients for each of the individual predictors.

The *omnibus significance test* provides similar information about the model as the overall F test in OLS regression. That is, it tests the null hypothesis that all regression coefficients (excluding the intercept) are equal to zero. In logistic regression, a likelihood ratio statistic is computed for the ratio of a model with the predictors to a model including only the intercept and no predictors. In most statistical software packages, a log-likelihood statistic is provided for both the intercept-only model and the model including the predictors. A transformation of the log-likelihood statistics for these two models results in test statistics that have an approximate chi-square (χ^2) distribution for hypothesis testing. The transformed log-likelihood statistic is referred to as the *-2 log-likelihood*. The -2 log-likelihood for the model including the predictors is then subtracted from the -2 log-likelihood for the intercept-only model. The resulting likelihood ratio statistic is approximately distributed as a χ^2 statistic with degrees of freedom equivalent to $k-1$, where k equals the number of parameters estimated in the model (including the intercept), and is commonly referred to as the *model chi-square*. A significant χ^2

indicates that one or more of the regression coefficients is significantly different from zero.[2]

Once it is determined that one or more of the predictors is significantly different from zero, examine the individual significance tests for each of the predictors controlling for other predictors in the model. For each of the predictors, an unstandardized regression coefficient is provided for the parameter estimate as well as a value for the standard error. Significance testing proceeds by dividing the unstandardized regression coefficient by its standard error and comparing this obtained ratio (distributed as a t statistic) to the critical value for a preestablished significance level of the t distribution with degrees of freedom equal to N-k. Alternatively, many statistical computing packages provide a Wald χ^2 test statistic for each individual predictor. Under asymptotic (large sample) assumptions, the Wald χ^2 is simply the square of the t statistic compared to a critical value of the chi-square distribution with one degree of freedom. The main difference between the two is that the t statistic allows for a test for either a one-tailed or two-tailed hypothesis, whereas the Wald χ^2 only allows for a test of a two-tailed hypothesis. Most statistical computing packages will provide either the t statistic or Wald χ^2 and the level of significance for each predictor.

Because the relation between the predictors and the logit is linear, we are able to interpret the individual regression coefficients by saying that a one unit change in a predictor is associated with a certain change in the *logit* or log odds of being a successful quitter, holding all other predictors in the model constant. The problem with interpreting the relations between predictors and the outcome in this manner is that it is difficult to understand what a change in the logit means. Fortunately, however, there is a simple transformation of the regression coefficient into a measure of the odds of an outcome occurring, and the relation of the predictors to a change in the odds of an outcome is much easier

[2]In addition to the model χ^2, how well your model fits the observed data can be tested. This is referred to as evaluating the goodness-of-fit of the model. For example, some statistical computing packages will report a statistic for which the log likelihood for the model is compared to that of a perfectly fitting model. The perfectly fitting model is one in which there are as many parameters to be estimated as data points, and is also called a saturated model. This obtained log-likelihood ratio is referred to as the *deviance*, and because the comparison model is a perfectly fitting model, a nonsignificant deviance statistic provides evidence that your model fits the data well. The deviance is similar in concept to the error sum of squares in OLS regression (Hosmer & Lemeshow, 1989). Thus, the ideal model is one in which the deviance is minimized.

to interpret than a change in the logit. With a dichotomous outcome variable, the odds are equal to the probability of an outcome occurring divided by the probability of an outcome not occurring, based on the value of a particular predictor variable. For example, the odds of successfully quitting smoking are calculated by dividing the probability of being a successful quitter by the probability of being an unsuccessful quitter, given the value of the predictor variable (e.g., beliefs about the likelihood of smoking in 1 year). If there is no relation between a predictor and the outcome, then the odds are equal to one because it is equally likely that a person will be in either group regardless of the value of the predictor variable ($P_1 = .50$ for being a successful quitter divided by $P_0 = .50$ for being an unsuccessful quitter). If greater beliefs that a person will be smoking in 1 year decreases the probability of being a successful quitter (i.e., a negative regression coefficient), the odds will be less than one. A positive relation between a predictor and the outcome will result in odds of greater than one because an increase in the predictor (e.g., value placed on a healthy lifestyle) increases the probability of the outcome occurring (e.g., being a successful quitter).

The relation of a predictor to the odds of an outcome occurring can be interpreted in terms of the odds ratio, which is the change in the odds of the outcome that occurs with a unit change in the predictor variable. In a model with more than one predictor, the odds ratio is conditional on the other variables in the model. Computing the odds ratio from the regression coefficient requires taking the base value of the natural log and raising it to the power of the regression coefficient. The odds ratio is commonly symbolized as $exp(\beta_k)$ and can range from zero to positive infinity. If there is no relation between a predictor and the outcome, then the odds ratio is equal to one [$exp(0)=1$] . If a predictor variable (e.g., greater beliefs that one will be smoking in 1 year) decreases the probability of being a successful quitter (i.e., a negative regression coefficient), the odds ratio will be less than one. A positive relation between a predictor and the outcome will result in an odds ratio of greater than one.

Consider an example, in which the results of a logistic regression analysis yield a negative relation of $\beta = -.37$ between a continuous variable, measuring an individual's perceptions of how likely it is that (s)he will be smoking in 1 year, and successful smoking cessation. This indicates that the more individuals perceive that they will be smoking in 1 year, the less likely it is that they will be successful quitters. The odds ratio is equal to *exp*(-.37) = .69. This is interpreted as "For a one unit increase in perceived likelihood of smoking in 1 year, the odds of being a successful quitter are *decreased* by a factor of .69, holding all other predictors in the model constant." On the other hand, suppose results indicate a positive relation between value placed on a healthy lifestyle and being a successful quitter (β = .20). The odds ratio for the relation between value placed on a healthy lifestyle and successful cessation is then *exp*(.20) = 1.22. Thus, for a one-unit increase in the value placed on a healthy lifestyle, the odds of being a successful quitter are *increased* by a factor of 1.22, holding all other predictors in the model constant.

Most statistical computing packages will compute the odds ratio for each of the predictors. However, it is important to keep in mind that, if there are categorical predictors, the way they are coded affects computation of the odds ratio. For instance, with a dichotomous predictor, most computer packages assume that the two groups are coded as 0 and 1, and the odds ratio that is provided is based on this assumption. However, if you use a different coding scheme such as unweighted effects coding where the two groups are coded as -1 and 1, then the odds ratio provided by the statistical program is incorrect. In the case of unweighted effects coding, the true odds ratio is not $exp\beta_k$, but instead is the base of the natural log raised to the regression coefficient multiplied by two [$exp(2\beta_k)$]. This is easy to see if you consider the definition of the odds ratio, which is the change in the odds for a one-unit change in the predictor variable. With a dichotomous predictor coded 1 and 0, the odds ratio can be thought of as the difference between the two categories (1-0), and the equation for the odds ratio can be expressed as $exp(\beta \times (1-0))$, which is equal to $exp(\beta)$. However, when a dichotomous variable is effect coded (1,-1), the equation for the odds ratio is: $exp(\beta \times (1-(-1))$, which is equal to $exp(2\beta)$. This difference in calculating the odds ratio for different dummy variable coding schemes again highlights the importance of knowing how the statistical computing package assumes that variables are coded (both predictors and the outcome), because the output that it gives may not be exactly what you need.

With the recent debate on significance testing leaning toward reporting of supplemental measures (beyond significance levels) of the relation between predictors and outcome variables (see Harlow, Mulaik, & Steiger, 1997), it is also useful to report confidence intervals. Although confidence intervals are not always presented in the output provided by statistical computing packages, they are relatively easy to compute. To calculate the lower limit of the 95% confidence interval, multiply the standard error for the regression coefficient by 1.96 (the *z* score value for p = .05). Then, subtract that value from the value of the regression coefficient. To calculate the upper limit of the 95% confidence interval, again multiply the standard error for the regression coefficient by 1.96, and then add that value to the value of the regression coefficient. For example, consider the standardized regression coefficient for the relation of perceived likelihood of smoking within 1 year and successful cessation (β = -.37) with an odds ratio = .69 and a standard error equal to .132. The regression coefficient for the lower limit of the 95% confidence interval would be (-.37) - (1.96)(.132), which equals -.63. The regression coefficient for the upper limit of the 95% confidence interval would be (-.37) + (1.96)(.132), which is equal to -.11. Again, in order to facilitate interpretation, the regression coefficients for the lower and upper limits of the 95% confidence interval can be transformed into odds ratios. Thus, the odds ratios for the 95% confidence interval are *exp*(-.63) = .53 for the lower limit and *exp*(-.11) =. 90 for the upper limit. What this means is that for a one-unit increase in the predictor, the decrease in the odds of the outcome being present is estimated with 95% confidence to be between .53 and .90.

Finally, there are useful summary statistics available for OLS regression such as R^2, the proportion of variance in the outcome variable that is accounted for by the predictors, that are popular among researchers. Unfortunately, because the relation of the predictors to the observed dichotomous outcome is nonlinear in logistic regression, there is no such simple way to summarize a model. When calculating R^2 for a linear regression model, an assumption is made that the mean and the variance of the outcome variable are independent of each other, which is the case when the observed outcome is continuous and the mean and variance can be obtained from the sample data. However, this is not the case for dichotomous outcome variables. As a result, traditional interpretation of the R^2 for the logistic regression model does not make much sense. Although some pseudo R^2 measures have been developed for the logistic regression model, they too are limited in their usefulness. There is no universal acceptance or use of a particular R^2 measure. For this reason, it is difficult to compare R^2 values for different models, and it remains unclear when an obtained R^2 is considered *good*. Because of these limitations, Aldrich and Nelson (1984) recommend that reporting pseudo R^2 be done with extreme caution, or even better, avoided altogether.

TESTING AND INTERPRETING INTERACTIONS

In many cases, it might be useful to examine whether predictors interact to influence the probability of an outcome occurring. That is, we might want to know whether the relation between a predictor and the probability of the outcome differs at different levels of another predictor. For example, in smoking behavior research, we often wonder whether the relation between predictors of smoking behavior vary as a function of certain demographic characteristics such as gender or education level. We might want to know, for instance, whether the relation between having friends who smoke and being able to successfully quit smoking differs for women compared to men or for those with differing education levels. Choosing the interactions to be tested should be done with care because a large number of interactions in a model can contribute to parameter instability (Hosmer & Lemeshow, 1989; Kleinbaum, Kupper, & Morgenstern, 1982). Hosmer and Lemeshow (1989) suggested that interactions be chosen based on theoretical guidance and specific research hypotheses. The procedure for testing and interpreting interactions in logistic regression parallels that for OLS regression.

The overall significance of interaction terms in the logistic regression model can be tested by first running a model with only main effect predictors. Then, run a model including both the main effects and the interaction terms. The first model with only the main effects is considered as being nested within the full model including the interaction terms. Thus, the model χ^2 for the main effect-only model can be subtracted from the model χ^2 for the full model including the interaction terms. The resulting χ^2 is then compared to a table of critical values for the χ^2 distribution with a predetermined significance level, and degrees of freedom equal to the degrees of freedom for the full model minus the degrees of freedom for the main effects model. A significant change in the χ^2 indicates that one or more of your interaction terms is significant.

Once it is determined that there are one or more significant interactions, their significance can be examined individually. Including a number of nonsignificant interaction terms in a model can inflate the standard errors of other predictors in the model, which makes it more difficult to detect significant effects (Aiken & West, 1991). So, finally, it is possible to test another model that retains only the significant interaction terms, and then subtract the model χ^2 for the model with fewer interactions from the model χ^2 including all the interaction terms with degrees of freedom equal to the degrees of freedom for the model including all interaction terms minus the degrees of freedom for the model including only the significant interaction terms. A nonsignificant change in the model χ^2 indicates that eliminating the nonsignificant interactions did not significantly reduce the predictive value of the model, and thus this more parsimonious model can be retained.

Once a model is identified and tested including the important interaction terms, the full nature of the interaction can be explored. As with the main effect predictors, a regression coefficient, standard error, *t* statistic, or Wald χ^2 statistic, and odds ratio is typically provided for each interaction term. It is important to remember that main effects of predictors involved in a significant interaction with other predictors should not be interpreted independent of the interaction. Furthermore, the regression coefficient for the interaction indicates only that the relation between a predictor and the outcome differs at different levels of another predictor, and the odds ratio associated with the interaction term cannot be interpreted. In our successful smoking cessation example, a significant interaction between gender and number of friends who smoke indicates that the relation between the number of friends who smoke and successful smoking cessation is different for men and women. What it does not do is show how the relations are different. In order to determine how men and women differ, slopes and odds ratios for men and women must be calculated separately at different levels of number of friends who smoke.

Essentially, testing interactions proceeds by determining the slopes, and associated odds ratios, for various combinations of the values of the predictors involved in the interaction. For example, if we determined that there was a significant interaction between gender and number of friends who smoke, we might want to explore the interaction by calculating the slopes and the odds of successful cessation, first for men at one standard deviation below the sample mean number of friends who smoke (-1 sd), at the mean, and at one standard deviation above the mean number of friends who smoke (+1 sd). Then, we would do the same for women. Cohen and Cohen (1983) suggested using these points (-1 sd, at the mean, +1 sd) as guidelines for testing interactions involving continuous predictors, but other values may be used depending on the specific research questions. However, it should be noted that, if there are extreme scores, using the highest and lowest values of the predictors may result in large changes in the odds that are not necessarily representative. Aiken and West (1991) referred to the individual slopes as simple slopes, and provided a detailed treatment of the procedure for interpreting interactions between categorical and continuous predictors (like our gender by number of friends who smoke example) and between continuous predictors. In addition, Hosmer and Lemeshow (1989) described a similar method for calculating simple slopes, standard errors, and odds ratios. Aiken and West (1991) also provided some simple SPSS statistical computing code for calculating the simple slopes and associated standard errors. This code can easily be modified to work with other statistical computing packages.

AN APPLICATION OF LOGISTIC REGRESSION IN SUBSTANCE USE RESEARCH

Cigarette smoking is the single largest preventable cause of premature death in the United States (U.S. Surgeon General, 1989). The positive health benefits of quitting smoking have been well documented, and many of those who currently smoke have made serious attempts to quit (Centers for Disease Control, 1994). However, the majority of these attempts end in failure. Thus, from a public health standpoint, it is critically important to identify predictors of successful smoking cessation.

The current example is drawn from a study examining prospective predictors of smoking cessation among participants from a long-term longitudinal study of the natural history of smoking from adolescence to adulthood (Rose, Chassin, Presson, & Sherman, 1996). Between 1980 and 1983, all consenting 6th through 12th graders in a midwestern county school system completed up to four annual, classroom-administered surveys (a total sample of 8,556 students were assessed at least once). Follow-up surveys were conducted in 1987 ("young adulthood") and 1993 ("adulthood") with retention rates of 73% of the original adolescent sample at each follow-up. For this study, we selected those who reported being regular smokers (at least weekly) in 1987, whose cohorts had graduated high school by 1987 (thus ensuring that participants could be considered "young adults"), and who in 1993 had reported making at least one serious attempt to quit smoking ($N = 530$). Among the 530 participants who had attempted to quit smoking, those who reported having not smoked a cigarette in over 6 months were coded as successful quitters ($Y = 1$; $N = 115$). Those who continued to smoke regularly despite having made a quit attempt were coded as unsuccessful quitters ($Y = 0$; $N = 415$).

All predictors of successful smoking cessation in 1993 were assessed in 1987 (young adulthood), with the exception of educational attainment (dichotomized as at least some college vs. no college education) and reasons for quitting smoking, which were measured in 1993. Additional predictors included gender, young adult smoking-related variables (e.g., likelihood of smoking in one year), smoking in the young adult's environment, smoking-related beliefs, motives for smoking, and young adult social role occupancy (for a more complete description of the measures see Rose et al., 1996). We also tested interactions of these predictors with gender, amount smoked in young adulthood (dichotomized as light smoking–less than 1 pack per day vs. heavy smoking-a pack or more per day), and smoking onset, which was dichotomized as early (before 12th grade) versus late (after high school) onset. All predictors were screened for nonnormality, collinearity, and outlying cases. Continuous predictors were centered (by subtracting the mean), and dichotomous predictors were effect-coded (-1,1 coding) to reduce mathematically caused multicollinearity (Aiken & West, 1991) in the multiplicative interaction terms.

In order to minimize the number of predictors in the models, we tested six separate models for each group of conceptually related predictors. In addition to these predictors, each model included educational attainment as a covariate, and the three moderator variables of interest (gender, amount smoked, and onset). We used the SPSS LOGISTIC REGRESSION program to conduct our analyses. Using the method described for testing the overall significance of the interaction terms, we first tested a model with main effects only, and then tested three additional models, each adding the two-way interaction terms associated with one of the moderators. If including the interaction terms did not significantly improve the model fit, then the main effects model was retained. If including the interaction terms significantly improved the model χ^2, they were examined for their individual significance level. When models with interactions were retained, nonsignificant interactions were eliminated as long as eliminating them did not significantly affect the model χ^2.

For example, consider the model testing reasons for quitting smoking. We first tested a model including only the main effects. The χ^2 value for the main effects model was χ^2 (7, N = 511) = 55.15, $p < .0001$. In the second step, we conducted three additional logistic regression analyses, each adding the interactions for one of the three moderators to the main effects model. The model with gender interactions significantly improved the model χ^2. The model χ^2 for the model including the gender interactions was χ^2 (10, N = 531) = 64.49, $p < .0001$. Subtracting the main effects model from the model including the interaction terms gave us a χ^2 difference equal to 9.34, with degrees of freedom for the χ^2 difference test equal to 3 (10-7). Consulting a table of critical values for the χ^2 distribution for 3 degrees of freedom indicated that the χ^2 difference value was significant at $p < .05$. The models including the amount smoked and smoking onset interactions did not significantly improve the model χ^2.

An examination of the Wald χ^2 test for the individual interaction terms indicated that only one of the interaction terms (gender x sensory reasons) was significant. A final *trimmed* model was tested in which the nonsignificant interaction terms were eliminated from the model. The χ^2 difference for the trimmed model (with degrees of freedom equal to the degrees of freedom for the model including all interaction terms minus the degrees of freedom for the trimmed model) was not significant, which indicated that eliminating the nonsignificant interaction terms did not affect the overall predictive value of the model.

Table 10.1 shows the results for each of the six final logistic regression models predicting successful smoking cessation. It also provides an example of what is important when presenting logistic regression results.

In the table, we present the model χ^2 and associated degrees of freedom. In addition, for each predictor we present the unstandardized regression coefficient, the standard error, the significance level, the odds ratio, and the 95% confidence interval for the odds ratio. In the first model including smoking variables, we found that education, amount usually smoked, and perceived likelihood of smoking in 1 year were all significantly related to successful smoking cessation. Gender and age of smoking onset were not significant predictors of successful smoking cessation. A positive regression coefficient for education indicated that a higher education level was associated with an increased likelihood (or probability) of being a successful quitter. Negative regression coefficients for amount usually smoked and likelihood of smoking in 1 year indicated that those who smoked more cigarettes per day and those who perceived a greater likelihood of smoking in the next year were less likely to be successful quitters.

TABLE 10.1

Logistic Regression Results for Predictors of Successful Smoking Cessation

Model and predictor	χ^2	*df*	*B*	*SE*	*OR*	*95%CI*
Smoking variables	43.48***	(5,N=496)				
Education			.41**	.128	2.27	(1.37, 3.75)
Amount usually smoked			-.53**	.170	0.35	(0.18, 0.68)
Gender			.07	.117	1.15	(0.68, 1.68)
Onset			.10	.129	1.22	(0.74, 2.02)
Likely will smoke in 1 year			-.37**	.132	0.69	(0.53, 0.90)
Smoking Environment	46.90***	(7,N = 466)				
Education			.28†	.143	1.75	(0.99, 3.06)
Amount usually smoked			-.61***	.183	0.30	(0.14, 0.60)
Gender			.01	.126	1.02	(0.62, 1.67)
Onset			.07	.139	1.15	(0.67, 1.98)
Parental smoking			-.18	.179	0.70	(0.35, 1.41)
No. of friends who smoke			-.18†	.095	0.85	(0.69, 1.01)
Perceived prevalence			.01	.008	0.99	(0.97, 1.01)
Smoking Beliefs	40.22***	(8,N=495)				
Education			.41**	.128	2.27	(1.37, 3.75)
Amount usually smoked			-.58***	.170	0.31	(0.16, 0.61)
Gender			.08	.118	1.17	(0.74, 1.86)

Table 10.1 continued

Table 10.1 continued

Model and predictor	χ^2	*df*	B	*SE*	*OR*	*95%CI*
Onset			.14	.127	1.32	(0.80, 2.18)
Personally relevant psychological beliefs			.03	.195	1.03	(0.82, 1.30)
Personally relevant health beliefs			-.05	.213	0.95	(0.64, 1.44)
General health beliefs			-.10	.205	0.90	(0.61, 1.35)
Value healthy lifestyle			.20*	.099	1.22	(1.01, 1.48)
Smoking Motives	53.21***	(12,*N*=482)				
Education			.38**	.137	2.14	(1.25, 3.66)
Amount usually smoked			-.74***	.254	0.23	(0.08, 0.62)
Gender			.03	.133	1.03	(0.63, 1.79)
Onset			.09	.139	1.09	(0.69, 2.05)
Addiction			-.14	.168	0.87	(0.63, 1.21)
Pleasure			-.10	.213	0.90	(0.66, 1.37)
Stimulation			-.30†	.158	0.74	(0.54, 1.01)
Sensorimotor			.83**	.304	2.29	(1.26, 4.16)
Habit			-.15	.172	0.86	(0.61, 1.21)
Affect control			.00	.172	1.00	(0.71, 1.40)
Gender X Stimulation			-.35*	.148		
Amount smoked X Sensorimotor			.81**	.295		

Table 10.1 continued

Table 10.1 continued

Model and predictor	$\chi 2$	*df*	*B*	*SE*	*OR*	*95%CI*
Reasons for Quitting[a]	60.99^{***}	(8,*N*=511)				
Education			$.37^{**}$	.131	2.10	(1.25, 3.50)
Amount usually smoked			$-.63^{***}$	.171	0.28	(0.64, 1.69)
Gender			.02	.123	1.03	(0.64, 1.69)
Onset			.19	.129	1.09	(0.88, 2.43)
Health reasons			-.18	.144	0.70	(0.40, 1.23)
Sensory reasons			$.46^{***}$	.128	2.51	(1.52, 4.15)
Social reasons			$-.32^{**}$	.126	0.53	(0.33, 0.85)
Gender X Sensory reasons			$-.29^{**}$	.123		
Social Roles	43.01^{***}	(9,*N*=431)				
Parental education			$.27^{\dagger}$	.139	1.72	(1.00, 2.26)
Amount usually smoked			$-.61^{***}$	.177	0.30	(0.15, 0.59)
Gender			.10	.123	1.22	(0.75, 1.98)
Onset			$.24^{\dagger}$	.131	1.62	(0.97, 2.73)
Age			-.07	.069	0.93	(0.81, 1.07)
Student			.25	.159	1.65	(0.88, 3.08)
Married			.23	.176	1.58	(0.79, 3.16)
Employed			$.30^{*}$	.149	1.82	(1.02, 3.27)
Live with children			$-.40^{*}$	.196	0.45	(0.21, 0.97)

Note. Ns vary as a result of listwise deletion of missing data. Wald chi-square was used to test significance of individual parameters. Gender was coded as -1 (women) or 1 (men). OR = odds ratio. 95% CI = 95% confidence limits for the odds ratio. [a]Trimmed models. $^{\dagger}p < .10$. $^{*}p < .05$. $^{**}p < .01$. $^{***}p < .001$.

The regression coefficient tells us only the direction of the relation between the predictors and the outcome, but the odds ratio provides more information about the magnitude of the relations. The output from SPSS LOGISTIC REGRESSION provides odds ratios for each of the predictors. However, the odds ratios for the dichotomous predictors that were provided by SPSS were calculated based on the assumption that the predictors' two categories were coded 0 and 1. We used unweighted effects coding (-1,1) for our dichotomous predictors, and as a result, the odds ratios provided in the output for these variables were incorrect. Consequently, we had to calculate the odds ratio using the equation *exp(2β)* instead of the equation *exp(β)*, the latter of which is appropriate for dichotomous variables that have 0,1 coding. Thus, in the model including smoking variables, the odds ratios for education and amount usually smoked were *exp*(.82) = 2.27 and *exp*(-1.06) = .35, respectively. The odds ratio provided by SPSS for the continuous predictor, likelihood of smoking in 1 year, was accurate and was *exp*(-37) = .69. Given the odds ratio for education, then, we could say that, in this study, having some college education increased the odds of being a successful quitter by a factor of 2.27 compared to those with no college education, holding all other variables constant. Similarly, being a heavy smoker (i.e., smoking a pack or more of cigarettes per day) decreased the odds of being a successful quitter by a factor of .35. Finally, for a one-unit increase in the perceived likelihood of smoking in the next year, the odds of being a successful quitter decreased by a factor of .69.

Alternatively, some researchers find it easier to express the relation in terms of the percent change in the odds ratio. This is done by applying the simple equation:

$$\textit{percent change} = 100(\textit{odds ratio}-1) \qquad (10.7)$$

where 1 is equal to the amount of change in the predictor variable (in this case 1 unit). For education, the percent change in the odds is 100(2.27-1), which is equal to 127%. Using the same formula, the percent change in the odds was 65% for amount smoked and 31% for likelihood of smoking in 1 year. Thus, you could say that in this study, having some college education was related to a 127% increase in the odds of being a successful quitter, whereas being a heavy smoker was related to a 65% decrease in the odds of being a successful quitter, holding all other variables constant. Furthermore, a one-unit increase in perceived likelihood of smoking in 1 year was related to a 31% decrease in the odds of being a successful quitter, holding all other variables constant.

We then calculated 95% confidence intervals for the odds ratios for each predictor. The regression coefficient for the lower limit of the 95% confidence interval for the relation of education to successful cessation was .41-(1.96)(.128) = .159, and the upper limit was .41+(1.96)(.128)=.661. Remember that, because we used unweighted effects coding for dichotomous predictors, we had to use *exp(2β)* to compute the odds ratio. The same is true for calculating odds ratios for the upper and lower limits of those variables. Thus, the odds ratio for the lower limit of the 95% confidence interval for education was *exp(2β)* = *exp*(.318) = 1.37. Likewise, the upper limit of the 95% confidence interval for education was *exp(2β)* = *exp*(1.322) = 3.75. So, the increase in the odds of successful cessation associated with having some college education is estimated with 95% confidence to be between 1.37 and 3.75. Notice that, for our nonsignificant predictors of gender and age of smoking onset, the 95% confidence interval for the odds ratio includes the value of one, indicating that we cannot be 95% confident that the estimate of the relation of gender and age of onset to successful cessation will be different from zero.

Finally, we had some significant interaction terms. For example, Table 10.1 indicates that there were significant interactions between amount smoked and sensorimotor motives and between gender and stimulation motives in the smoking motives model. In addition, there was a significant interaction between gender and sensory reasons in the reasons for quitting model. Although sensorimotor motives, stimulation motives, and sensory reasons were significant main effect predictors of successful smoking cessation, these main effects need to be interpreted within the context of the interaction. That is, the relation of sensorimotor motives to successful cessation differs for light and heavy smokers. Likewise, the relation of stimulation motives for smoking, and sensory reasons for quitting, to successful cessation differs for men and women.

Note that, in the table, we do not present an odds ratio for the interaction terms. The interaction term means only that there is a significant difference between light and heavy smokers (and men and women) on the relation of smoking motives to successful cessation and between men and women on the relation of sensory reasons to successful cessation. It does not indicate how light and heavy smokers (and men and women) differ on these relations, and therefore, the odds ratio does not provide any information for interpreting these relations. To find out how these groups differ, we need to calculate simple slopes and odds ratios for the relation of sensorimotor motives to successful cessation separately for light and heavy smokers, and we need to calculate simple slopes for the relations of stimulation motives and sensory reasons to successful cessation separately for men and women.

We used Cohen and Cohen's (1983) guidelines to calculate simple slopes and associated odds ratios for light and heavy smokers separately at one standard deviation below the mean, at the mean, and at one standard deviation above the mean for sensorimotor motives (low, medium, and high levels of sensorimotor motives). We used the same procedure to explore the relations for men and women smokers separately for stimulation motives. Sensory reasons was a dichotomous variable indicating whether or not participants endorsed any reason for quitting associated with sensory aspects (e.g., quitting to improve senses of taste and smell). Therefore, we calculated simple slopes and associated odds ratios for men and women separately for each of the two values of reasons for quitting. We used SPSS LOGISTIC REGRESSION to compute the simple slopes. Because the procedure for calculating simple slopes is clearly outlined by Aiken and West (1991), it will not be covered in detail here. However, for researchers interested in exploring interactions, it is recommended that they refer to Aiken and West (1991) to obtain a greater understanding of interpreting interactions and to obtain the computer code to calculate simple slopes.

The odds ratios for the significant interactions are shown in Table 10.2. The odds ratios show that the relation between sensorimotor motives (e.g., enjoy holding a cigarette) and the likelihood of successful cessation varied with the amount typically smoked. Among light smokers, sensorimotor motives were unrelated to successful cessation, that is, the odds of successful cessation did not change regardless of the level of sensorimotor motives (odds ratios = .27, .28, and .28 for low, medium, and high levels of sensorimotor motives). However, among heavy smokers, the odds of successful cessation were .02, .06, and .23 for low, medium, and high levels of sensorimotor motives, indicating that those with high levels of sensorimotor motives were more likely to be successful quitters. In addition, the relation between stimulation motives (e.g., smoking to "perk up") and the likelihood of successful smoking cessation differed for men and women. For women, stimulation motives were unrelated to

TABLE 10.2

Estimated Odds Ratios for Significant Interactions

Predictor	Odds Ratio		
	Low (-1 SD)	*Medium (Mean)*	*High (+1 SD)*
Sensorimotor Motives			
Light Smokers	0.27	0.28	0.28
Heavy Smokers	0.02	0.06	0.23
Stimulation Motives			
Men	0.24	0.14	0.08
Women	0.12	0.13	0.13
	Sensory Reasons		*No Sensory Reasons*
Sensory Reasons for Quitting			
Men	0.56		0.09
Women	0.25		0.16

successful cessation. The odds of successful cessation for women were .12, .13, and .13 at low, medium, and high levels of stimulation motives, indicating no change in the likelihood of successful cessation regardless of the level of stimulation motives. For men, however, the odds of successful cessation for men at low, medium, and high stimulation motives were .24, .14, and .08, respectively. Finally, we explored the significant interaction between gender and sensory reasons for quitting smoking. For both men and women, those who reported sensory reasons for quitting were more likely to successfully quit smoking than those who did not. However, the relation was stronger for men. For men, the odds of successful cessation were .56 for those reporting sensory reasons compared to .09 for those who did not. For women, the difference in the odds was much smaller. The odds of successful cessation for women who reported sensory reasons were .25 compared to .16 for women who did not report sensory reasons for quitting.

CONCLUSION

Logistic regression is an ideal method for analyzing dichotomous outcome variables. Although it closely parallels the linear regression model in many ways, it is able to overcome the problems that arise when linear regression with OLS estimation is used to analyze dichotomous outcome variables. This can be done by thinking of the observed dichotomous outcome as having an underlying, unobserved continuous outcome variable that is linearly related to the predictors. However, despite the greater flexibility of the logistic regression model, many of the same assumptions of the linear regression model (normality, homoscedasticity, and independence) apply to the logistic regression model. Furthermore, because a maximum likelihood estimation method is necessary, logistic regression tends to require larger sample sizes than are required for linear regression with OLS estimation. Finally, the logistic regression model is somewhat more difficult to interpret than the linear regression model in the sense that more steps are required to fully understand the relations of the predictors to the outcome, and simple summary statistics such as R^2 are not easily available or ideal.

Although the limited scope of this chapter precludes an in-depth discussion of regression diagnostics, it should be noted that further assessing the fit of the model through regression diagnostics is an important step after finalizing a predictive model. This can help to identify empty cells or cases that might be exerting too much influence on model results. Logistic regression diagnostics typically consist of generating plots of leverage, influence, and residuals. Most statistical computing packages, such as SPSS and SAS, include options to save a number of regression diagnostic statistics as variables that can be plotted. One example of a diagnostic plot would be to look at the change in χ^2 for the predicted probabilities generated by the model in order to test the influence of individual patterns of values of predictors on the overall fit of the model. Patterns that do not fit the model well are indicated by points that are found in the top left or top right corners of the plot. In addition, the influence of predictor patterns on the estimated values of the regression coefficients can be examined by plotting the change in the estimated regression coefficient against the predicted probabilities. Large changes in the estimated regression coefficients at small values of the predicted probabilities indicate predictor patterns that are exerting a strong influence on the values of the estimated regression coefficients.

Methods for handling patterns of predictor values that are exerting an unusual influence on the results depend a good deal on the goals of the analysis. If a pattern is theoretically plausible, it might make sense to retain those cases, especially if they do not affect the conclusions that are drawn from the model. In general, influential patterns should be examined carefully for their effects on the results and on the conclusions drawn. Patterns that make little theoretical sense, but greatly influence the results, should be viewed with suspicion. A detailed discussion of logistic regression diagnostics can be found in Hosmer and Lemeshow (1989). Furthermore, Cook and Weisberg (1982) provided a detailed treatment of general regression diagnostics.

In this chapter, only dichotomous outcomes were discussed. However, this method can be extended to analyze ordinal outcome variables (the ordered logit model) or, more generally, nominal outcome variables (the multinomial logit model). Both are considered extensions of the binary logit model discussed in this chapter. Similar to dichotomous outcome variables, ordinal outcome variables can also be viewed as having an underlying continuous variable, to which the observed variable is linked by several thresholds. Clark, Rakowski, Kviz, and Hogan (1997) used ordered logit modeling to identify predictors of stages of readiness to quit smoking. Multinomial logit modeling of nominal outcomes can be viewed as simultaneously testing a series of binary outcomes. Hasin, Van Rossem, McCloud, and Endicott (1997) used multinomial logit modeling to identify baseline diagnosis predictors of no DSM-IV diagnosis of alcohol dependence, alcohol dependence, and alcohol abuse. A good discussion of both the ordered and multinomial logit models that includes real data applications of the models can be found in Long (1997).

REFERENCES

Aiken, L. S., & West, S. G. (1991). *Multiple regression: Testing and interpreting interactions.* Newbury Park, CA: Sage.

Aldrich, J. H., & Nelson, F. D. (1984). *Linear probability, logit, and probit models.* Newbury Park, CA: Sage.

Centers for Disease Control. (1994). Cigarette smoking among adults–United States, 1993. *Morbidity and Mortality Weekly Report, 43,* 925–930.

Clark, M. A., Rakowski, W., Kviz, F. J., & Hogan, J. W. (1997). Age and stage of readiness for smoking cessation. *Journals of Gerontology: Series B: Psychological Sciences and Social Sciences, 52B,* S212–S221.

Cohen, J., & Cohen, P. (1983). *Applied multiple regression/correlation analyses for the behavioral sciences* (2nd ed.). Hillsdale, NJ: Lawrence Erlbaum Associates.

Cook, R. D., & Weisberg, S. (1982). *Residuals and influence in regression.* New York: Chapman Hall.

Harlow, L .L., Mulaik, S. A., & Steiger, J. H. (Eds.). (1997). *What if there were no significance tests?* Mahwah, NJ: Lawrence Erlbaum Associates.

Hasin, D. S., Van Rossem, R., McCloud, S., & Endicott, J. (1977). Differentiating DSM-IV alcohol dependence and abuse by course: Community heavy drinkers. *Journal of Substance Abuse, 9,* 127–135.

Hosmer, D. W., & Lemeshow, S. (1989). *Applied logistic regression.* New York: Wiley.

Kleinbaum, D. G., Kupper, L. L., & Morgenstern, H. (1982). *Epidemiological research: Principles and quantitative methods.* New York: Van Nostrand Reinhold.

Long, J. S. (1997). *Regression models for categorical and limited dependent variables.* Thousand Oaks, CA: Sage.

Peduzzi, P., Concato, J., Kemper, E., Holford, T. R., & Feinstein, A. R. (1996). A simulation study of the number of events per variable in logistic regression analysis. *Journal of Clinical Epidemiology, 49,* 1373–1379.

Rose, J. S., Chassin, L., Presson, C. C., & Sherman, S. J. (1996). Prospective predictors of quit attempts and smoking cessation in young adults. *Health Psychology, 15,* 261–268.

U.S. Surgeon General. (1989). *Reducing the health consequences of smoking: 25 years of progress.* Washington, DC: U.S. Government Printing Office.

11

Summary and Future Directions

Clark C. Presson
Laurie Chassin
Arizona State University

Steven J. Sherman
Jennifer S. Rose
Indiana University

We began this volume by noting some of the changes in drug abuse research that have occurred in the 20 years since the publication of NIDA's monograph on multivariate techniques for analyzing substance abuse data (Bentler, Lettieri, & Austin, 1976). Compared to studies conducted 20 years ago, current research on substance use and abuse focuses on questions of change over time, and routinely generates complex longitudinal data sets with multiple waves of assessments. Researchers are interested in trajectories of substance use and substance use-related impairment both to describe the natural history of drug use, and to evaluate the results of prevention and treatment interventions. Moreover, research has moved beyond simple description to an interest in theory testing and in identifying the processes underlying changes in substance use and abuse over time. Thus, in addition to describing trajectories of use and changes in use because of intervention, current research seeks to understand etiological processes as well as the processes underlying effective interventions.

These research interests create a need for testing longitudinal models of process, including processes of mediation and moderation. Adding another dimension of complexity, however, current theory and research in substance use recognizes that these processes encompass multiple levels of analysis. That is, it is not sufficient to consider only individual factors because individuals are embedded within multiple social contexts including families, classrooms, neighborhoods, treatment groups, cultures, and so on. The need to analyze complex, longitudinal data sets, testing models of mediation and moderation, with multiple levels of analysis creates a host of data analytic challenges. In addition, these kinds of studies raise several important methodological issues that require novel, creative analytic strategies. For example, because of the difficulty of tracking subjects longitudinally, there are commonly problems of missing data. Moreover, the outcomes of interest in drug use research are often non-normally distributed and categorical in nature—for example, researchers are interested in abstinence as an outcome or clinical diagnosis of substance abuse/dependence as an outcome.

The chapters in this book are offered to illustrate some of these data analytic challenges, and to provide examples of recent approaches to solving them using data from studies of substance use and abuse. Latent curve analysis, multilevel modeling, and latent transition analysis all offer methods for modeling trajectories of substance use over time (see the chapters by Curran, Hyatt & Collins, Khoo & Muthén, and Muthén). Techniques are offered for testing mediational models (see the chapter by MacKinnon), moderator models (see the chapter by Wood), and epidemic models (see the chapter by Rowe, Rodgers, & Gilson). Techniques are offered for analyzing multilevel data both with hierarchical linear models and latent curve models (see the chapter by Khoo & Muthén). Approaches are providing for handling missing data (see the chapter by Hedeker & Rose) and for analyzing categorical outcomes (see chapters by Hyatt & Collins, and Rose, Chassin, Presson, & Sherman). Finally, an approach is provided for applying latent variable models to the analysis of drug prevention data in a Solomon Four Group Design (see the chapter by Ullman, Stein, & Dukes). As these chapters illustrate, researchers in the field of substance abuse have benefitted tremendously from recent advances in quantitative methods. These methods not only provide more appropriate tests of research questions, but have actually influenced the questions that are posed. For example, now that techniques are available to evaluate mediation and moderation, hypotheses concerning mediating and moderating relations have become cornerstones of our theoretical models.

Of course, these chapters are only examples of many techniques that are currently under development that will bring new tools for researchers in substance use, and there are many data analytic challenges that have not yet been answered. In fact, many of the aspects of the methods illustrated in this volume have not been fully developed. In thinking about these unanswered questions and what the future of quantitative methodology has in store for substance abuse researchers, we polled the authors of the chapters in this volume. We asked them what new tools might be realistically available to drug abuse researchers in the next 5 years. Many of the responses centered on the ways in which the modeling techniques illustrated in this volume would become even more flexible. Techniques are expected to better handle violations of assumptions of multivariate normality and model mis-specification (Bentler & Dudgeon, 1996; Bollen, 1996), to allow the use of continuous predictors in latent transition analysis (Reboussin, Reboussin, Liang, & Anthony, 1998), to further expand the capacity to model clustered data (Hedeker & Mermelstein, 1998), to further expand the capacity to model categorical latent variables (Muthén, 1999), to further expand the capacity to model ordered categorical outcomes by quantifying thresholds that can indicate a change in classification (Hedeker, Mermelstein & Weeks, in press), to better address problems of missing data (Schafer & Olsen, 1998; Schafer, 1997), and to accommodate small-sample research (which is sometimes characteristic of drug abuse studies, Hoyle, in press). Furthermore, there is ongoing development of statistical software to make these complex techniques more accessible and easier to implement (e.g., TETRAD II, which assists researchers in correctly specifying structural equation models; Scheines, Spirtes, Glymour, Meek, & Richardson, 1998; Mplus for latent variable random coefficient growth modeling; Muthén & Muthén, 1998).

In addition to expanding the flexibility of the models, many aspects of these techniques are not yet fully developed, and important advances are in progress. Methods for testing mediational models in latent curve analysis or hierarchical linear models are not yet readily available, but are currently under development (see e.g., Krull, 1998; Krull & MacKinnon, in preparation). Research has not yet fully described distributions of goodness-of-fit indices, often making evaluation of these models difficult to accomplish (e.g., for latent transition analysis see Hyatt & Collins, this volume; for categorical outcomes in hierarchical linear modeling, see Bryk, Raudenbush, & Congdon, 1996). Moreover, ways of estimating power of model tests in latent curve analyses, latent transition analysis, or hierarchical linear models are not yet readily available, although some methods are beginning to emerge (Brown & Liao, 1996; Brown, Wang, Muthén, & Dagne, 1998; Hedeker, Gibbons, & Waternaux, 1999; Muthén & Curran, 1997; see also Lashley & Kenny, 1998, for power estimation in dyadic research). Methods for estimating the power of these models are obviously important for designing studies of substance use and abuse.

In addition, recent work in quantitative methods is starting to address some of the complex issues involved in modeling longitudinal growth processes. For example, a long-standing problem in longitudinal studies has been the equivalence of measures of a construct at different developmental stages or ages. Longitudinal analyses are only meaningful if there is a stable meaning of a variable across time. If the underlying construct changes with maturation or has different ways of manifesting itself across development, then readministering a single instrument to measure this construct over the study will not be adequate. For example, drug use researchers are often interested in the construct of conduct problems. An item such as "gets in trouble in school" might be an extremely important indicator of conduct problems, but will be an age-specific indicator that is appropriate only for those of school age, so that the same measure may not be able to be administered at all time points. Educational researchers have been applying vertical equating of overlapping test forms with increasing difficulty levels to address this problem, but similar techniques are just beginning to be applied to psychological data (Khoo, 1998). Similar issues of measurement equivalence arise from the presence of distinct subgroups in the data (e.g., does an instrument assess a construct in the same way for men and women or for individuals of various ethnicities?). Researchers are currently providing methods to evaluate this equivalence (Knight, 1994; Millsap, 1995, 1997). Finally, assuming that measurement issues can be resolved, the presence of distinct subgroups in the data may produce different functional forms of growth within the sample. Methods are currently being developed to simultaneously analyze these different functional forms of growth (Muthén & Shedden, in press, see Muthén this volume, Neale, 1998) and to allow modeling of nonlinear growth (see Muthén, this volume; Hedeker & Mermelstein, in press). Finally, important etiological questions might be focused on ways in which multiple trajectories of growth (i.e., growth in multiple domains) are interrelated. For example, researchers may be interested in the interrelations between trajectories of conduct problems and trajectories of substance use. Methods development is underway to allow modeling of multiple trajectories (see Muthén, this volume, Collins, 1997).

Although some future advances will come from further methods development, there was also some consensus among the authors of this volume that the advantages and abilities of these models as they currently exist have not been fully realized. Largely, this is because these modeling techniques have been applied to existing data rather than to data gathered explicitly to be analyzed by these approaches (e.g., overlapping cohort designs, repeated baseline assessments, and additional repeated measures over time). New studies are now being designed that are capable of taking maximal advantage of these techniques, and design recommendations for such longitudinal studies are now available (Willett, Singer, & Martin, 1998). These new studies will be able to more fully capitalize on the techniques that are illustrated in this volume. One example can be found in the expansion of the numbers of repeated measures taken over time. Recent improvements in experience sampling methods (e.g., gathering multiple data points naturalistically through ambulatory monitoring and/or hand-held computer technology) now allow for collecting longitudinal data with great frequency. These methods allow for a more microanalytic look at behavior in its natural context—capturing time series effects that occur during short time lags rather than the longer lags of typical longitudinal designs. For studies of very proximal determinants of behavior (e.g., predictors or outcomes of drug craving) such a microanalytic perspective is critical. These designs combined with advances in time series analysis, such as studying intraindividual change using dynamic latent variable models, allow tremendous advances in studying the individual as a functioning system by fully exploring the structure of relations among variables obtained from numerous repeated measures on a single individual (Molenaar, 1994; Wood, 1994). In essence, the information obtained from multiple repeated measures on a single individual represents a rich data set by itself, for which the analysis of dynamic latent variable models provides an in-depth exploration of short-term fluctuations and the pattern of relations among variables within the individual over time.

Another new direction that is in its relative infancy in terms of application to the psychological phenomena of substance use is the modeling of Nonlinear Dynamical Systems (NDS). These NDS models (see the Rodgers & Rowe chapter on epidemic modeling for one simple version) introduce whole new levels of both conceptual and mathematical complexity, allowing for relations of feedback between independent and dependent variables and coaction among variables, nonlinear oscillators, and even "catastrophes"–points of discontinuity at which the underlying relations that are modeled can totally shift. NDS models are often inspired by chaos theory in the mathematical and physical sciences (see Baker & Gollub, 1990, for an introduction to chaos theory), and address the interaction of environmental influences and the individual's adaptive and self-regulatory reactions to these influences on individual change (Boker & Graham, 1998). Boker and Graham (1998) provide a substance use example of this approach, in which they show how differential equations can relate scores on measures of substance use behavior to both the first derivative of these measures (the rate of change) and the second derivative (how quickly the change in substance use behavior is accelerating or decelerating) to model change in adolescent smoking and drinking behavior.

To this point, we have been considering the advantages and opportunities that these sophisticated, multivariate techniques provide for researchers in the field of substance use and abuse. This is not to say that these methods are without difficulties for the applied researcher, and no presentation of these techniques is complete without a discussion of these potential difficulties. First, there is real danger that, with these sophisticated analytic tools, research can become technique-driven rather than theory-driven. Research questions must always be developed from the underlying conceptualization of the problem rather than from the techniques that are available for analyzing the data. Although researchers should ensure that their designs are suitable for the data analytic technique that they wish to employ (Willett et al., 1998), researchers must take care to match the technique to the research question rather than to match the research question to the technique. Second, the complexity and abstractness of the models often draw researchers far from their data. A label on a latent variable makes it all too easy to forget what you have actually measured, and makes it easy to reify constructs. A beautifully drawn model can tempt researchers to believe that they have correctly identified the underlying processes without considering that competing alternative models may provide an even better fit to the data. Thus, these techniques place high demands on researchers to examine the underlying data set (as well as examining their underlying theoretical models).

Such an examination can reveal possible limiting conditions of effects, moderators of effects, problems in distributions that can produce misleading estimates of effects, and suggestions of underlying mechanisms of effects that may not have been previously considered.

As a very simple example, before testing a multivariate theoretical model predicting growth in substance use, an important first step is to simply plot the observed changes over time in the outcome variable to know what the trajectory looks like and what form of growth you should be modeling. A linear growth model is limited if the actual form of growth is quadratic. However, in practice, researchers may sidestep these preliminary examinations because the model conveys an illusion of reality, and because software programs do not demand it. As software becomes increasingly user-friendly, drifting from the data set becomes even more likely. Thus, because these modeling techniques can be implemented with the researcher far from his/her data, there is a great potential for error. An analysis can produce a well-fitting model, with hypothesized significant effects, but a closer look might reveal evidence of mis-specification. If researchers do not look closely and carefully for indicators of problems, they will publish happily without being aware of errors. Finally, the cost of using these complex, multivariate techniques can be measured in both time and money involved. From the design point of view, considerable time and expense is involved in collecting the large sample size and large number of data points that are required. The data analytic steps require considerable effort and time as well as sophisticated software and hardware. Perhaps most daunting is the investment of training that is required to use these techniques intelligently.

Given the intensive resources that are required for the use of these techniques, it is worth considering what steps are necessary to disseminate them more widely into substance use research. Perhaps most important are significant changes that are needed in the structure of research teams. Given the intensive training that these techniques require, it is unlikely that a single investigator will have the expertise to both design and implement the drug abuse questions of interest and the analyses of the data. Parallel with other changes that have pushed research efforts to be broader and to incorporate more interdisciplinary collaboration, the advances and dissemination of these complex, multivariate techniques call for further broadening of research teams to include the skills of quantitative methodologists. As described by Muthén (1989) and also by Aiken et al. (1990) in their assessment of human capital issues in quantitative methods, the field requires training and nurturing of both "developers," who will innovate and create new techniques and methods and "bridgers," who will work with content researchers to apply these methods appropriately to create new knowledge about the particular research questions of interest. Methodologists must be fully integrated into the collaborative team from issues of initial study design through publication of the results. Moreover, for this to be accomplished, universities will need to invest substantially in quantitative methods so that advanced training opportunities are available to graduate students, postdoctoral fellows, and faculty (Aiken et al., 1990).

Of course, making all this a reality requires considerable tangible resources and financial support as well. Funding agencies (and the reviewers who evaluate grant applications) may need to re-examine the time that they have been willing to support for data analytic purposes. Too often funds are available for data collection, but not for the more extended periods that are required by these complex data analytic methods. Moreover, funding agencies play key roles in providing support for the development and dissemination of these techniques, and it is important that resources be made available for methods development studies as well as for research on content questions. In this regard, the field of drug abuse research has benefitted tremendously from investments in the development of quantitative methods. As illustrated by the chapters in this volume, important advances in quantitative methods for drug abuse research have come from projects supported by NIAAA (see chapters by Muthén, by Khoo & Muthén, and by Wood, this volume), NIDA (see chapters by Hyatt & Collins; by Curran; by MacKinnon, and by Ullman et al., this volume), NICHD (see chapters by Hedeker & Rose, by Rose et al., and by Rowe et al., this volume), and NIMH (see the chapter by Hedeker & Rose). Such investments have both made possible the development of new quantitative methods for the analysis of drug use data and have facilitated the application of new and innovative techniques to existing data sets in drug use research.

CONCLUSION

We hope that this volume has illustrated both the complex problems involved in designing and analyzing drug use research studies, as well as the innovative solutions that are already available and some that are on the horizon. We hope also that these examples help to bring together researchers pursuing questions about substance use and abuse with researchers working to develop and innovate quantitative methods. These collaborations will benefit both fields—as content questions stimulate the development of new methods, and the introduction and dissemination of new methods changes the way that we think about our theories and research questions. The result will be a richer perspective on substance use and abuse etiology, prevention, and treatment.

REFERENCES

Aiken, L. S., West, S. G., Sechrest, L., & Reno, R. R. (1990). Graduate training in statistics, methodology, and measurement in psychology. *American Psychologist, 45*, 721–734.

Baker, G. L. & Gollub, J. P. (1990). *Chaotic dynamics: An introduction*. New York: Cambridge University Press.

Bentler, P. M. & Dudgeon, P. (1996). Covariance structure analysis: Statistical practice, theory and directions. *Annual Review of Psychology, 47*, 563–592.

Bentler, P. M., Lettieri, D. J., & Austin, G.A. (1976). *Data analysis strategies and designs for substance abuse research*. Rockville, MD: National Institute on Drug Abuse.

Boker, S. M., & Graham, J. W. (1998). A dynamical systems analysis of adolescent substance use. *Multivariate Behavioral Research, 33*, 479–508.

Bollen, K. A. (1996). An alternative two stage least squares (2SLS) estimator for latent variable equations. *Psychometrika, 61*, 109–121.

Brown, C. H., & Liao, J. (1996, July 12). *Principles for designing randomized trials in mental health* [WWW document]. URL http://yates.coph.usf.edu/research/psmg/ Power/schoolpower.html

Brown C. H., Wang W., Muthén B. O., & Dagne G. (1998, May 4). *Power calculations on the WWW for longitudinal intervention trials* [PostScript file]. URL http://yates.coph.usf.edu/research/psmg/Power/growthpower.html

Bryk, A., Raudenbush, S., & Congdon, R. (1996). *Hierarchical and nonlinear modeling with the HLM/2L and HLM/3L programs*. Chicago, IL: Scientific Software International.

Collins, L. M. (1997). *Modeling risk of drug abuse with an expanded latent transition analysis procedure* [On-line abstract]. URL http://methcenter.psu.edu/abstracts.html

Hedeker, D., Gibbons, R. D., & Waternaux, C. (1999). Sample size estimation for longitudinal designs with attrition: Comparing time-related contrasts between two groups. *Journal of Educational and Behavioral Statistics, 24*, 70–93.

Hedeker, D., & Mermelstein, R. J. (1998). A multilevel thresholds of change model for analysis of stages of change data. *Multivariate Behavioral Research, 33*, 427–456.

Hedeker, D., Mermelstein, R. J., & Weeks, K. A. (in press). The thresholds of change model: An approach to analyzing stages of change data. *Annals of Behavioral Medicine*.

Hoyle, R. H. (in press). *Statistical strategies for small sample research*. Thousand Oaks, CA: Sage.

Khoo, S.-T, (1998, June). Measuring change, changing measurements: Issues in longitudinal development. Presented at the annual meeting of the Society for Prevention Research, Park City, Utah.

Knight, G. P., Virdin, L., & Roosa, M. (1994). Socialization and family correlates of mental health outcomes among Hispanic and Anglo-American children: Consideration of cross-ethnic scalar equivalence. *Child Development, 65*, 212–224.

Krull, J. L. (1998, March). Mediation in multi-level models: A prevention intervention example. Presented at the Conference on Mediational Models in Prevention Research, Arizona State University, Tempe, AZ.

Krull, J. L. & MacKinnon, D. P. (in preparation). Multilevel mediation modeling in group-based intervention studies.

Lashley, B. R., & Kenny, D. A. (1998). Power estimation in social relations analyses. *Psychological Methods, 3*, 328–338.

Millsap, R. E. (1995). Measurement invariance, predictive invariance, and the duality paradox. *Multivariate Behavioral Research, 30*, 577–605.

Millsap, R. E. (1997). Invariance in measurement and prediction: Their relationship in the single factor case. *Psychological Methods, 2*, 248–260.

Molenaar, P. C. M. (1994). Dynamic latent variable models in developmental psychology. In A. von Eye & C. C. Clogg (Eds.), *Latent variables analysis: Applications for developmental research.* Newbury Park, CA: Sage.

Muthén, B. (1989). The future of methodological training in educational psychology: The problem of teaching students to use new sophisticated techniques. In M. Wittrock & F. Farley (Eds.), *The future of educational psychology* (pp. 181–189). Hillsdale, NJ: Lawrence Erlbaum Associates.

Muthén, B. (1999). Second generation structural equation modeling. To be presented at the annual meeting of the AERA.

Muthén, B. O., & Curran, P. (1997). General longitudinal modeling of individual differences in experimental designs: A latent variable framework for analysis and power estimation. *Psychological Methods, 2*, 371–402.

Muthén, L., & Muthén, B. (1998). *Mplus user's guide.*

Muthén B. O., & Shedden, K. (in press). Finite mixture modeling with mixture outcomes using the EM algorithm. *Biometrics.*

Neale (n.d./1998). *Mx homepage* [WWW document]. URL http://griffin.vcu.edu/mx

Olsen, M. K. & Schafer, J. L. (in press). Analysis of incomplete data by multiple imputation. *Multivariate Behavioral Research.*

Satorra, A., & Bentler, P. M. (1988). Scaling corrections for chi-square statistics in covariance structure analysis. *Proceedings of the American Statistical Association* (pp. 308–313).

Schafer, J. L. (1997). *Analysis of incomplete multivariate data.* London: Chapman & Hall.

Scheines, R., Spirtes, P., Glymour, C., Meek, C., & Richardson, T. (1998). The TETRAD project: Constraint based aids to causal model specification. *Multivariate Behavioral Research, 33*, 65–117.

Willett, J. B., Singer, J. D, & Martin, N. C. (1998). The design and analysis of longitudinal studies of development and psychopathology in context: Statistical models and methodological recommendations. *Development and Psychopathology, 10*, 395–426.

Wood, P. K. & Brown D. (1994). The study of intraindividual differences by means of dynamic factor models: Rationale, implementation, and interpretation. *Psychological Bulletin, 116*, 166–186.

Author Index

Subject Index

L

M